FLOODS

FLOODS

Editor

Klemens Oelberg

scitus
academics

Floods

Edited by **Klemens Oelberg**

This Book is Printed on Acid-Free Paper. ♾

ISBN: 9781681178820
Library of Congress Control Number: 2017948909

SCITUS Academics LLC,
616, Corporate Way, Suite 2, 4766,
Valley Cottage, NY 10989

www.scitusacademics.com

Notice

Printed in United States of America.

Preface

Flooding may occur as an overflow of water from water bodies, such as a river, lake, or ocean, in which the water overtops or breaks levees, resulting in some of that water escaping its usual boundaries, or it may occur due to an accumulation of rainwater on saturated ground in an aerial flood. While the size of a lake or other body of water will vary with seasonal changes in precipitation and snow melt, these changes in size are unlikely to be considered significant unless they flood property or drown domestic animals. Flood impact is one of the most significant disasters in the world. More than half of global flood damages occur in Asia. Causes of floods are due to natural factors such as heavy rainfall, high floods and high tides, etc., and human factors such as blocking of channels or aggravation of drainage channels, improper land use, deforestation in headwater regions, etc. Throughout the last century flooding has been one of the most costly disasters in terms of both property damage and human casualties. Major floods in China, for example, killed about 2 million people in 1887, nearly 4 million in 1931, and about 1 million in 1938. The 1993 flood on the upper Mississippi River and Midwest killed only 47 people, but the U.S. Army Corps of Engineers estimates the total economic loss at between 15 and 20 billion dollars. A flood can devastate homes, commercial buildings, agricultural and pastoral lands, public goods, and other physical properties. However, during the flood and its aftermath, there are also threats to one's health and safety. In the flood occurred in Bangladesh in 1988, diarrhea was found to be most common illness and a major cause of death amongst the population affected by the flood that helped spread the disease. Prevalent respiratory infection induced by the flood was also blamed for the high morbidity and death toll – 46,740 patients and 154 deaths – in the devastated areas. In developing regions of the world, acute sanitation problems and various water-borne diseases – such as diarrhea, dysentery, cholera and typhoid – threaten disaster affected populations, especially the poor and vulnerable, due to lack of access to safe drinking water, medicine and hygienic food. Problems become more critical due to more severe and frequent flooding likely caused by climate change, socio-economic damage, population affected, public outcry and limited funds.

This volume 'Floods' focuses on flooding, but in particular as it seeks to highlight the breadth and complexity of flood risk management and new ideas, approaches and innovations in flood risk management. It presents the coverage of loss prevention and mitigation includes structural flood control measures such as construction of dams or river dikes and non-structural measures such as flood forecasting and warning, flood hazard and risk management, public participation and institutional arrangement, etc. It describes concepts, policy, plan and operation on integrated urban flood disaster and risk management. Examples on urban flood disaster and risk management are illustrated and discussed. This compilation is intended to students, practitioners, researchers, policy makers, and civil engineers.

Table of Contents

Chapter 1	The Reality of Life During Floods 1
Chapter 2	Flooding and Flood Risk Reduction in Nigeria: Cardinal Gaps 23
Chapter 3	Low-salinity flooding in a selected carbonate reservoir: experimental approach 53
Chapter 4	Sensitivity of Coastal Flood Risk Assessments to Digital Elevation Models 77
Chapter 5	Technologies to Support Community Flood Disaster Risk Reduction 93
Chapter 6	Assessment of Prolonged Unsafe Conditions through Tidal Flood and Rain water Flood at Lower Bengal Basin, India 105
Chapter 7	Saving Lives: Timely Flash Flood Warnings in the UK 121
Chapter 8	The 2011 Brisbane Floods: Causes, Impacts and Implications 139

Chapter 9 Dilemmas and Pathways to Dealing with Flood Problems in Twenty-First Century China 169

Chapter 10 Momentum Attraction by Flood Plains in Compound Channel 193

Chapter 11 Effective and Sustainable Flood and Landslide Risk Reduction Measures: An Investigation of Two Assessment Frameworks 209

Chapter 12 Study and Analysis of Chennai Flood 2015 Using GIS and Multicriteria Technique 241

Chapter 13 The Sustainable Procedure Framework for Disaster Risk Management: Illustrated by the Case of the EU Floods Directive in Sweden 259

Chapter 14 Coastal Flooding in the Solent: An Integrated Analysis of Defences and Inundation 283

Chapter 15 Response and recovery measures for two floods in north China during the nineteenth century: a comparative study 319

Chapter 16 Flood Hazard Mapping of Lower Indus Basin Using Multi-Criteria Analysis 337

Chapter 17 Flood Management in California 349

Chapter 18 Framework of National Non-Structural Measures for Flash Flood Disaster Prevention in China 365

Chapter 19 **Managing the Limbe Floods: Considerations for Disaster Risk Reduction in Cameroon** **379**

Index **401**

CHAPTER 1

The Reality of Life During Floods

Luke Kuźmiński[1*], Luke Szałata[2] and Jerzy Zwoździak[3]

[1]*Wroclaw University of Economics, Ul. Komandorska 118/120, 53-345 Wroclaw, Poland*

[2]*Wroclaw University of Technology, Pl. Grunwald 9, 50-377 Wroclaw, Poland*

[3]*IMGW-PIB Warsaw Branch, Ul. Podleśna 61, Warsaw, Poland*

ABSTRACT

The aim of the article is to assess the variability of the risk of flood risk with the use of distributions maximum values. The copyright research is hydrological data in the form of daily water levels from the period 1981-2011. The collected data are derived from the hydrological station Bystrzyca Klodzka for the Nysa Klodzka river and the hydrological station Lesna for the Kwisa river. For the purpose of estimating flood risk from the collected data has been selected quarterly maxima of daily water levels. As a measure of the flood risk authors accepted the likelihood of exceeding the alarm condition in the test sections of the rivers. This risk is calculated by using the theoretical cumulative distribution distribution of quarterly highs water levels. The study was used Gumbel distribution. At the same time the article was paid significant attention to the possibility of using the solutions for integrated flood risk management process in accordance with applicable national legislation and European.

Keywords: Block method; Environmental risk management; Flood; Flood risk; Gumbel distributnion; Odra basin

1. INTRODUCTION

In order to more broadly understand the complex nature of the course of floods in some cases a closer description of the Kłodzka land as an area in south-western Poland in Central and Eastern Sudetes was made. It is characterized by varying morphology of the terrain and by its hydrographic network. The main drainage river for the Kłodzka Area is the Nysa Kłodzka together with its larger tributaries: the Biała Lądecka, the Bystrzyca Dusznicka and the Ścinawka. Due to large declines of river banks and river beds as well as high density of valley floors the reception the basin of the Nysa Kłodzka quickly responds to heavy rain with violent floods and high flows. Major floods do not last longer than two or three days and small ones - from a few to several hours [1].

1.1 Purpose and scope of the article

The primary objective of this article is to analyse the dynamics of the flood risk in the function of time and space on selected receptor basin areas. Given this, the authors of this article ask themselves a question in terms of the reality of life in floods and identify effective operational tools of technical and non-technical nature which will enable a more realistic prediction of the flood phenomena and contribute to the effective reduction of the socio-economic impact of the flood risk occurrences.

1.2 Characteristics of Flooding in the Oder Basin

The Oder river basin is highly developed and extremely asymmetrical. The reception basins of the left-side tributaries, the source areas of which are located in the Sudetenland and Sudeckie as well as Olza which flows from the Silesian Beskid, are all categorised as mountain-lowland rivers. There are three main types of landscape: the mountain and piedmont, upland and lowland. The differentiation of the environment of this area affects not only the amount of rainfall, but also the speed of the water run-off and the retention capabilities of the reception basin.

July and August are the months with a high risk of flooding on the Oder. This is confirmed by the hydrological data available in the database of the Institute of Meteorology and Water Management. To confirm the rank and scale of the flood risk and in accordance with historical data for the years 988-1774, chronicles recorded 36 major floods in the Oder river basin. In the nineteenth century, catastrophic floods occurred in the years: 1813, 1854, 1855 and 1888. However, in the twentieth century a series of severe floods took place, among others in the years: 1903, 1915, 1924, 1938, 1940,

1947, 1958, 1960, 1963, 1964, 1965, 1970, 1972, 1977, 1980, 1985, and 1997. The biggest flood of unprecedented size, exceeding the most catastrophic assessment, occurred in July, 1997. The second that flood in magnitude happened also in July, but in 1903 and was considered the biggest flood of the twentieth century until July 1997. Events of 2010 became the confirmation of the reproducibility of flood events. It should also be noted that the percentile ratio of the length of the Oder river in relation to the length of the individual tributaries is confirmed by the significant preceding of the culmination of the tributaries in contrast to the Oder observed in the event of natural water swells. The smallest preceding of the culmination is observed in the Nysa Kłodzka river. This preceding slightly reduces the amount of water in the Otmuchów and Nysa reservoirs, which is caused by the prolonging of a part of the falling wave on Nysa Kłodzka.

The likelihood of catastrophic floods occurring in the upper, middle and lower Oder is very small. In the lower reaches of the Oder river, there are usually spring floods, often related to the procession of ice. They can also occur in the time of not favourable storms on the Baltic Sea that hinder the flow of the flood water into the sea, which is confirmed by recorded peak flows during floods on the Oder river and the levels.

To make the flood risk more real, in 1598, torrential rains caused a significant landslide of the Bardzka Mountain slopes in Bardo and a partition of Nysa Kłodzka which flooded the city and then changed its river-bed. In 1997, during the "Flood of the millennium" a 9 meter high wave swept through the river and flooded Kłodzko. Storage reservoirs did not protect also Nysa from flooding. Throughout the Kłodzko Valley the rivers belonging to the Nysa Kłodzka basin caused huge damages. Long-term average flow at the mouth of the Nysa Kłodzka flowing into the Oder is 50 m^3/s, the flow that does not cause losses in the infrastructure below the dam is dependent on the swell of the Oder and varies between 150-190 m^3/s, while in July 1997 this flow was over 1500 m^3/s. Then, the emergency dump of Nyskie Lake flooded, among others: Nysa, Lewin Brzeski as well as numerous villages in the basin of the river and contributed to a significant increase of the water levels in the Oder, which flooded Wrocław [1]. In this paper, the authors focused their attention especially on the Upper Oder River Basin, taking into account the available water levels for Kwisa and Nysa Kłodzka rivers.

1.3 The Occurrence of the Flood Risk in terms of Water Management

The directions of water management, including the flood protection system as well as the reduction of the flooding system are defined by the national and European legislation on the broadly understood environmental protection. Appropriate references can be found in the program documents laid down in the Ministry of the Environment, such as The National Environmental Policy or environmental protection and sustainable development strategies. The directions of cooperation in the field of water management are also determined by international agreements concluded by the government or the Minister of the Environment both, the ratified ones as well as the bilateral agreements that do not require ratification. Offices of government together with lower level offices support the implementation of tasks in so far as they relate to their competences. By definition, "the socio-economic development, in which in order to balance access opportunities of individual communities or their citizens - both contemporary and future generations - to the environment, a process of integration of political, economic and social actions takes place maintaining the natural balance and the permanence of basic natural processes" and the area of operation of sustainable development shows that there is a need for the conservation of nature and the improvement of the natural environment's state. Continuing the development of a given region requires constant raising of social awareness in this respect, particularly, indicating the interdependence between the sustainable development and the economic interests [2].

The principle of sustainable development is written down in the Article 5 of the Polish Constitution. The sustainable development and the environmental protection constitute an important element of social and economic policies and the spatial development of the country [2].

The water management, in accordance with the Polish Water Law, as well as the Water Framework Directive and the Flood Directive, takes place in the hydrographic system, that is, in the river basins and water regions. The socio-economic aspect, in which the activities related to the natural balance and to the permanence of basic natural processes are integrated in order to guarantee the possibility of satisfying the basic needs of individual communities or citizens of both the present and future generations plays a very important role in achieving sustainable development. The Water Law of 18 July 2001 requires conducting water management in accordance with the principle of sustainable development concerning, in particular, the development of water resources protection as well as water use and water management. Water management is conducted according to the principle of rational and holistic treatment of surface and groundwater resources, including their quantity and quality, respecting the principle of common interests. It is implemented by the cooperation of the public administration, water users and representatives of local communities, so as to obtain the maximum social benefits, including economic and ecological ones. The application of the principle of sustainable development in water management and flood protection systems means striving for such a way of meeting the needs associated with the use of water resources so as not to curtail the access to them for future generations, while, at the same time, protecting the aquatic and water-dependent ecosystems in order to protect and preserve the stability of natural processes. Currently, the process of implementing these principles by the provisions of the Water Law, which covers the regulations of the Water Framework Directive and Flood Directive in this respect, is taking place. The actions being undertaken lead to the spread of an integrated approach to water management as well as the implementation of the flood protection system and methods for flood risk mitigation. While discussing the socio-economic functioning of the water management one should also pay special attention to the measurable socio-economic impact of the flooding and flood risk. At this point the following activities should be mentioned:

- Banning the construction of public facilities such as hospitals, schools, kindergartens, nursing homes and others of a similar nature.
- Banning the placing of facilities, such as sewage treatment plants, municipal waste landfills, toxic waste landfills, and chemical warehouses etc., which, in the case of flooding, pose a secondary risk.
- Allowing the construction of residential and public buildings, provided that their utility premises (e.g. residential premises) are located above the reliable flood level (i.e. the hundred year flood level), e.g. on an embankment, or that they are secured by encircling them with protective constructions or embankments (www.wroclaw.rzgw. gov 2016).

At this point, the description of this chapter should focus on the activities related to avoiding the social or economic indicators of the system's vulnerability such as the awareness of the flood risk or the degree of preparation for the coming of a flood all of which could be a source of significant damages. For example, according to German data, preparing the lower storey of a detached house to flooding reduces the damage by, between 5-30%, while according to the Flood Hazard Research Centre from the US, reducing the time of warning the residents of impending floods results in reducing the damages by up to 10%, with a range of reduction depending on the depth of flooding.

Modelling the flood risk by the use of probability distributions of selected hydrometric parameters as risk measures constitutes a practical tool for the evaluation of the actual financial risk in the insurance field for floodplains.

In summary, flood damages refer to a wide range of influences during the course of a flood in terms of their impact on:

- People, their health and property,

- Municipal infrastructure (water supply systems, waste disposal systems, transportation systems, power systems, etc.),
- Cultural heritage,
- Ecological systems,
- Industrial production,
- The competitiveness of the flood-affected businesses.

Depending on the manner of categorising, they are divided into material damages - expressed in monetary units and non-material damages - expressed in natural units, e.g. The number of casualties or the amount of square meters of the ecosystem flooded with sewage.

Therefore, the evaluation and assessment of hazards and flood risks and a proper reaction as well as the reduction of the potential flood damages in the described aspects is given a high priority in this respect.

1.4 Applications of Extreme Value Theory in Hydrology - A Literature Review

Historically, the year 1709 marks the beginning of working on the analysis of the extreme values problems. Then, Nicolas Bernoulli led reflections on the average largest distance between given n data points spread randomly on a straight line with a fixed length'ť' [3]. Very rich and comprehensive bibliography of the literature on the theory of extreme value distributions and their applications consists of more than 1,100 positions counting until the beginning of the 21st century. There is no way to present them all as it would require a separate multi-volume monograph devoted solely to the subject study. Such a vast literature indicates a great interest in this field of science as well as its wide application. Therefore, in this chapter, only selected items will be presented that, in the author's opinion, had a significant impact on the development of the theory and that are closely related to the issues raised in the article. Obviously, such a rich literature in one area has also its drawbacks. The main problem is the lack of coordination between researchers in this field and the inevitable duplication (and even tripling) of the results appearing in various publications around the world. Probably, the first one to use the extreme values in studying floods in his article was Fuller in 1914). The systematic development of a general theory of extreme values, however, is associated with the work of Bortkiewicz, which concerned the distribution range in a random sample from a normally distributed population. This work is very important, since the author introduced and clearly defined the concept of distribution of the highest value there for the first time [4]. Gumbel first drew the attention of engineers and statisticians on the possibility of use the formal theory of extreme values for certain distributions that were previously regarded as empirical. He applied the distribution of extreme value to the analysis of stream flows in the USA in 1941 [5]. In subsequent works he continued his research and discussions on the estimation of extreme stream flows and flood forecasts [6-8]. In the course of his research, Teodorovic acquired the observed frequencies N(T), meaning the number of days in a period which was T days long, when the water flow in the Greenbrier river in West Virginia exceeded 17 000 cubic feet. The period of his observation took 72 years, from 1896 to 1967. He then compared the observed frequency with the theoretical Poisson distributions. In the results it could be seen that the discrete observations N(T) for the studied river and for the given climate can be very well modelled with Poisson distributions [9]. In the 70s - 90s of the twentieth century, a lot of papers were written on the subject of applying elements of the extreme value theory to solve problems associated with floods. Pericchi and Rodriguez-Iturbe conducted research based on data on daily water flows in the Feather river in Oroville, California, USA. The data collected came from the years 1902-1960. From these data, they selected to annual flow peaks and fitted their empirical distribution t Gumbel distribution. In addition to that, in their work they proposed schedules such as: gamma (Person type III), gamma-log (log - Pearson type III) and log - normally for the analysis of selected peaks. In their research they also suggest the use of distribution function for overflows and gambling function for flood risk analysis [10]. The use of probability distributions for the flood frequency estimation was also illustrated in Greis' and Wood's work [11]. Shen applied the probability distributions to forecast flood events [12]. Rossi proposed a two-component extreme value distribution to analyse the frequency of flooding [13]. In subsequent years, Smith, Jain and Singh as well as Ahmad brought a discussion on the application of type I extreme value distribution to analyse the frequency of flooding [14,15]. At the end of the twentieth century, after the great flood in the USA, which caused huge losses in the Midwest of the country, Hipel presented the use of extreme value theory in the analysis of flood events in his work. He accurately presented the analysis of emergency level exceedings over the span of 100 years in the context of the flood of 1993 [16].

The beginning of the 21st century is also rich in terms of studies of hydrological and meteorological phenomena using the extreme value theory.

In their article, Katz with co-authors presented a comprehensive study using distributions of extreme values on hydrological data collected in Fort Collins, Colorado, USA [17]. Engeland, Frigessi and Hisdal presented the analysis of flood and drought risks using the generalized extreme value distributions and Pareto. They conducted their research on data concerning streamflows on the Ha river in south-west of Norway [18]. In their work,

Bordi, along with coauthors, analysed wet and dry periods in Sicily. For this purpose they applied monthly rainfall maxima [19]. Yurtal and others compared in their work the method of maximum likelihood to weighted method of moments for estimating the parameters of hydrological data distributions probability obtained from measuring stations on the Ceyhan river in southern Turkey [20]. After a great number of floods in the Czech Republic, Holičky and Sykora used a log-normal distributions and Persona III in their research to estimate the flood risk for cultural heritage [21]. Nachabe and Paynter conducted research using generalized distribution of extreme values on hydrological data from the selected lakes in the south-west of Florida [22]. Chaibandit and Konyai, in their studies, analysed hydrological data obtained on a monthly basis from 6 stations on the Yom river. The study used the distributions of extreme values, normal distribution and log-normal distribution as well as the return period method [23]. Arns and others, in their studies, estimated the flood risk by estimating the probability of achieving a certain water level in rivers [24]. Charon, along with other scientists in their work compared a very large number of probability distributions used to model wind speeds. The data came from 9 meteorological stations in The United Arab Emirates [25].

The above presented overview of the applications of the extreme values theory constitutes only selected items that relate to the phenomena of floods. Additionally, it should be mentioned, that there exists a very large number of publications dealing with the use of extreme value theory to model many other meteorological phenomena as well as phenomena in the field of economics, engineering and many others. Due to the limited scope of this publication this point will mark the end of the chapter on application literature review.

1.5 Selected Issues of the Theory of Extreme Values Distribution

This chapter contains selected issues concerning the distribution of extreme values, which were used to carry out the research for this study. Issues covered in this chapter include: selected probability distributions of the maximum values along with the methods of parameter estimation and compliance tests, methods for visual representation of empirical distributions and the block method used to select the maximum values of specified data blocks.

2. MAXIMA

We assume that the y_i observations are the maxima, i.e., that,

$$y_i = \max\{x_{i1}, \ldots, x_{im}\}, i = 1, \ldots, n, \quad (1)$$

Where x_{ij} may not be observable. In the case where x_{ij} are observable, the selection of certain maxima from certain sets with m number of elements is a form of selection of the upper extreme values from a data set. This method is called the block methodor the Gumbel method [26].

The block maxima method requires defining the time horizon (the block), and calculating the maxima of the tested variable for the said horizon. Most commonly, blocks of one year, half a year, quarter, month or smaller size are used depending on research needs. For data in the form of hydrometric parameters blocks of the mentioned above size are used. The block size cannot be too small to prevent the occurrence of the relationship between the maximum values of the neighbouring blocks of time. 10 day period is considered to be the minimum limit value of the size of the time block for which the independence of neighbouring maxima can be accepted.

There can also be cases when, during the long-lasting floods, there may occur a risk of a dependence even between the maxima of adjacent blocks of time. In such situations, when such a relationship between the variables under consideration occurs it is necessary to apply the cumulative distribution of extreme values for dependent random variables for the analysis of the distribution of the maximum values [27].

At this point, one more fact deserves attention, namely, that the observations y_i are the embodiments of the random variable $M_m$defined by the formula:

$$M_m = \max\{X_1, \ldots, X_m\} \quad (2)$$

2.1 Probabilistic models of maxima values

In the probabilistic studies of maxima distributions for hydrometric data it is suggested to first apply the Gumbel distribution, which is one of the three types of extreme value distributions [28]. The [IACWD (US Interagency Advisory Committee on WaterData - Hydrology Subcommittee) 1983] report recommends the Persona III distribution with the log-normal transformation for long-term data to predict flood events as well as the log-normal distribution.

According to the theorem concerning the types of extreme value distributions [29] the distributions of extreme values are one of three distribution functions described by the following formulas:

$$\textit{Gumbel (EV0 lub Typ I)}: \quad G_0(x) = \exp\left(-e^{-x}\right), \quad -\infty < x < \infty$$

$$\textit{Frechet (EV1 lub Typ II)}: \quad G_1(x) = \exp\left(-x^{-\alpha}\right), \quad \textit{dla pewnego } \alpha > 0,\ x > 0$$

$$\textit{Weibull (EV2 lub TypIII)}: \ G_2(x) = \exp\left(-(-x)^{\alpha}\right), \ \textit{dla pewnego } \alpha > 0,\ x \le 0. \tag{3}$$

The family of distribution functions of the maxima distributions specified in the (3) formula can be extended by introducing the parameterization to distribution function's formulas expressed with the above formula. The location parameter μ and the scale parameter σ are introduced into the formulas of the distribution functions. The transformation based on adding the above-mentioned parameters takes place according to the Theorem 1.

THEOREM 1: If the random variable X has distribution function F, then the random variable (μ + σX) has the distribution function $F_{\mu,\sigma}(x) = F((x-\mu)/\sigma)$ [26].

After the parameterization according to Theorem 1, distribution functions expressed with the formula (3) take the following form:

EV0Model: $G_{0,\mu,\sigma}(x) = \exp\left(-e^{-(x-\mu)/\sigma}\right)$

EV1 and EV2 Models:

$$G_{i,\alpha,\mu,\sigma}(x) = G_{i,\alpha}\left(\frac{x-\mu}{\sigma}\right), i = 1, 2. \tag{4}$$

The above parameterization of the family of distribution functions of the maximal values significantly broadens the spectrum of possibilities related to modelling of the distributions of maximal values of various random variables. By utilizing the parameterized distribution functions of maximal values one can very precisely choose a theoretical distribution function which satisfactorily describes the distribution of the value of the given maximal characteristics.

Concluding the topic of the family of distribution functions of extreme values it should be mentioned that all distribution functions presented in formulas (3) and (4) as members of one family of distributions described by the formula: [18,30].

$$G_{\gamma,\mu,\sigma}(x)=\begin{cases}\exp\left\{-\left[1+\gamma\left(\frac{x-\mu}{\sigma}\right)\right]^{-\frac{1}{\gamma}}\right\} & if\ \gamma\neq 0\\ \exp\left\{-\exp\left(\frac{x-\mu}{\sigma}\right)\right\} & if\ \gamma=0\end{cases}$$

(5)

In examining the empirical distributions of the maximal values of hydrological characteristics one should be aware of the normal distribution. There are situations when empirical distributions of the maximal values of the studied characteristics are more consistent with the Gaussian distribution then with the distributions expressed with the formula (4).

In the research that was conducted for this article a tool in the form of empirical density function of the probability distribution, also called kernel density, was used to visualise empirical distributions of the maximal values of specific hydrological characteristics. This precise form of an empirical histogram is described with the formula [27]:

$$f_{n,b}(x)=\frac{1}{nb}\sum_{i\leq n} k\left(\frac{x-x_i}{b}\right),$$

(6)

Where k is the function (kernel), such that $\int k(y)dy=1$ and b>0 is the desired bandwidth. The functionalkhas several forms defined which are used. In this study, the functional $k(x) = 0{,}75(1- x^2)I(-1\leq x \leq 1)$ was used

3. ESTIMATION METHODS AND TESTS OF SIGNIFICANCE

In this study, for the purpose of the conducted research, the highest reliability method was applied to estimate the distribution parameters of the maximal values of the family of distributions described with the formula (4). This method gives effective results in specific cases. These cases concur with the cases considered in this study [30,31]. The parameter γestimator exists for $\gamma > -1$ and for $\gamma > -0.5$ the variance has asymptotically normal distribution [14].

In order to verify the hypotheses concerning the compatibility of the studied empirical distributions with the selected theoretical distributions of maximal values described in the paper the following compliance tests were applied: chi - square test, Kolmogorov - Smirnov test and Anderson - Darling test.

In addition, within the family of distributions of extreme values described by the formula (5) the credibility quotient test was used to verify the hypothesis H0: $\gamma = 0$, that the tested distribution is better described by Gumbel distribution, against an alternative hypothesis H_1: $\gamma \neq 0$, that the tested distribution is better tested by other distributions of this family. The test statistics for this test is described with the formula: [26].

$$T_{\mathrm{LR}}(x) = \frac{\prod_{i \le n} g_{\hat{\gamma},\hat{\mu},\hat{\sigma}}(x_i)}{\prod_{i \le n} g_{0,\tilde{\mu},\tilde{\sigma}}(x_i)} \tag{7}$$

From $(\hat{\gamma},\hat{\mu},\hat{\sigma})$ and $(\tilde{\mu},\tilde{\sigma})$ representing the sets of the maximal likelihood estimators in the EV0, EV1 and EV2 models described by the formula (4). Since the sets of parameters are 2- and 3-dimensional, it is known that the test statistic has an asymptotically chi - square with 1 degree of freedom. Consequently, p - value has the formula:

$$p_{\mathrm{LR}}(x) = 1 - \chi_1^2(T_{\mathrm{LR}}(x)) \tag{8}$$

The significance level is achieved with higher accuracy using the Bartlett's adjustment (Hosking 1984), which consists in replacing the TLR test statistics with the statistics given by the formula T_{LR} / (1+2.8/n). In this case, the p – value has the formula (9).

Modelling the Distributions of Quarterly Hydrometric Parameters Maxima for the Studied Areas

4. HYDROMETRIC DATA

The study in this paper utilizes one of the hydrometric parameters, namely, the level of water [32]. For the purposes of the study, daily water level measurements were collected for the Nysa Kłodzka river taken from a water gage in the hydrological station located in Bystrzyca Kłodzka in the period from 01.01.1981 to 31.12.2011 which gives a sample size of n = 11 322 and for the Kwisa river measured at the water gage in the hydrological station in the village of Leśna from the period from 01.01.1982 to 31.12.2011 which, in this case, gives a sample size of n= 10 957. The studied periods have been divided into three parts. In the case of data for the Kwisa river the period I spans over the years 1982 - 1991, the period II spans over the years of 1992 - 2001 and the period III spans over the years 2002 to 2011. In contrast, for the data from the Nysa Kłodzka river the period I spans the years 1981 - 1990, the period II spans over the years of 1991 -

2000 and the period III spans over the years 2001 to 2011. As can be noted the period III for the Nysa Kłodzka river is an 11 year period. The remaining periods for both surveyed locations are 10 year periods.

By using the block method, as described in the previous section, the quarterly highs of the daily water levels collected in two hydrological stations were selected.

Consecutive quarters in a calendar year have an uneven number of days: 90, 91 and 92. For the purpose of the study, a simplification was made by assuming that each quarter has the length of 91 days. The differences arising from the simplification introduced into the probabilistic analyses are statistically insignificant.

The formula (1) for all three periods studied in the Leśna village and for the first two periods in Bystrzyca Kłodzka takes the form:

$$y_i = \max\left\{x_{i1}, \ldots, x_{i91}\right\}, i = 1, \ldots, 40 \tag{10}$$

While for the data from the third period in Bystrzyca Kłodzka it takes the form:

$$y_i = \max\left\{x_{i1}, \ldots, x_{i91}\right\}, i = 1, \ldots, 44 \tag{11}$$

Empirical and theoretical probability distributions for the quarterly maxima of the daily water levels

The empirical function of probability density given by the formula (6) with optimally matched functional k and parameter b was used to graphically depict the empirical distributions of quarterly maxima of the studied characteristics in the three periods for each of the two areas. Additionally, each graph presents the function of probability density of the theoretical distribution optimally matched to the empirical distribution of the studied maxima. The charts referred to above are presented in **FIGURES 1**and **2**.

By using the highest likelihood method the parameters of theoretical distributions optimally suited to the empirical distributions of quarterly daily water level maxima for the three analysed periods in the two measurement points: Bystrzyca Kłodzka and Leśna. The values of the μ position and σscale parameters for the theoretical distributions for the data from Bystrzyca Kłodzka and Leśna are presented in **TABLE 1** and **TABLE 2** respectively.

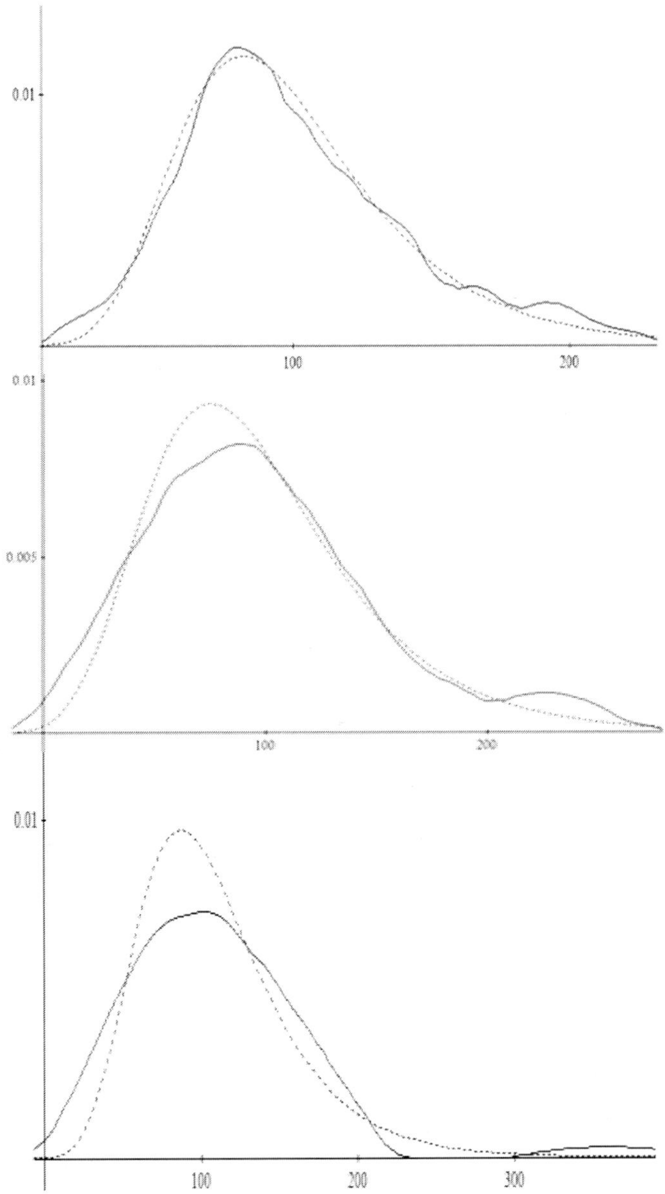

Figure 1: The empirical functions of distribution density of the quarterly water level maxima for the data from Bystrzyca Kłodzka for the periods I - III (solid line) and optimally matched theoretical distributions for quarterly maxima (dotted line). The chart on the left - the period I, the graph in the middle - the period II and the graph on the right - the period III. (Source: own materials).

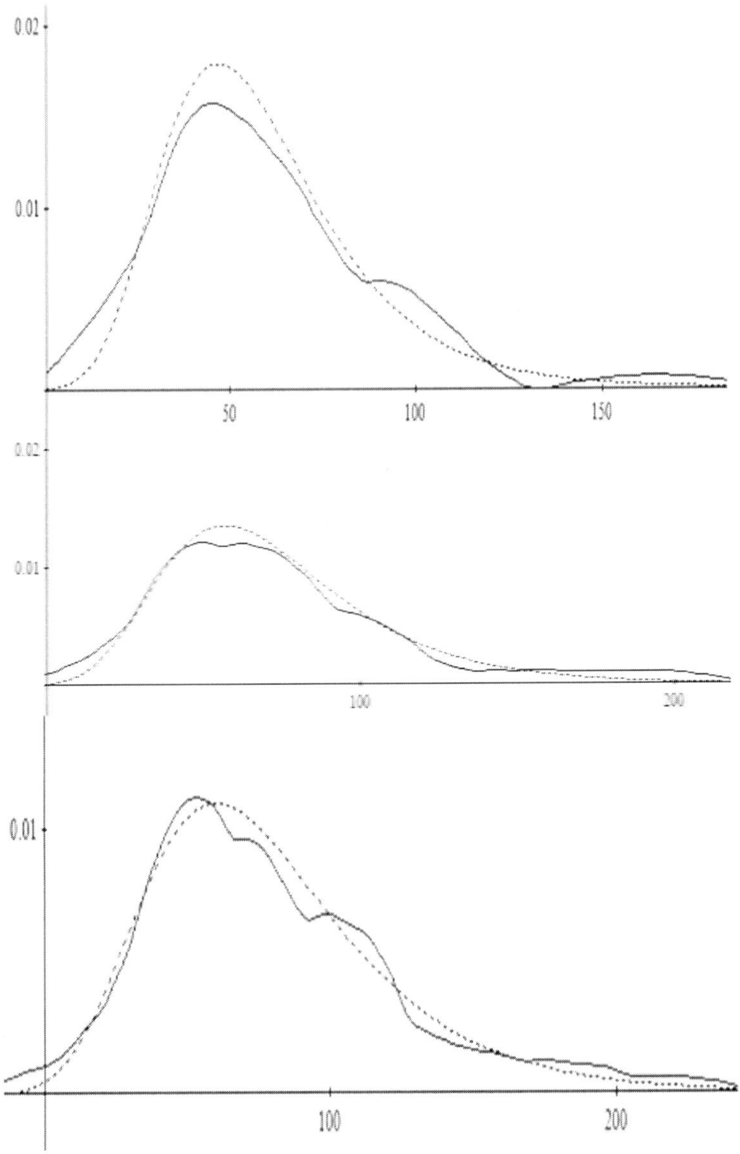

Figure 2: The empirical functions of distribution density of the quarterly water level maxima for the data from Leśna for the periods I - III (solid line) and optimally matched theoretical distributions for quarterly maxima (dotted line).The chart on the left - the period I, the graph in the middle - the period II and the graph on the right - the period III. (Source: own materials).

Table 1: The values of the μ, σ parameters estimates of the theoretical distributions for the Nysa Kłodzka river.

Periods	The estimator values
1981 – 1990	$\hat{\mu}_I = 82.02 \; \hat{\sigma}_I = 31.87$
1991 – 2000	$\hat{\mu}_{II} = 87.14 \; \hat{\sigma}_{II} = 37.9$
2001 – 2011	$\hat{\mu}_{III} = 75.29 \; \hat{\sigma}_{III} = 45.41$
Source: Own materials.	

Table 2: The values of theμ, σ parameters estimates the theoretical distributions for the Kwisa river.

Periods	The estimators values
1982 – 1991	$\hat{\mu}_I = 46.92 \; \hat{\sigma}_I = 20.59$
1992 – 2001	$\hat{\mu}_{II} = 56.9 \; \hat{\sigma}_{II} = 27.34$
2002 – 2011	$\hat{\mu}_{III} = 60.17 \; \hat{\sigma}_{III} = 33.67$
Source: Own materials.	

Presented below are the formulas of distribution functions for the distributions of the quarterly water level maxima for each of the three studied periods in the cases of the Nysa Kłodzka and the Kwisa. After the research, for all three periods studied, at the measurement point in Leśna and at the measuring point in Bystrzyca Kłodzka, the EV0 distribution (the Gumbel model) was proven to be optimally matched theoretical distribution, whose cumulative distribution is represented by the formula (4). The formulas of the distribution functions for the measurement point in Bystrzyca Kłodzka including the parameter estimators from **TABLE 1**:

1981 - 1990 period: $G_{0,\mu,\sigma}(x) = \exp\left(-e^{-(x-82.02)/31.87}\right)$

1991 - 2000 period: $G_{0,\mu,\sigma}(x) = \exp\left(-e^{-(x-87.14)/37.9}\right)$

2001 - 2011 period: $G_{0,\mu,\sigma}(x) = \exp\left(-e^{-(x-75.29)/45.41}\right)$

The formulas of the distribution functions for the measurement point in Leśna including the parameter estimators from **TABLE 2**:

1982 - 1991 period: $G_{0,\mu,\sigma}(x) = \exp\left(-e^{-(x-46.92)/20.59}\right)$

1992 - 2001 period: $G_{0,\mu,\sigma}(x) = \exp\left(-e^{-(x-56.9)/27.34}\right)$

2002 - 2011 period: $G_{0,\mu,\sigma}(x) = \exp\left(-e^{-(x-60.17)/33.67}\right)$

The charts of empirical density functions of probability distributions of the quarterly maxima for the analysed periods and locations along with the probability density functions of the corresponding theoretical distributions show a very good match of some theoretical distributions (**FIGURES 1** and **2**).

In order to confirm the goodness of fit of the empirical distributions with the matched theoretical distributions resulting from a visual assessment of the probability distribution function graphs two commonly used tests were performed: Anderson-Darling test and Kolmogorov-Smirnov test. The test results for the distribution of the three periods of two locations in the form of p-value shown in **TABLES 3** and **4** confirm a very high goodness of fit of the selected theoretical distributions with the corresponding empirical distributions.

Table 3: The p – value of the probability distributions, compatibility tests of the quarterly water level maxima in the three periods studied for the Nysa Kłodzka river.

Test Period	K-S	A – D
1982 – 1991	$p_v = 0.4032$	$p_v = 0.7107$
1992 – 2001	$p_v = 0.2662$	$p_v = 0.5857$
2002 – 2011	$p_v = 0.4281$	$p_v = 0.7331$
Source: Own materials.		

Table 4: The p – value values of the probability distributions, compatibility tests of the quarterly water level maxima in the three periods studied for the Kwisa river.

Periods	p_v NysaKłodzka	Periods	p_v Kwisa
1981 – 1990	0.7468	1982 – 1991	0.5897
1991 – 2000	0.3244	1992 – 2001	0.3628
2001 – 2011	0.9195	2002 – 2011	0.904
Source: Own materials.			

In addition, to confirm a very high goodness of fit of the theoretical distributions selected on the basis of visual charts with the empirical and theoretical probability distributions a likelihood ratio test was performed within the family of distributions for extreme values described in detail in Section 3.3. The values of p-value calculated using the formula (9) for all distributions in the examined periods and points of measurement are included in **TABLE 5**. The test results confirm the validity of the use of the Gumbel distribution to model the distribution of quarterly maxima of the studied hydrometric parameter in each of the cases.

Table 5: The p-value values for the likelihood ratio test in the family for the family of extreme values distributions.

Periods	p_v NysaKłodzka	Periods	p_v Kwisa
1981 – 1990	0.7468	1982 – 1991	0.5897
1991 – 2000	0.3244	1992 – 2001	0.3628
2001 – 2011	0.9195	2002 – 2011	0.904
Source: Own materials.			

5. ANALYSIS OF THE FLOOD RISK DYNAMICS ON THE KWISA AND THE NYSA KŁODZKA RIVERS

This section consists of two main parts. In the first part, flood risk measures will be calculated in the periods studied for the two measuring points. The second part of the chapter is solely devoted to the analysis of the dynamics of the flood risk in the analysed periods.

Defining the concept of risk proves to be a difficult time every time. Providing a clear, precise definition is impossible. Risk is defined on the basis of various branches of knowledge and theories, including economics, behavioural sciences, legal sciences, psychology, statistics, insurance, probability theory and others. According to the authors, the following two definitions of risk are most suitable to determine the flood risk. The first one treats risk as the possibility or likelihood of loss, e.g. Due to flooding [33,34]. The second definition assumed risk to be the probability of a system failure or the failure of its element p_f, which, in the particular cases, may be equated with flooding [35].

In this paper, based on the aforementioned two definitions, the probability of exceeding a certain water level (u) by the maximal daily water level from the time horizon assumed in the study was adopted as the measure of the flood risk on a given area. In the case in this work, a quarter (91 days) was adopted as the time horizon. The water level, which was used to calculate the flood risk measures in the periods studied in both areas is the emergency level set for the studied hydrological stations [36,37].

The emergency levels for the Nysa Kłodzka river in the hydrological station in Bystrzyca Kłodzka and for the Kwisa river in the hydrological station in the Leśna village are as follows: $u_{NK}(A)$ = 180 cm and $u_K(A)$ = 100 cm.

According to the formula (2) the maximal quarterly water level is a random variable marked as M_{91}. For the measurement point on the Nysa Kłodzka river this variable is marked as $M_{NK,91}$ while for the measurement point on the Kwisa river as $M_{K,91}$. **TABLES 6**and **7** present the results of calculations of the risk measure or the probability of exceeding the emergency level by the random variables $M_{NK,91}$ and $M_{K,91}$ for all the studied periods.

Table 6: Measures of flood risk for the Nysa Kłodzka river.

Risk measure Periods	$P(M_{NK,91} > u_{NK}(A))$	Dynamics indexes $I_{t/t-1}$	Percentile changes +/-
I: 1981 – 1990	0.0452	-	-
II: 1991 – 2000	0.0827	1.8296	82.96
III: 2001 – 2011	0.0949	1.1475	14.75
Source: Own materials.			

Table 7: Measures of flood risk for the Kwisa river.

Risk measure Periods	$P(M_{K,91} > u_K(A))$	Dynamics indexes $I_{t/t-1}$	Percentile changes +/-
I: 1982 – 1991	0.0731	-	-
II: 1992 – 2001	0.1876	2.5663	156.63
III: 2002 – 2011	0.2639	1.4067	40.67
Source: Own materials.			

Additionally, **TABLES 6** and **7** contain the indexes of flood risk changes dynamics in the studied period in the two study areas. The results show that the trend is clearly increasing in both studied locations. On the Nysa Kłodzka river, in the measurement station in Bystrzyca Kłodzka, the risk of flooding during the period II was 82.96% higher than in the period I, while in the period III the risk was higher by 14.75% than in the period II. On the Kwisa river, in the studied period, the risk is characterized by higher dynamics. During the period II, the flood risk was even 156.63% higher than in the period I, while in the period III the risk was 40.67% higher than in the period II.

In the two studied points, the dynamics of changes in flood risk varies. At the measurement point on the Kwisa river, in the hydrological station in the Leśna village, this dynamics is clearly higher. There are many factors influencing this situation, but this issue will not be addressed in this article. Despite the differences in the risk dynamics in both locations during the studied period proportionality of changes of the flood risk can be seen in both areas. In both studied points the increase during the period II in comparison to period I is significantly higher than in period III compared to the period II [38-41].

6. CONCLUSIONS

The temporal and spatial analysis of the change dynamics of the flood risk in the studied measurement points indicates growth trends of the flood risk in the described time horizons. At the measuring point for the Kwisa river at the gage Leśna (km of the river according to RZGW - 89 + 230, km according to IMGW-86+600) as well as for the gage Bystrzyca Kłodzka on the Nysa Kłodzka river (in km of the river according to RZGW -151+438, km of the river according to IMGW -147+800) the risk dynamics in 1991-2000 period compared to the 1981-1990 period was higher than in comparison with the 2001-2011 period. As a result of the assumed train of thought and the conducted analyses, the authors conclude that the hard and soft actions to minimize the flood risks until now have been insufficient. The current actions should therefore be carried out on the basis of the risk measures based on the results of the monitoring network. On the basis of the risk measures the authors suggest the development of indicators that should be included in such programs as ISOK and in dynamic and static flood models. Spatial planning must take the levels of the flood risk into account by defining the technical conditions for buildings, security systems for people and institutions.

REFERENCES

1. Łach J (2009) The history of floods on the ground Kłodzko and their influence on the direction of the transformation dendollnnych Nysa Kłodzka and White Lądecka, Nature Science Technology. Volume 3:3.
2. Nowak Z (2001) Environmental management, part-I. Gliwice: Silesian University of Technology.

3. Gumbel E (1954) Statistical theory of extreme values and some practical applications, National Bureau of Standards, Applied Mathematics Series, Vol. 33.
4. Bortkiewicz L (1992) Mean variation and error. Sitzungsber Berli MethGoose-5.
5. Gumbel E (1941) The return period of flood flaws, Ann Math Statist- 5.
6. Gumbel E (1944) On the plotting of flood discharges, Trance. Amer. Geophys. Union 25: 699-719.
7. Gumbel E (1945) Floods estimated by probability methods. Engrg News-Record 134: 97-101.
8. Gumbel E (1948) The statistical forecast of floods. Water Resources Board Bulletin No. 5.
9. Todorovic P (1978) A probabilistic approach to analysis and prediction of floods. Proc. 43rd ISI.
10. Pericchi L, Rodriguez-Iturbe I (1985) On statistical analysis of floods, In: Celebration Statistics, A. Fienberg, S. Atkinson, (eds.) 511-541.The ISI Centenary Volume.
11. Greis N, Wood E (1981) Regional flood frequency estimation and network design. Water Resources Res 1167-1177.
12. Shen H, Bryson M, Ochoa I (1980) Effect of tail behavior assumptions on flood, Water Resources Re. 16: 361-364.
13. Rossi F, Fiorentino M, Versace P (1986) Two - component extreme value distribution for flood frequency analysis. Water Resources Res. 22: 1551-1564.
14. Smith R (1987) Estimating the upper tail of flood frequency distributions, Water Resources Res. 23: 1657-1666.
15. Ahmad M, Sinclair C, Spurr B (1998) Assessment of flood frequency models using empirical distribution function statistics. Water Resources Res. 5.
16. Hipel K (1994) Extreme values: Floods and droughts, Proceedings of International Conference on Stochastic and Statistical Methods in Hydrology and Environmental Engineering Vol 1. Kluwer, Dordrecht.
17. Katz R, Parlange M, Neveau P (2002) Statistics of extremes in Hydrology. Advances in Water Resources: 8-12.
18. Engeland K, Frigessi A, Hisdal H (2005) Practical extreme value modeling of hydrological floods and droughts: A Case Study. Extremes 7: 5 - 30.
19. Bordi I, Fraedrich K, Petitta M, Sutera A (2006) Extreme value analysis of wet and dry periods in Sicily. Theor Appl Climatol 5.
20. Dogan A, Haktanir T, Seckin S, Yurtal R (2010) Comparison of probability weighted moments and maximum likelihood methods used in frequency analysis for Ceyhan river basin. Arabian J Sci & Eng 35 Issue 1B: 49-69.
21. Holičky M, Sykora M (2010) Assessment of flooding risk to cultural heritage in historic sites. Journal of performance of constructed facilities: 432-438.
22. Nachaben M, Paynter S (2011) Use of generalized extreme value covariates to improve estimation of trends and return frequencies for lake levels. Journal of Hydro-informatics 13.1: 13-24.
23. Chaibandit K, Konyai S (2012) Using statistics in hydrology for analyzing discharge of the Yom river, APCBEE Procedia,1.
24. Arns A, Wahl T, Jensen IJ, Pattiaratchi C (2013) Estimating extreme water level probabilities: A direct comparison of the methods and recommendations for best practice. Coastal Eng 81: 51-66.
25. Charon C (2015) Probability distributions of wind speed in the UAE. Energy Conversion and Manage 93 : 414-434.
26. Thomas M, Reiss R (2007) Statistical analysis of extreme value with applications to insurance, finance, hydrology and other fields. Basel: Birkhauser.
27. Kuźmiński L (2013) Limit distribution function of extreme values for dependent random variables, Econometrics 2 (29), Publishing House of the University of Economics in Wroclaw.
28. Yue S, Bobe B, Legendre P, Bruneau P (1999) The Gumbel mixed model for flood frequency analysis, J. Hydrol., Pages 88-100.
29. Leadbetter M, Lindgren G, Rootzen H (1983) Extremes and related properties of random sequences and processes, New York, Springer-Verlag.

30. Kotz S, Nadarajah S (2005) Extreme value distributions. Theory and applications. London: Imperial College Press.
31. Smith R (1985) Maximum likelihood estimation in a class of non-regular cases. Biometrika, pages 69-90.
32. Bajkiewicz- Grabowska, E, Mikulski Z (2011) Hydrology general.Warsaw: PWN.
33. Jedynak P (2001) Limit distribution function of extreme values for dependent random variables, Insurance economy. Selected elements of theory and practice.
34. Kunreuther H, RothR (1998) Paying the price, Washington, Joseph Henry Press.
35. Yen B (1988) stochastic methods and reliability analysis in water resources. Advances Water Resources, Vol. 11.
36. Beran, M, Hoskin J, Arnell N (1996) Comment on two - component extreme value distributio n for flood frequency analysis. Fabio Rossi, Mauro Fiorentino, Pasquale Versace, (eds.) Water Resources Res. 5.
37. Fuller W (1914) Floods flows. Trance Amer Soc, Civil Engineers 77: 564-583.
38. Hosking J (1984) Testing whether the shape is a zero in the generalized extreme value distribution, Biometrika 71: 367-374. http://mapy.isok.gov.pl/imap/
39. Jain D, Singh V (1987) Estimating parameters of EV1 distribution for flood frequency analysis, Water Resources Res. 5. https://wroclaw.rzgw.gov.pl/en/index

CHAPTER 2

Flooding and Flood Risk Reduction in Nigeria: Cardinal Gaps

Nkwunonwo UC[1,2*], Whitworth Malcolm[2] and Baily Brian[3]

[1]*Department of Geoinformatics and Surveying, University of Nigeria, Enugu Campus, Nigeria*

[2]*School of Earth and Environmental Sciences, University of Portsmouth, UK*

[3]*Department of Geography, University of Portsmouth, UK*

ABSTRACT

This study reflects on critical issue relating to flooding in Nigeria such as causes, impacts and remedies. Flooding which arguably has been more damaging for Nigeria has worsened recently due to a number of possible factors including rapid population growth, urbanization, poor urban planning and climate change especially in increased frequency and intensity of rainfall. Attempts to tackle the hazard in Nigeria appear to be limited by lack of flood data and other remote causes which are yet to be identified. In view of this background, the present study reviews the widespread flooding in Nigeria and efforts to tackle it. Over the period 1985 to 2014, flooding in Nigeria has affected more than 11 million lives with a total of 1100 deaths and property damage exceeding US$17 billion. Lagos state has experienced most of the floods while more frequent floods are recorded in Niger, Adamawa, Oyo, Kano and Jigawa states possibly due to the influence of rivers Niger, Benue, Ogun and Hadeja. It is argued that more robust and scientific approaches to flood risk reduction such as: flood modeling and vulnerability assessment are lacking. To align the focus of flood risk reduction in Nigeria with the objectives of such a task in more developed countries (such as the United States, United Kingdom and the Netherlands) which is among three fundamental issues to be addressed in Nigeria, the present study makes pivotal recommendations.

Keywords: Flooding; Developing countries; Nigeria; Flood risk; Climate change; Flood modeling; Flood vulnerability assessment

1. INTRODUCTION

Arguably, the rate of flooding occurrence in recent times has been unprecedented. With 70 million people globally exposed to flooding every year, and more than 800 million living in flood prone areas [1], climate change with more frequent and severe rainfall events, sea level rise, rapid population growth and urbanization, the rate of development on floodplains, the level of awareness of flood risk and the ineffectiveness of efforts towards tackling flooding in many places are factors of concern within the global [2,3]. The economic consequences of flooding reported in the last two decades amount to tens of billions of US dollars [4]. Over 3700 flood disasters globally are recorded in the EM-DAT database, covering the period 1985 to 2014 [5]. These events were responsible for hundreds of thousands of deaths mainly in Asia (most notably China, Thailand and Bangladesh) and adversely affected billions of people mostly through homelessness, spread of diseases, physical injuries, mortality (mainly through drowning) and psychological conditions mainly depression, anxiety and posttraumatic stress [6-10]. In the US, 32.9% of the total natural disasters in 2012 were hydrological with floods accounting for the most part, affecting more than 9 million people and causing about US$ 0.58 billion worth of damage [5]. The same source shows, for that year, more than US$4.7 billion worth of damage recorded for Europe, and about US$0.83 billion and US$19.3 billion damage for Africa and Asia respectively resulting from flooding. Four different floods that hit cities in UK in 2012 caused a total loss of $2.9 billion, with hundreds of people who were affected [4]. In many African countries for example Nigeria, flooding has impoverished hundreds of thousands of people through displacement from homes and loss of tangible properties [11].

In Nigeria, flooding and means of addressing its challenges are critical issues [12]. Evidently, the country has experienced devastating floods which affected millions people and caused fiscal losses amounting to billions of US dollars [13]. These hazards were generally linked to poor urban planning and climate change especially in increased frequency and intensity of rainfall [11,14-16]. The impacts of floods in Nigeria include mortality, physical injuries, widespread infection and vector-borne diseases, social disorders, homelessness, food insecurity, economic losses (mainly through destruction of farmlands, social and urban infrastructure) and economic disruption (most notably in oil exploration in the Niger delta, traffic congestion in many cities in Nigeria, disruption in telecommunication and power supply) [17-19]. In 2012, Nigeria experienced the worst flooding in more than 40 years as a result of heavy rainfall that lasted for many days. The incidence affected 32 states with 24 considered severely affected [13]. The floods extended from July to October that year and affected 7.7 million people with more than 2 million others reckoned as internally displaced (IDPs). More than 5000 people were physically injured along with over 5900 houses which were destroyed.

Historically, flooding in Nigeria which dates back to the early 1950' are fluvial, coastal and pluvial in nature and have been a major cause of concern for rural areas and cities within the country [20,21]. Fluvial and coastal flooding both of which affected mainly coastal environments were influenced by seasonal interruption of major rivers and water overtopping their natural and artificial defences and overflowing areas not typically submerged [22]. Fluvial floods account for the majority of the flood threats experienced in locations along the plains adjoining major rivers in the country, including rivers Niger, Benue and Hadeja. The states in Nigeria mostly affected by such floods are Adamawa, Kano, Niger, Jigawa, Kaduna, Cross River and Kebbi [23,24]. The worst fluvial flood in Nigeria was the Kaduna flood disaster of 2006 which affected hundreds of thousands of human lives with economic loss worth millions of US dollars [25,26]. Coastal floods in Nigeria affect the low-lying areas in the southern part of the country (comprising for examples Lagos, Oyo, Ondo, Akwa-Ibom and Bayelsa states). The impacts of such floods have been severe due to the number of human populations exposed following the attractions of coastal areas for economic and social reasons [27]. Nigeria is globally ranked with the top 20 countries whose present population and future scenarios in the 2070s (including climate change and socio-economic factors) are exposed to coastal flooding [28].

Pluvial floods usually occur annually during rainy seasons (between July and October) and affect mainly the urban areas in Nigeria. Such floods which are arguably unprecedented in recent times are caused by more frequent and severe rainfall which overwhelms the efficiency of drainage systems and soil infiltration capacity [29,30]. The significance of urban areas in the economic and political development of Nigeria is generally acknowledged [31]. However, urbanization is a critical anthropogenic influence on climate change and hydrological cycle in the country given that much impervious surfaces increase surface water runoffs and reduce soil infiltration capacity [32-35]. Along with urbanization is the rapid population increase in many Nigerian cities which is also a global concern within the context of flooding in urban areas [36]. It is estimated that more than half of the world's population has been residing in cities since the last 6 years and by 2030 the number of people living in urban areas (with urban areas of developing countries (DCs) accounting for the most part) will grow to 5 billion (i.e. 60% of the world's population) [37-39]. Regrettably, a major challenge with rapid population growth and urbanization in Nigeria which also seems to influence the risk of flooding in the country both presently and in the future (if not addressed) has been poor urban planning (in particular inadequate drainage system and the range of poorly serviced urban utilities) [14]. The authors argue that urbanization in Nigeria has not to be accompanied by corresponding strategies to support humanitarian needs and anthropogenic activities. Apparently, this concern has not received adequate attention in the literature, especially with regards to the implications of future urban scenarios on environmental sustainability. Moreover, the reality of widespread flooding in Nigeria coupled with the notion that floods are inevitable phenomena which can never be fully constrained within the natural environment appears to overwhelm efforts towards finding a solution. Thus, it is rather easier to feel some developing attitudes of anticipating increasing human vulnerability within the country as the years succeed each other, and the authors perceive that "*there seems to be fear on the horizon*".

However, various levels of government, the community and other stake holders have been active with measures to tackle flooding in Nigeria [40]. These measures have been criticized as ad-hoc, nongeneralizable and not well established [12]. In the light of 'best practices' in flood risk reduction and 'lessons learned' from other countries' experiences of flooding, it can be argued that such stake holders' efforts are at best limited most probably due to lack of quality data, which among other things are needed to systematically tackle flooding, poor perception of flooding among the general public, lack of funds and improved technology as well as poor political will power. Best practices in flood risk reduction ideally is based on "living with floods and not fighting them" idea, which dominates key environmental risk research themes (for examples: Disaster Risk Reduction (DRR) and Climate Change Adaptation (CCA)) and integrates structural and nonstructural measures to reduce the impacts of flooding on social systems and to achieve key requirements in risk management which are prevention/ mitigation, preparedness, emergency response, recovery and lessons learned [41-47].

Arguably, given the growing number of flood victims and the constrained sustainable development caused by flooding within the country, researches into solutions to widespread flooding in Nigeria are weak towards flood risk reduction. Recent studies focus attention on the impacts of flooding and its remedies in local communities, geopolitical regions and various states within the country [27,48-52]. It is argued that a more comprehensive understanding of the widespread flooding in Nigeria requires wide ranging discussions and accounts which few studies presented [12,20,53-58]. The influence of climate change and anthropogenic activities, poor urban planning and environmental management on the widespread flooding in Nigerian cities has been extensively discussed [14,50,54,59]. While the lack of definite measures and capacity to radically address the challenges of flooding within the country has been arguably overwhelming, concerted efforts in the form of environmental and infrastructural planning, policy directives, social responses, physical intervention and enhanced public enlightenment programmes, vital to tackling the prevailing flood hazard in the country have been considered [20,21,24]. Other vital tools needed to address flooding in Nigeria are community based early warning systems [56], humanitarian aids from government and private sectors [54,60] and appropriate level of preparedness and capacity building [57]. The need for science and technology to embrace environmental education in Nigeria is posited [58]. In [27,48,61,62], flood hazard mapping and assessment of vulnerabilities of lives and properties to flooding are important procedures factors which play key roles towards building community resilience to flooding. Few studies considered modeling of flood by means of hydrodynamic, GIS, cellular automata and statistical techniques [63-72]. The importance of reinforcing present strength and capacities of all agencies, including local communities within Nigeria to deal with flood hazard situations underlined [12]. The study investigated existing institutional approaches that deal with flood challenges in Nigeria with a view to more efficient, robust and satisfactory flood intervention strategies in the country.

question: "what is the remedy to the recurrent flooding in Nigeria?" remains unanswered. The lack of flood data and other ancillary data which is a major setback towards tackling flooding in Nigeria was not addressed. Attention has solely rested on general knowledge of the causes, impacts and remedies of flooding, suggesting that the general view of the situation in these studies has been lop-sided and sloppy. The need for more scientific approaches such a flood modeling which drives flood risk management in more developed countries was not highlighted. A general critique, which should provide a nuanced understanding of the strengths and limitations of present efforts to addressing the threats of flooding in the country, is lacking and gaps between increasing flood occurrences and vulnerabilities of local communities were not identified. It is well known that the success of flood risk reduction depends to a large extent on knowledge-based decision, robust institutional framework and flood risk communication [3,18] but these factors are missing in Nigeria and where they exist they are poorly addressed. Knowledge-based decision uses available information relating to flooding to draw conclusions on possible strategies to be adopted for tackling flooding. Flood risk communication aims at creating awareness of flooding and its impacts in stake holders and the general public. Institutional framework that tackles flooding in Nigeria refers to government response procedures which include policies, regulations, guidelines and agencies engaged in planning and managing flood emergency conditions or in helping victims to cope with and recover speedily from extreme flooding events [12]. Many flood risk based studies argue that these requirements are fundamental to information relating to flood hazard and its consequences and may be sourced from public opinions, expert knowledge, research findings and flood risk/hazard maps of an area [73-75]. It has also been argued that various flood modelling approaches are pivotal components of flood risk reduction because they are capable of quick, continuous and routine simulation of flood data (most notably flood water depth, inundation extent and duration as well as water flow velocity) required for flood hazard / risk assessment [30,76,77].

Despite the lack of these critical data for tackling flooding in Nigeria, flood modelling is poorly carried out [69]. Although the acquisition of these datasets (particularly the flood risk/hazard maps) have been a daunting task for Nigeria as they require critical understanding of the drivers of flood hazard/risk. However, it is important to acknowledge the specific roles flood modelling can play in this regard and in general towards tackling flooding in Nigeria. Over the recent past and in many flood management policies such as the EU commission directive on flood, the United States flood control policy, national flood insurance program (NFIP) and other regionally-based flood risk management policies, the relevance and provision of flood data highlight the significance of flood modeling [30,45]. For Nigeria, besides the general roles of flood modelling, the technique can assist in provision of critical information for strategic planning and effective flood risk management within the country. It will equally drive an improved understanding of flood phenomena and prompt improvements into more robust flood management approaches such as flood forecasting, flood early warning system and flood damage estimation.

This need for flood modelling was emphasized by the DG of Nigerian Hydrological Services Agency [78] in a recent mission statement:

"...in view of flooding in Nigeria, governments at all levels should create awareness on the need for communities to relocate to safer terrain. Moreover, while the current trends in climate variations prevail, the need to develop flood modeling and early warning systems cannot be overemphasized... There is also need to carry out a comprehensive flood hazard mapping for all areas considered at risk of flooding in the country..."

It is argued that whilst this statement recognizes flooding situation in Nigeria and the relevance of flood data for flood risk reduction, it laments the dearth of flood data for the country and the peculiar challenges that arise from utilizing commercial flood modelling codes to simulate such data. Moreover, the fact that Nigeria is one of the most populated countries of the world with population size estimated at over 170 million people compounds the situation [79].

The theory that future population growth will drive future flood risk highlights the importance of urgency in finding means of preparing human population in the country to adapt to floods. It will be a welcome development for Nigeria to implement flood modelling since within European Union framework, such a technique is operationalized towards flood hazard/risk mapping of the constituting States [73]. In the Netherlands and the US, flood modeling, among other roles, supports investigation into estimation of damage caused by flooding [80- 83]. Many Asian countries, notably China, Vietnam and Bangladesh although having 'not too well' established flood management systems and methodologies utilize flood modeling methodologies for flood risk assessment and mitigation [38,84].

Thus, to assist human populations towards adapting to flooding and to support stake holders' efforts towards tackling flooding in Nigeria, the present study is an attempt to bridge several gaps in flood risk reduction in Nigeria. The authors synthesized and analyzed relevant literature and available historical flood data spanning the period 1985 to 2014, to identify these gaps and to make relevant recommendations towards building flood resilient communities and more effective remedies to the threats of flooding. It is argued that promoting flood modelling in Nigeria will underpin current efforts at tackling the hazard.

This study in general and the recommendations in particular are driven by three key issues, which are: firstly, to demonstrate the roles more robust and scientific techniques such as flood modeling can play in flood risk reduction within the context of Nigeria, secondly, to align the focus of flood risk reduction in Nigeria with the objectives of such a task in more developed countries such as the US, the Netherlands and United Kingdom, and finally, to promote flood risk awareness in the general public as well as to delineate safer terrains for relocation of human populations during flooding in Nigeria.

The methodology and data for the research are discussed in section 2 while the study area is described in section 3. Section 4, the result, discusses efforts towards tackling flooding in Nigeria and cardinal gaps which have been identified to undermine such efforts.

Recommendations relevant to strengthening present efforts of tackling flooding in the study area are presented in section 5. Section 6 gives a general conclusion of the paper.

2. METHOD AND DATA

A search process to identify the body of literature relevant to flooding and efforts towards addressing its threats in Nigeria was undertaken. Combination of terms such as “flooding and management in Nigeria”, “flooding and human health in Nigeria”, “flooding and modeling in Nigeria” and “flooding and climate change in Nigeria” was applicable to the search. Overall, 429 publications were identified of which 17 focused on the causes of flooding in Nigeria, 132 addressed the impacts, 181 discussed the remedies, 54 looked at climate change issues, 14 discussed public perception of flooding while 31 addressed urban management and planning (**TABLE 1**). The scientific quality of these papers was assessed based on the publishing journal. This is consistent with academic standard and regulations. Although locally published articles provided most of the information to establish the case in the present study, however, the greater weight was given to articles published by Elsevier, Science Direct, Taylor and Francis, Wiley and sons, ASCE, Nature, Sage, Springer and Copernicus publishers and on International conferences. The data that provided much of the evidence regarding the prevalence of flooding in Nigeria was sourced mainly from EM-DAT database and Nigerian ministry of Environment. These findings are fundamental to discussions presented in this study.

Table 1: The results of literature search of issues relating to flooding in Nigeria. From these results, whilst remedies have received ample attention, improving public perception of flooding is poorly addressed. This is a reason for concern knowing that the success of present approaches of flood risk reduction depends on public participation which is driven by public perception of flooding.

S. No.	Issues relating to flooding in Nigeria	Literature search
1.	Flooding causes – causes	17
2.	Flooding – impacts	132
3.	Climate change issues	54
4.	Mitigation activities	181
5.	Public perception	14
6.	Urban planning and management	31

2.1 Description of the study area

Nigeria, a sub-Saharan West African country, is on the Gulf of Guinea, east of the Greenwich and north of the equator. The country, made up of 36 states including the federal capital territory (FCT), Abuja, lies between latitudes 4° and 14°N, and longitudes 2° and 15°E, with a total land area of 923,768 km2 (**FIGURE 1**), and borders with Republics of Benin and Niger, Chad, and Cameroon. It maintains a large expanse of coastline, over 853 km in magnitude, with hydrological features which includes the rivers Niger and Benue, both of which confluence at Lokoja, and flows further southwards passing through the Niger Delta to empty into the Atlantic ocean.

Figure 1: Map of Nigeria showing the 36 states and Rivers Niger and Benue. Inset is Africa showing Nigeria's location. Source: Drafted by the authors.

The 2006 census confirmed over 140 million people in Nigeria, but this population has grown steadily, and is presently estimated at more than 170 million people, making the country the seventh most populous country in the world [12,85]. According to United Nations projections, Nigeria is one of the eight countries expected to account collectively for half of the total population increase in the world from 2005-2050, and will by 2100, record a population amounting between 505 million and 1.03 billion people [86].

3. RESULT AND DISCUSSION

First and foremost, the present study argues that the impacts of flooding in Nigeria continue to trigger concerns for food security, vulnerability of the general public and local communities within the country, humanitarian needs and services, primary health delivery, environmental management, solid waste management, urban development, professionalism in journalism practice and the dynamism or lethargy of Nigerian democracy and political system [12,21,27,60,87- 89]. Whilst investigating these factors vis-à-vis flood risk reduction in Nigeria is vital, key features of flooding which influence its level of impacts in the country include flood water depth, inundation extent and duration as well as velocity of flood flow.

Figure 2: Evidence of flooding impacts in Nigeria: (a) Major public access road affected by flood water (b) street lanes submerged by flood water, (c) community center submerged by flooding, and (d) residential house submerged by flooding.

Respectively, flood width, height, annual frequency and duration in Nigeria can measure up to 700 m, 11 m, 10 and 25 days respectively [53]. It is shown from EMDAT database that most floods in Nigeria lasted up to 79 days. Thus based on these characteristics, the dangers posed to human lives and properties by flooding in Nigeria can be appreciated (**FIGURE 2**).

Paradoxically, relevant flood data which reveal widespread flood threats in Nigeria and how the situation differs from other places are lacking. Arguably, such dearth of flood data not only constrains investigations into gaining a better understanding of the spatial and temporal distribution of flooding across the states of Nigeria, but it also undermines efforts towards tackling the hazard. Although there is a large body of literature that highlights the gravity of flooding situation in the country, inconsistency of flood narratives in Nigeria has been overwhelming [59,90]. Undoubtedly, this inconsistency of flood narratives will undermine the general flood awareness which underpins more proactive ways of tackling the hazard in Nigeria.

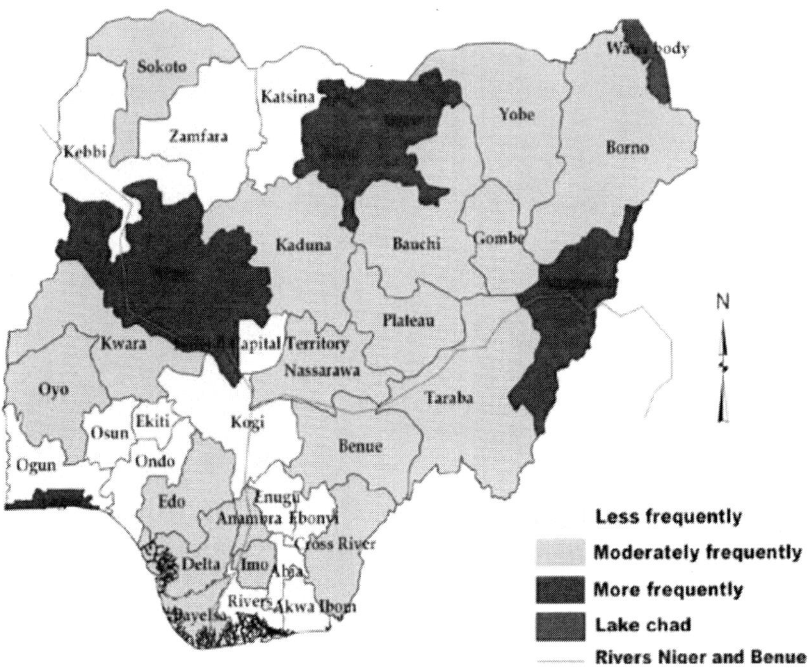

Figure 3: Spatial distribution of flooding across the states of Nigeria from 1985 to 2014. Here it is shown that Lagos, under coastal influence along with under states under the influence of Rivers Niger, Benue, Hadeja and Ogun have experienced more frequent floods in the country.

However, based on flood data from EM-DAT, CRED and Dartmouth Flood Observatory (DFO) databases, present investigations (which extend studies by [12,24,25,54,91]) relating to flood threats in Nigeria have been made.

Based on data on historical flooding in Nigeria from 1985 till 2014, the present study reveals that flooding is a common phenomenon for various states in Nigeria. Although Lagos state appears to have experienced most of the floods in the country, more frequent flooding are recorded in Niger, Adamawa, Kano, Oyo and Jigawa states. Whilst Lagos state flooding can be attributed to urban growth and coastal influence among other key factors, the influence of rivers such as Niger, Benue, Ogun and Hadeja may account for the rest of the states with more frequent floods (**FIGURES 3 AND 4**).

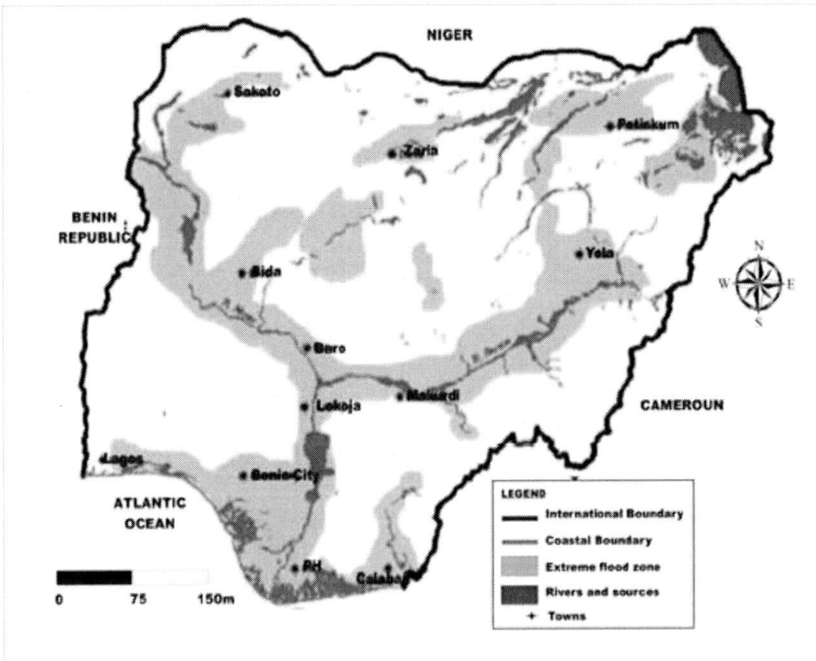

Figure 4: Spatial distribution of areas affected by extreme floods in Nigeria between 2000 and 2012. Source: Federal Ministry of Environment [92].

To understand where Nigeria stands in global and regional perspective since 1985, the 2012 flooding in Nigeria reputed as the worst in 40 years is compared with the most devastating floods in the world between 1985 and 2014. The result of this comparison shows that within the period under consideration, Nigeria may be ranked as third in the world, following peoples Republic of China and Soviet Union (**TABLE 2**). In Africa, the study shows that the country ranks topmost (**TABLE 3**), overtaking Mozambique and Algeria in terms of economic loss. This reality highlights the need for more proactive efforts towards tackling flooding in the country [92].

Table 2: 20 most devastating floods in the world between 1985 and 2014. Source: Complied by authors

S.No.	Country	Year	Mortality	Economic loss (billion US$)	Affected Human Population (millions)
1.	Soviet Union	1991	N/A	60	0.04
2.	China	1998	3656	20	240
3.	Nigeria	2012	363	16.9	7.7
4.	Korea (P Dem Rep)	1998	405	15	5.2
5.	United States	1993	50	15	N/A
6.	Russia	2002	167	0.9	0.3
7.	Italy	1994	N/A	6	N/A
8.	Poland	1997	N/A	4.3	0.24
9	United Kingdom	2007	30000	3.2	N/A
10.	Ecuador	2008	N/A	1	0.289
11.	Bangladesh	1998	3836	2	30
12.	Netherland	1998	N/A	5.3	0.02
13.	Argentina	1985	N/A	1.3	1
14.	Vietnam	1999	702	0.78	1.7
15.	Philippines	2012	1268	2.19	0.45
16.	Pakistan	1993	3084	1	6.184
17.	Afghanistan	1992	3000	N/A	N/A
18.	Colombia	2011	450	1.2	4.98
19.	Nepal	1993	1048	N/A	0.53
20.	Brazil	2008	246+	0.75	1.5

Apart from China which presently reputes as the most flood prone country in the world, characterized by recurrent perennial floods due to among other things, the influence of population growth and mainly the River Yangtze [93]. The fact that other countries with known extreme flooding experience (for examples: Netherlands, the US, Brazil, United Kingdom and many other European countries) may be ranked below Nigeria in terms of devastating floods within the period considered suggest among other things the significance of flood risk reduction measures which are presently in place in those countries.

Table 3: 10 most devastating floods in Africa since 1985.

S/No	Country	Year	Mortality	Economic loss (billion US$)	Affected Human Population (millions)
1.	Nigeria	2012	363	16.9	7.7
2.	Algeria	2001	921	0.3	0.045
3.	Mozambique	2000	800	0.419	4.5
4.	Ethiopia	2006	705	0.32	N/A
5.	South Africa	1987	506	0.765	0.065
6.	Burkina-Faso	2009	N/A	0.15	0.15
7.	Malawi	1991	472	0.024	0.15
8.	Kenya	2006	114	N/A	0.723
9	Egypt	1994	600	0.14	0.16
10.	Sudan	2007	N/A	0.3	0.56

The Netherlands with more than half of the country at or below sea level experienced a severe flood in 1953 that devastated majority of the nation's economic and human infrastructure. The estimated impact of the flood was 1835 deaths and 1 billion Dutch guilders (US$ 558 million). That flood challenged various stake holders, particularly the general public and local communities and Dutch government towards more effective strategies of mitigating the threats of flooding.

The reduced impacts of flooding in the country in recent times validates these efforts. Similarly, recent floods in the US have lesser impacts compared to the flood of 1972 which caused 238 deaths, 357 injuries and destroyed about 1335 homes with estimated fiscal loss of over 800 million US$.

In the UK, although there have been devastating recent floods, the 1947 floods were considered the worst in recent history with overall impact estimated at about £4.5 million (USD$ 6.81 million) at current value, with millions of displaced human populations, farm animals and agricultural products [94]. For Brazil, the floods of 2010 have reduced impacts compared to the floods of 1967 which was considered the deadliest in the country's history, claiming about 610 lives with economic loss amounting to about US$ 1.2.

3.1 Efforts towards tackling flooding in nigeria

The threats of flooding in Nigeria have been tackled through a number of specific and general actions including physical intervention, legislation and policy formulation, creation of awareness of flooding, urban renewal and development, engineering structures such as dams, bridges and drainage systems, relocation of human populations during flooding and assisting flood victims with basic humanitarian needs [40,78,95]. To simplify discussion regarding these efforts, the present study simplifies and groups these actions into institutional approach, actions by local communities and the general public, action by humanitarian agencies and actions by research institutions and the media.

3.2 Institutional approach

Institutional approach towards addressing the threats of flooding in Nigeria dates back to the early 1960's with the establishment of federal and state ministries of works [96]. However, the increasing frequency and severity of floods across the country prompted the establishing of the Federal Environmental Protection Agency (FEPA) as a unit in the Federal Ministry of Works and Housing in 1988 [26] and the Federal Ministry of Environment (FME) in 1999 [92]. Among other things, the key roles of FME towards flooding risk reduction in Nigeria is to assess the flooding potentials as well as design, determine, develop and/ or authorize the development of appropriate flood reduction measures for the country [92].

With the FME comes various ministries and agencies for tackling flooding in Nigeria which include: Federal Emergency Management Agency (FEMA), National Emergency Management Agency (NEMA), State Emergency Management Agency (SEMA), Local Emergency Management Agency (LEMA), National Orientation Agency (NOA), National Commission for Refugees (NCR), National Environmental Standards and Regulations Enforcement Agency (NESREA) which by 2009 Nigerian Acts supersedes the FEPA, Nigerian Meteorological Agency (NIMET), Nigeria Hydrological Services Agency (NIHSA), NEST (Nigeria Environmental Study/Action Team) and Building Nigeria's Response to Climate Change (BNRCC).

NEMA is basically a coordinating body for disaster management in Nigeria. Actions towards addressing the threats of flooding which the agency coordinates include but not limited to policy formulation, leasing with and assessing the state of preparedness of all other relevant agencies, data collation from relevant agencies, education of the general public on flooding and interaction with SEMA towards the distribution of relief materials to disaster victims within states and local government areas. Recently, a memorandum of understanding was signed with NESREA and NOA to intensify efforts towards flood risk management in Nigeria [13].

The protection and development of the environment, a measure that helps keep check flooding, is the key role of NESREA, the agency which enforces all environmental laws, guidelines, policies, standards and regulations in Nigeria, as well as enforcing compliance with provisions of international agreements, protocols, conventions and treaties on the environment to which Nigeria is a signatory

Whilst the provision of humanitarian needs such as shelter, clothing and floods for internally displaced persons is anchored by NCR, the importance of local communities being aware of flooding and actively participating in discussions and decisions which might increase their resilience and adaptability to the hazard highlights in the roles of NOA, which re-orientates and keeps Nigerians informed about ways of taking part in issue that affects them.

The poor perception of flooding in the country should be the concern of this agency.

NIHSA provides reliable and high quality hydrological and hydrogeological data on a continuous basis for the for the purposes of assessing the status and trends of the nation's water resources including its location in time and space, extent, dependability, quality and the possibilities of its utilization and control. Since 2013, the agency has been creating awareness of flooding through the "flood outlook" initiative. Other activities of NIHSA include; provision professional advice to various levels of government in Nigeria on all aspects of hydrology, collaborates with NIMET to issue flood forecast and contributes towards creating awareness of flooding among local communities.

NIMET furnishes the country with weather report, and other meteorological information, issues alerts and early warning and forecast on impending flood disasters within the country.

Issues relating to flood insurance are coordinated by FEMA, the agency which makes federally funded insurance protection policy available for property owners in Nigeria. Policies relating to assisting flood victim at state and local government levels are coordinated by SEMA and LEMA.

As climate change is complicit with other factors that influence flooding in Nigeria, BNRCC's key role is to collaborate with other agencies to promote the capacities of the generality of human populations within the country to cope with all effects of climate change. NEST undertakes continuous research required for enhancing decisions and robust measures towards addressing flooding in Nigeria [97,98].

3.3 Action by local communities and the general public

The peculiar attitude of Nigerians towards issues they probably have no solution to is to find a possible natural means to adapt. Although such attitude has cost many lives and properties, however, it is arguably a significant potential for Nigerians and has severally favored them in emergency situations. Families in Nigeria co-habit and this offers a comparative advantage in the event of flood disasters. In many flooding incidences in Nigerian cities, the general public has often offered assistance to victims, assisted in evacuation of those displaced and in protecting property from further damage. Many IDPs easily find shelter and other humanitarian needs from families and friends while awaiting intervention by authorities. Although anti-social behavior, such as looting and sexual harassment of some of the internally displaced victims often arise, however the civilized attitudes of the general public, which may be comparable to those in the developed world such as the US following the hurricane Katrina of 2005 [99] has been commendable.

3.4 Action by humanitarian agencies

Humanitarian response to flooding in Nigeria dates back to the early 1980s following the Ogunpa, Sokoto-Rima, Banguada flood disasters [100]. Almost in all cases of flooding in Nigeria victims have received humanitarian supports with most notably the International Federation of Red Cross (IFRC), United Nations, World Bank, Foreign countries including UK, the United States, China, Japan, France as well as religious organizations including the Catholic, Anglican and Pentecostal churches and missionary societies. Although this study has not provided an empirical evidence of supports received so far in view of flooding in Nigeria, recent records by OCHA [60] indicate that such supports have been considerable. The 2012 flooding saw humanitarian response amounting to over US$70 million [60]. These supports which seem to indicate solidarity for Nigeria as a whole and the people affected by flooding within the country in particular have been evidently considerable.

3.5 Action by research and the media

Arguably, much of what is known about flooding in Nigeria can be attributed to efforts by research and media. In the literature, undoubtedly, considerable attention has been given to flooding in Nigeria. Presently there are more than five hundred publications indexed in the that relates to flooding and means of tackling it in Nigeria. In addition, the country has a number of countrywide research-based groups such as the NEST (Nigeria

Environmental Study/Action Team), BNRCC (Building Nigeria's Response to Climate Change) and university-based research groups that focus on flooding within Nigeria. Similarly, the Nigerian media has been given the credit of providing qualitative information regarding the widespread flooding in Nigeria [59].

3.6 Gaps in Flood Risk Reduction in Nigeria

The fundamental idea of flood risk reduction is to build the resilience of human populations to flooding [44]. In the US, UK and Netherland, it can be argued that this idea is underpinned by much profound measures for tackling flooding. For Nigeria, there are potential efforts for tackling flooding but in the light of best practices in flood risk reduction, there are gaps and limitations which the present study identifies and discusses as follows:

3.7 Poor attention to flood modelling and assessment of vulnerability to flooding

The crucial gap in flood risk reduction in Nigeria is poor attention to flood modelling and assessment of vulnerability to flooding. The present study has shown that flood risk reduction under the idea of living with floods integrates structural and nonstructural approaches which are underpinned by flood modelling. Flood data is fundamental to tackling flooding in Nigeria, but the dearth of such data is overwhelming. Similarly, assessment of vulnerability to flooding aides the understanding of variations in human sensitivities, exposures and lack of capacities to cope with floods. These approaches are poorly considered in efforts at tackling flooding in Nigerian.

3.8 Weak institutional framework

Despite the seeming organized nature of Nigeria's institutional approach to flood risk reduction, the approach has been acutely flawed [12,24,97]. Although key roles within the approach has consisted of facilitating the evacuation of victims affected by floods and providing them with urgent humanitarian needs, the level of dissatisfaction and agitations from large numbers of the flood victims, especially the IDPs, queries the effectiveness of institutional approach in Nigeria. Although it is unjustifiable to claim that the weakness of this institutional approach in Nigeria probably leads to more frequent flooding in the country, however, it can be argued that whilst that such measures have not improved the country with regards to the idea of "living with floods" is clearly acknowledged [27,35,98].

Arguably, institutional framework with a complex chain of action which characterizes Nigeria is not ideal for a country with urgent needs to address the threats of flooding. The widespread flooding in Nigeria requires less complex framework with reduced chain of action and increased authority and responsibilities towards flooding and ways of managing its threats, as is the case in the United States, UK and Netherlands where a centralized authority such as FEMA, Environment Agency (EA) and *Rijkswaterstaat* respectively operates within the framework of institutional approach with more abundant resources towards addressing the challenges of flood risk.

3.9 Poor human responsibility

The vulnerabilities of local communities to flooding in Nigeria may indicate among other factors an overwhelming level of irresponsibility towards flooding and ways of addressing its challenges. Although a number of actions by the people towards flooding is commendable, some actions which characterize a cross section of Nigerian such as the failure to comply with environmental laws and regulations and to adhere to weather warnings and alerts are possible situations where lack of responsibilities of the general public is highlighted [53]. Poor perception of flooding among local communities is a major issue which underscores the current activities of NOA within Nigerian institutional framework [18,62]. The indifference towards research questionnaires and surveys most likely constrain research effort at tackling flooding in Nigeria. In Nigeria there seems to be a gap between efforts by the government and activities by the people such that it appears difficult to know the peoples' responsibilities towards addressing the challenges of flooding. The government claims the people are sabotaging its efforts. Whilst this claim is hard to substantiate, the attitude of a considerable population of Nigeria towards environmental management, regulations and rules in the country shows irresponsibility towards the threats of flooding.

3.10 Lack of detailed plan

We argue that the lack of detailed plan and strategy for disbursing funds and inaccurate information relating to those who have been affected by flooding most probably undermine humanitarian support in Nigeria. The disbursement of these funds and relief materials, although unjustifiable to argue that it highlights the level of corruption and financial mismanagement in Nigeria has been argued to indicate the weakness of institutional framework within the country [12]. Humanitarian actions in Nigeria are generally for post-disaster and emergency situations suggesting some limitation based on what can be achieved through financial support. Given that most local communities in Nigeria consist of poor human populations, it is arguably a topbottom approach to concentrate humanitarian supports in the event of crises rather than prioritizing improving the living condition of these people which will not only ultimately reduce their chances of being vulnerable to flooding and assist in minimizing financial mismanagement, but also it will boost the credence of humanitarian supports towards natural disasters in general and flooding in particular in Nigeria.

3.11 Insufficient research and skewed focus of local education and publications

Considering what has been done globally in research (mostly in China, India, UK, the US, Bangladesh and Netherlands) the specific situation of Nigeria regarding research in terms of flooding is arguably weak. The study of science and technology in Nigeria are yet to embrace environmental education and this has been identified [58]. Whilst flooding and climate change are passive subjects in education curriculum of studies in Nigerian schools, current key issues in flood research such as flood modeling, uncertainty analyses, early warning systems and flood forecasting, vulnerability assessment and climate change models are lacking in the country. From the literature search conducted in this study, it is clear that perception of flooding in Nigeria has only received little attention, and this seems to undermine possible efforts as assessing the level of perception and public awareness of flooding in Nigeria. A number of researches relating to flooding in Nigeria seem to recycle issues that are well known such as causes and impacts of flooding. More scientific investigations are few probably due to lack of funds and the indifference of political leaders.

3.12 Poor media reporting

The inconsistency in media reporting in Nigeria which was identified in [90] highlights some disconnect between the media and agencies tackling flooding in the country particularly the NEMA. The standard of media reporting in Nigeria following recent flood incidences has been very poor. This is compounded by the irregular power supply in the country which overwhelms the chances to take advantage of information gadgets such as radio and TV. A large number of the population seems to rely on Facebook pages, which also has its own challenges based on internet accessibility. This situation seems to have caused a general lack of confidence in the media generally such that a growing parlance among researchers in recent times regarding flooding is "of what relevance is what the media have to say?" Whilst this situation undermines the professionalism in Nigerian media industries in particular, a key question that lingers with regards to widespread flooding in the country is "how would the media industry contribute to addressing the challenges of flooding Nigeria?"

It is possible political factors and market competition to a large extent influence media reporting, but it also appears the importance of piloting the perception and awareness of flooding in the general public and local communities as well as sensitizing them on issues relating to flood alerts and early warning has been compromised.

4. RECOMMENDATIONS

Based on the findings of this study, lessons learned from other countries' experiences of flooding and "best practices" in flood risk reduction [100,101], the authors argue that the following key issues are fundamental to the success of efforts towards addressing the challenges of flooding in Nigeria. Firstly, the understanding and demonstration of the roles more scientific approaches such as flood modeling, can play in flood risk reduction within the context of Nigeria. Secondly, the need to align the focus of flood risk reduction in Nigeria to the objectives of such task in more developed countries such as the US, UK and the Netherlands, and thirdly, the need to promote awareness of flooding among local communities and the general public and to delineate more suitable locations for relocation of human populations during flooding events. For this reasons, the study makes the following recommendations:

1) Following the European Union framework, which requires all constituting States to prepare flood hazard/risk maps [102], a strong legislation that requires each state of Nigeria to produce a flood hazard/ risk map, is recommended for Nigeria. This will to a large extent strengthen existing institutional framework and stimulate increased responsibility towards flood risk reduction among the states in the country.

2) The Environment Agency in United Kingdom is adequately empowered to coordinate, among other things, research which drives current approaches of addressing the challenges of flooding [83]. Thus, it is recommended that Nigeria's NEST be reviewed in terms of its research activities, how relevant they have been and how they can be further strengthened to promote more scientific researches towards ways of addressing the challenges of floods and other environmental challenges in Nigeria.

3) The importance of a centralized institution having all the powers and resources to manage the challenges of flooding highlights the success of flood risk reduction in the Netherlands, the US and UK [103]. The roles

of NEMA in Nigeria are acknowledged, but improvement is required. Lack of adequate resources especially accurate data clearly undermines the effectiveness of the agency in disaster management. Thus, we recommend improvement in resources allocation to NEMA, but also acquisition of data should be the agencies top priority.

4) Hurricane Katrina, among other major disasters in the US showed the unique roles of security personnel in managing emergencies and securing victims of disasters. In Nigeria, the victims of flooding live in apprehension of theft and sexual abuse. These dimensions which if addressed can restrict the impacts of flooding on human populations, are often overlooked by security personnel in the course of duty. For this reason, we recommend training for rescue and evacuation operations during for the civil defense corps and other military and paramilitary personnel, including the police, immigration and road safety commission.

5) Flood risk reduction under the "living with floods" idea is multi-disciplinary indicating that various industries can assist in reducing the impacts of flooding. In UK, evidences of collaboration from various companies and institutions towards addressing flood challenges are undisputable [83,101,103]. Thus, the need for multinationals and banking industries in Nigeria to sponsor research and promote sustainable development within Nigerian cities, as well as augment humanitarian supports to improve the living standards of local communities whilst reducing their vulnerabilities and building their resilience is should not be ignored.

6) Flood insurance is a non-structural approach which many property owners have benefitted from in developed countries following flood disasters. To support the roles of flood insurance in Nigeria, it is recommended that the role of FEMA in this regard should be extended to states and whilst encouraging insurance companies to commence sensitization exercises for properties owners to take positive step in this direction.

7) Enforcement of environmental standard and laws is often a key factor towards containing adverse effects of climate change including flooding. Indiscriminate waste disposal, production of Carbon monoxide, construction along flood plain, felling of trees and indiscriminate car parking, among other anthropogenic activities which influence flooding in Nigeria are illegal. In view of addressing these matters, NESREA should embark on arrest, prosecution and proportionately fine Nigerian who violate these rules.

8) The flooding of 1953 is historic in the Netherlands, which with its half at or below sea level is reputed as the top in the world in terms of flood

management [81]. Such activities in the Netherlands described as imperfect but pragmatic is built on a strong commitment to resist any attempt of a repeat of history. The people are committed, and so is the government implying that collective efforts underlie success towards addressing the challenges of flooding. The old English adage "*God created the world, but the Dutch created Netherlands*" is often used to highlight the commitment and responsibilities of people in the Netherlands towards flooding and its challenges. It is claimed that flood defense in the Netherlands cost each person a few hundred Euros each year and the people rarely flinched at the responsibility [81]. The high level of adherence to regulations and rules shown by British citizens is highlighted in the conservation of nature and high environmental standards within the country. Such positive attitude is also exhibited towards weather reports, disaster warning and alerts informing a significant level preparedness which appears to influence reduced damage following flooding event. Against this background, Nigerians need to known that the role of protecting the country is a collective one. Thus a change of attitude towards flooding its management in which the people should participate in matters relating to flooding which most largely affects their lives. This can be done by asking relevant questions, seeking to know and willing to adapt to individual actions which can potentially influence flood risk reduction within the country. Individuals in politics should ensure that laws which underlie the enforcement of environmental standards and regulations are made. Equally, the general public and local communities in Nigeria should support research through positive and accurate responses to questionnaire and surveys.

9) Arguably, it appears research globally is proportional to success towards tackling flooding. From a routine search, literature relating to flooding in Nigeria appears insignificant compared to those of the US, Netherlands, UK, etc. This is a strong pointer and indicator that more research is required for the country. Equally, the universities and research agencies should be empowered to include in their curriculum studies and programmes tailored towards improving flood awareness and management. Flood alert and flood early warning systems should be introduced and utilized at all levels of flood disaster management. More research should be directed towards developing bespoke hydrologic and hydraulic flood models for simulating flood hazard and other hydrological parameters in Nigeria.

The government should play more active parts in this regards. Research is capital intensive and many Universities and research agencies are constrained by this factor. Indeed, many universities have benefitted from the Tertiary Education Intervention (TETFUND). However, due to limited funds available, accessibility by a number of academics has been poor. Thus we recommend improvement on funding and acquisition of high accuracy input data like LiDAR (Light Detection and Ranging) and SAR (Synthetic Aperture Radar) for flood modeling within the country.

4. CONCLUSION

Flooding has been devastating for Nigeria with large scale threats on people, critical infrastructure and economic activities. The hazard which has been generally linked to climate change and poor urban planning has received the attention of government at all levels, local communities, humanitarian agencies and research. However, efforts so far at tackling the hazard seem limited mostly due to lack of data relating to flooding and other factors which are yet to be identified. Although flood hazard is widespread in the country, its knowledge in the wider population is poor.

To attempt towards solving the challenges of flooding in Nigeria, the present study reviews flooding and ongoing efforts at tackling the hazard in the country over a period 1985 till 2014, the present study reveals considerable impacts of flooding, arguing that such impacts have not been accompanied by corresponding efforts to address them. A number of limitations and gaps which needs to be addressed were identified with those ongoing efforts and they include poor attention to flood modelling and assessment of vulnerability to flooding, weak institutional framework, poor human responsibility, etc.

It is argued that flood modeling and vulnerability assessment, have both been relevant to all aspects of flood risk reduction [39,104], as they provide information about flooding and also serves as the bases for evaluating the susceptibilities of social systems to the hazard. Efforts towards addressing the challenges of flooding in most places for example the US and Europe include these more robust operations to achieve the three-tier risk management cycle, which includes mitigation/ prevention, preparedness, response and recovery. Unfortunately, the implementation of these operations in Nigeria has been inadequate.

Across the world, there are many predictions of the worsened threat of flooding in the future, which in turn is stimulating research into finding solutions. There seems to be some increased interest in preparing for future uncertainties with regards to exposure and vulnerabilities of social systems flood risks. For example in Europe, the EU adopted a directive in 2007 which requires all member States to carry out flood hazard and risk mapping in their territory, in order to support the transition from traditional flood defense strategies to a flood risk management approach at the basin scale [76]. In the US, despite the high level of flood events, based on available records, it can be argued that the country seems to have contained the degree of impacts on social systems at least for the present [4]. This may be as a result of various on-going initiatives especially in flood modeling (for example: HEC-1, HEC-RAS, and HEC-HMS). Given the proliferation of models existing at present for simulating flood hazard along with recent development in remote sensing and computer technology, it is now possible to undertake flood forecasting and flood early warning systems in many places [105].

However, looking at these developments vis-à-vis Nigeria, one wonders and asks "how prepared are the government and the people in this present and future fight against flooding?" If perhaps the situation with Nigeria is that of unpreparedness and limitations with regards to addressing the threats of flooding, them much apprehension should abound bearing in mind that the country has one of the fastest growing urban areas and population density in the world [9], and knowing that the risks of flooding will capitalize on those factors leaves more to be desired. It is not productive to attribute blame as to why flooding occurs, yet responsibility is required to mitigate the extent of damage and disruptions to economic activities. It is in view of this responsibility, which is often defined by awareness, and positive action that the present study has made a number of recommendations hoping that Nigeria will join her counterparts in various parts of the world to become more proactive in finding lasting solutions to the threats of flooding.

ACKNOWLEDGEMENTS

This paper results from a PhD research funded by the UNN/TETFUND. Some grant was equally provided by the Surveyors Council of Nigeria (SURCON). Most of the analyzed flood data were sourced from the center for research in epidemiology of disasters (CRED/EM-DAT), Nigeria's ministry of environment and Nigeria Environmental Study Action/Team (NEST). A number of previous studies which were consulted and the anonymous reviewers of this paper are acknowledged.

REFERENCES

1. Peduzzi P, Dao H, Herold C, Mouton F (2009) Assessing global exposure and vulnerability towards natural hazards: the disaster risk index, Natural. Hazards Earth System. Science 9: 1149-1159. https://www.art.man.ac.uk/planning/cure
2. Raaijmakers R, Krywkow J, van der Veen A (2008) Flood risk perceptions and spatial multi-criteria analysis: an exploratory research for hazard mitigation. Natural hazards46: 307-322.

3. Guha-Sapir D, HoyoisPh, Below R (2013) Annual Disaster Statistical Review 2012: The Numbers and Trends. Brussels: CRED. https://www.emdat.be/
4. Ahern M, Kovats RS, Wilkinson P, FewR, Matthies F (2005) Global health impacts of floods: epidemiologic evidence. Epidemiologic reviews27: 36-46.
5. Tapsell SM, Tunstall SM (2008) "I wish I'd never heard of Banbury": The relationship between 'place' and the health impacts from flooding. Health & place14: 133-154.
6. Hunter PR (2003) Climate change and waterborne and vector-borne disease. Journal of Applied Microbiology94: 37-46.
7. Tapsell SM, Penning-Rowsell EC, Tunstall SM, Wilson TL (2002) Vulnerability to flooding: health and social dimensions. Philosophical Transactions of the Royal Society of London. Series A: Mathematical. Physical and Engineering Sciences360: 1511-1525.
8. Tunstall S, Tapsell S, Green C, FloydP, George C (2006) The health effects of flooding: social research results from England and Wales. J Water Health4: 365-380.
9. ActionAid (2006) Climate change, urban flooding and the rights of the urban poor in Africa: Key findings from six African cities. London: Action Aid International.
10. Obeta CM (2014) Institutional Approach to Flood Disaster Management in Nigeria: Need for a Preparedness Plan. British Journal of Applied Science & Technology 4: 4575-4590.
11. NEMA (Nigeerina Emergency Manageent Agency) (2013) Report on flood disasters in Nigeria. Abuja: Government Press.
12. Adeloye AJ, Rustum R (2011) Lagos (Nigeria) flooding and influence of urban planning. Urban Design and Planning 164 (DP3): 175-187.
13. Cline WR (2007) Global Warming and Agriculture. Centre for Global Development, Peterson Institute for International Economics, Washington, DC, USA.
14. www.infoo@climatechange.org
15. Ogunbodede EF, Sunmola RA (2014) Flooding and traffic management in Akure (Nigeria) metropolitan environment. International Journal of Innovation and Scientific Research7: 121-130.
16. Ologunorisa TE, Adeyemo A (2005) Public perception of flood hazard in the Niger Delta, Nigeria. Environmentalist25: 39-45.
17. Fadairo G, Ganiyu SA (2010) Effects of flooding on the built environment in Akure, Nigeria. In West Africa built environment research (water) conference. pp. 281.
18. Bashir O, Oludare H, Johnson O, Aloysius B (2012) Floods of fury in Nigerian cities. Journal of Sustainable Development 5: 69-79.
19. Douglas I, AlamK, Maghenda M, Mcdonnell Y, McLean L, et al. (2008) Unjust waters: climate change, flooding and the urban poor in Africa. Environment and Urbanization20: 187-205.
20. Akintola EO (1994) Flooding phenomenon. Ibadan Region. Ibadan: Rex Charles Publication, pp.244-255.
21. Iloje NP (2004) A New Geography of Nigeria. Lagos, Longman publishers.
22. Agbola BS, Ajayi O. Taiwo OJ, Wahab BW (2012) The August 2011 flood in Ibadan, Nigeria: anthropogenic causes and consequences. International Journal of Disaster Risk Science 3: 207-217.
23. Adebayo AA, Oruonye ED (2013) An Assessment of the effects of the 2012 Floods in Taraba State, Nigeria. Paper delivered at the Annual National Conference, organized by the Association Hydrological Science at University of Agriculture, Abeokuta, Ogun, state. Nigeria 13-18.
24. Obeta MC (2009) Extreme river flood events in Nigeria: A geographical perspective of Nigerian. Journal of Geography and the Environment 1: 170-179.
25. Adelekan O (2010) Vulnerability of poor urban coastal communities to flooding in Lagos, Nigeria. Environment and Urbanization 22: 433-450.
26. Nicholls R, Hanson JS, Herweijer C, Patmore N, Hallegatte S, et al. (2008) Ranking of Port cities with high exposure and vulnerabilities to climate extremes. OECD Environment Working Papers, No.1. University of Southampton, United Kingdom: OECD Publishers.
27. Houston D, Werritty A, Bassett D, Geddes A, Hoolachan A, et al. (2011) Pluvial (rain-related) flooding in urban areas: the invincible hazard. York, UK: Joseph Rowntree Foundation.

28. Merz B, Thieken AH, Gocht M (2007) Flood risk mapping at the local scale: concepts and challenges. In Flood risk management in Europe. Springer Netherlands.
29. Ujoh MF, Isa D, Olarewaju O (2009) Understanding urban sprawl in the Federal Capital City, Abuja: Towards sustainable urbanization in Nigeria. Journal of Geography and Regional Planning 3: 106-113.
30. Mujumdar PP (2001) Flood wave propagation: The St. Venant equations. Resonance 6: 66-73.
31. Shuster WD, Bonta J, Thurston H, Warnemuende E, Smith DR (2005) Impacts of impervious surface on watershed hydrology: a review. Urban Water Journal 2: 263-275.
32. Kalnay E, Cai M (2003) Impact of urbanization and land-use change on climate. Nature 423: 528-531.
33. Seto KC, Shepherd JM (2009) Global urban land-use trends and climate impacts. Current Opinion in Environmental Sustainability 1: 89-95.
34. UNSTTPUNDA (United Nations System Task Team on the Post-2015 UN Development Agenda). (2012) Disaster and Resilience. United Nations. New York.
35. UN (United Nations) (2007) World Population Prospects: the 2006 revision - executive summary. Department of Economic and Social Affairs, population division, United Nations. New York.
36. Huong HTL, Pathirana A (2011) Urbanization and climate change impacts on future urban flood risk in Can Tho city, Vietnam. Hydrology and Earth System Sciences Discussions 8: 10781-10824.
37. Li Y, Wang C (2009) Impacts of urbanization on surface runoff of the dardenne creek watershed, St. Charles county, Missouri. Physical Geography 30: 556-573.
38. Olorunfemi FB (2011) Managing flood disasters under a changing climate: lessons from Nigeria and South Africa.NISER Research Seminar Series, NISER, Ibadan.
39. BalbiS, Giupponi C, Gain A, Mojtahed V, Gallina V, et al. (2012)A Conceptual Framework for Comprehensive Assessment of Risk Prevention Measures: The KULTURisk Framework (KR-FWK). Available at SSRN 2184193.
40. Di Baldassarre G, Uhlenbrook S (2012) Is the current flood of data enough? Atreatise on research needs for the improvement of flood modelling. Hydrological Processes 26: 153-158.
41. DEFRA (Department for Environment Food and Rural Affairs) (2013). Desktop review of 2D hydraulic modelling packages. Environmental Agency. Bristol. https://www.unisdr.org/files/657_lwr1.pdf
42. EC (European Commission) (2004) Flood risk management - Flood prevention, protection and mitigation. Communication from the Commission to the Council, the European Parliament, the European Economic and Social Committee and the Committee of the Regions.
43. ZhuD, Zhou N, Jiang S (2011) Research overview of runoff model for urban rainwater. Journal of Water Resources and Water Engineering 22: 132-137 (in Chinese).
44. https://scholarship.kentlaw.iit.edu/cgi/viewcontent.cgi?article=1772&context=fac_schol
45. Ologunorisa TE (2004) An assessment of flood vulnerability zones in the Niger delta,Nigeria. International Journal of Environmental Studies 61: 31-38.
46. Ojigi ML, Abdulkadir FI, Aderoju MO (2013) Geospatial Mapping and Analysis of the 2012 Flood Disaster in Central Parts of Nigeria. 8th National GIS Symposium. Dammam. Saudi Arabia 1067-1077.
47. Aderogba KC, Martins OM, Oderinde S, Afelumo T (2012) Challenges of Poor Drainage Systems and Floods in Lagos Metropolis, Nigeria. International Journal of Social Science and Education 2: 412-427.
48. Ogwuche JA, Abah IA (2014) Assessment of flood disaster vulnerability for flood insurance programme in part of Makurdi floodplain, Benue state, Nigeria. Donnish Journal of Ecology and the Natural Environment 1: 1-5.
49. Nwilo PC, Olayinka DN, Adzandeh AE (2012) Flood modelling and vulnerability assessment of settlements in the Adamawa state floodplain using GIS and cellular framework. Global Journal of Human Social Science 12: 10 20.

50. Aderogba KA (2012a) Qualitative studies of recent flood and sustainable growth and development of cities and towns in Nigeria. International Journal of academic research in Economics and management science 1: 11-25.
51. Adeoye NO, Ayanlade A, Babatimehin O (2009) Climate change and menace of floods in Nigerian cities: Socio-economic implications. Advances in Natural and Applied Sciences 3: 369-377.
52. Ali D, Hamidu S (2014) Environmental hazard: climate change and flooding, the impact on the built environment in Nigeria. Journal of Environmental Sciences and Resources Management 6: 136-144.
53. Agbonkhese O, Agbonkhese EG, Aka EO, Joe-Abaya J, Ocholi M, et al. (2014) Flood Menace in Nigeria: impacts, remedial and management strategies. Civil and Environmental Research 6: 32-40.
54. Adedeji OH, Odufuwa BO, Adebayo OH (2012) Building capabilities for flood disaster and hazard preparedness and risk reduction in Nigeria: need for spatial planning and land management. Journal of Sustainable Development in Africa 14: 45-58.
55. Terungwa UC, Torkwase IC (2013) Current issues in flood disaster: challenges and implications for science and technology to enhance environmental education. Academic Journal of Interdisciplinary Studies 2: 61-65.
56. Aderogba KA (2012b) Global warming and challenges of floods in Lagos metropolis, Nigeria. Academic Research International 2: 448-468.
57. https://www.unocha.org/
58. Adeaga O (2008) Flood Hazard Mapping and Risk Management in Parts of Lagos N. E.
59. Ajibade I, McBean G, Bezner-Kerr R (2013) Urban flooding in Lagos, Nigeria: Patterns of vulnerability and resilience among women. Global Environmental Change 23: 1714-1725.
60. Pinho J, Ferreira R, Vieira L, Schwanenberg D (2014) Comparison between two hydrodynamic models for flooding simulations at river Lima basin. Water Resources Management 1-14.
61. Thompson JR, Hollis GE (1995) Hydrological modelling and the sustainable development of the Hadejia-Nguru wetlands, Nigeria. Hydrological Sciences Journal40: 97-116.
62. Adewale PO, Sangodoyin AY, Adamowski J (2010) Flood routing in the Ogunpa River in Nigeria using HEC-RAS. Journal of Environmental Hydrology 18.
63. Onwuteaka J (2014) GIS Modeling of Flooding Exposure in Nigerian Coastal Areas from Sea Level Rise. Journal of Environment and Earth Science4: 81-94.
64. Nwilo PC, Olayinka DN, Adzandeh AE (2012) Flood Modelling and Vulnerability Assessment of Settlements in the Adamawa State Floodplain Using GIS and Cellular Framework Approach. Global Journal of Human Social Science.
65. van de Sande B, Lansen J, Hoyng C (2012) Sensitivity of Coastal Flood Risk Assessments to Digital Elevation Models. Water4: 568-579.
66. Nkwunonwo UC, Whitworth M, Baily B, Inkpen R (2014) The development of a simplified model for urban flood risk mitigation in developing countries. In Vulnerability, Uncertainty, and Risk@ Quantification, Mitigation, and Management. ASCE.
67. Wolski P, Savenije HHG, Murray-Hudson M, Gumbricht T (2006) Modelling of the flooding in the Okavango Delta, Botswana, using a hybrid reservoir-GIS model. Journal of Hydrology331: 58-72.
68. Garba H, Ismail A, Ibrahim A, Ahmed S, Faustinus B (2013) Hydrological Modeling of the impact of climate change on a tropical perennial river flooding. Research Inventy: International Journal of Engineering and Science. 3: 30-35.
69. Adegbola AA, Jolayemi JK (2012) Historical rainfall-runoff modeling of river Ogunpa, Ibadan, Nigeria. Indian Journal of Science and Technology5: 2725-2728.
70. Sayers P, Li Y, Galloway G, Penning-Rowsell E, Shen F, et al. (2013) Flood risk management: a strategic approach. Paris, UNESCO.
71. Sayers PB, Hall JW, Meadowcroft IC (2002) Towards risk-based flood hazard management in the UK. Proceedings of ICE Civil Engineering 150: 36-42.
72. Smith K (2013) Environmental hazards: assessing risk and reducing disaster sixth edition. Routledge. London.

73. deMoel H, van Alphen J, Aerts JC (2009) Flood maps in Europe – methods, availability and use. Natural Hazards and Earth System Sciences 9: 289-301.
74. de Almeida GA, Bates P, Freer JE, Souvignet M (2012) Improving the stability of a simple formulation of the shallow water equations for 2-D flood modeling. Water Resources Research 48: 1-14.
75. https://www.nihydro.gov.ng/wp-content/uploads/2012/08/AMENDED-REPORT-OF-The-Nigerian-Hydrological-Services-Agency.pdf
76. World Bank (2013) The world Bank: Working for a world free of poverty, Population (Total). Washington, DC: World Bank Group.
77. Jonkman SN, Bockarjova M, Kok M, Bernardini P (2008) Integrated hydrodynamic and economic modelling of flood damage in the Netherlands. Ecological Economics66: 77-90.
78. Vis M, Klijn F, De Bruijn KM, Van Buuren M (2003) Resilience strategies for flood risk management in the Netherlands. International journal of river basin management1: 33-40.
79. Neal J, Schumann G, Fewtrell T, Budimir M, Bates P, et al. (2011) Evaluating a new LISFLOOD-FP formulation with data from the summer 2007 floods in Tewkesbury, UK. Journal of Flood Risk Management4: 88-95
80. EA (Environment Agency) (2010) Working with natural processes to manage flood and coastal erosion risk. London, Environment Agency.
81. Renyi L, Nan L (2002) Flood area and damage estimation in Zhejiang, China. Journal of environmental management66: 1-8.
82. National Population Commission (NPC) (2007) 2006 Population and Housing Census: National and State Population and Housing Tables: Priority Tables I-IV. FCT, Abuja: Federal Government of Nigeria.
83. United Nations (2004) World Population to 2300. New York: United Nations Department of Economic and Social Affairs, Population Division.
84. Clement AR (2013) An application of Geographic Information System in mapping flood risk zones in a north central city in Nigeria. African Journal of Environmental Science and Technology7: 365-371.
85. Uzochukwu CE, Ekwugha UP, Marion NE (2014) Media coverage of the environment in Nigeria: issues and prospects. International Journal of Social Sciences and Humanities Review4: 1-10.
86. Ochuko MO (2014) Solid waste management in Obantoko area of Abeokuta, Nigeria. Journal of Emerging Trends in Engineering and Applied Sciences5: 111-115.
87. https://www.e-ir.info/2013/02/04/the-inconsistency-of-the-flood-narrative-in-nigeria/
88. https://www.Figurenet/pub/fig2011/papers/ts06j/ts06j_etuonovbe_5002.pdf
89. FME (Federal Ministry of the Environment) (2012) Bulletin on Ecological Disasters, Abuja, Nigeria. FME.
90. Zhang Q, Liu C, Xu CY, Xu Y, Jiang T (2006) Observed trends of annual maximum water level and stream flow during past 130 years in the Yangtze River basin, China. Journal of Hydrology324: 255-265.
91. EA (Environment Agency). (1997) Tyne Estuary Project 1996. Report of the Tyne Estuary Group. Newcastle.
92. Odunuga S (2008) Urban land use change and the flooding in Ashimowu watershed, Lagos, Nigeria. University of Lagos, Nigeria.
93. Ibitoye M (2007) The need for planning of peri-urban growth in south western Nigeria: the surveyors' perspective. Symposium organized by the Nigerian Institution of Surveyors (NIS). University of Lagos, Lagos.
94. Kolawole OM, Olayemi AB, Ajayi KT (2011) Managing flood in Nigerian cities: Risk analysis and adaptation options–Ilorin city as a case study. Archives of Applied Science Research3: 17-24.
95. Akintola FO, Ikwuyatum GO (2012) Issues in sustainable flood management in Nigeria. In Ivbijaro FA, Akintola FO. Editors; Sustainable Environmental Management in Nigeria. Ibadan, Oxford University Publishers.

96. Wilson S, Temple B, Milliron M, Vazquez C, Packard M, et al. (2007)The lack of disaster preparedness by the public and its effect on communities. The Internet Journal of Rescue and Disaster Medicine. 7.
97. Anih SC (2004) Effective survival measures against natural hazards in settled area. In L. Mbah, U. Uchegbu, & L. Muoghalu, Management of environmental problems and hazards in Nigeria. England: Ashgate publishing limited.
98. Water UK (2008) Lesson Learned from Summer Floods 2007-Phase 1 report-Emergency Response.Review Group on Flooding.
99. EA (Environment Agency) (2003) Risk Assessment for Flood & Coastal Defence for Strategic Planning R&D Technical Report W5B-030/TR1, Authors Sayers PB, Hall JW, Meadowcroft, IC, Dawson, R, Deakin R, Rosu C.
100. Pitt M (2008) Lessons from the 2007 floods. London: Pitt Review.
101. Fewtrell L, Kay D (2008) An attempt to quantify the health impacts of flooding in the UK using an urban case study. Public Health122: 446-451.
102. Bates PD, Dawson RJ, Hall WJ, Horritt MS, Nichollos RJ, et al. (2005) Simple two-dimensional numerical modelling of coastal flooding and example applications. Coastal Engineering 52:793-810.

CHAPTER 3

Low-salinity flooding in a selected carbonate reservoir: experimental approach

Hazim H. Al-Attar, Mohamed Y. Mahmoud, Abdulrazag Y. Zekri, Reyadh Almehaideb, Mamdouh Ghannam

Chemical and Petroleum Engineering Department, United Arab Emirates, University Al Ain, United Arab Emirates

ABSTRACT

Low-salinity waterflooding has been used to improve oil recovery for many decades. Several theories regarding the mechanism of low-salinity flooding have been discussed in the literature including interfacial tension reduction, wettability alteration, change in pH value, emulsion formation, and clay migration. This work presents the results of flooding tests on selected carbonate core samples taken from Bu Hasa field in Abu Dhabi using sea water and two field injection waters, Um-Eradhuma (UER) at 197,357 ppm and Simsima at 243,155 ppm. These results were used to evaluate the effects of brine salinity and ionic composition on the possible interactions of limestone rock/brine/and oil system and to identify the oil recovery mechanism. The field injection waters were diluted to salinities of 5,000 and 1,000 ppm and the optimum salinity was determined and then modified by varying the sulfate and calcium ion concentrations. Wettability alteration was determined by contact angle measurements. Interfacial tension measurements of the studied systems were also performed in an attempt to evaluate the flow mechanism with low-salinity flooding. The experimental results revealed that a significant improvement in the oil recovery can be achieved through alteration of the injection water salinity.

Reducing the salinity of UER water from 197,357 to 5,000 ppm resulted in an improvement of oil recovery from 63 to 84.5 % of OOIP and the latter salinity was used to evaluate the impact of changing the sulfate and calcium ion concentrations on oil recovery. Results also indicated that sulfate concentration has a significant effect on the flooding process and that increasing the sulfate concentration beyond some optimum concentration of 46.8 ppm resulted in a negative effect on the flooding process. Contact angle measurements indicated that lowering the solution salinity could shift the wettability of the system towards intermediate wettability levels and that the UER water exhibits higher shift toward intermediate wettability compared to other waters. Results also indicated that there is no clear correlation between the improvements in oil recovery and interfacial tension and the pH of the studied systems. The results of this work are useful for people working in this field.

Keywords: Low-salinity flooding, IFT EOR, Wettability, Oil recovery

1. INTRODUCTION

Low-salinity waterflooding was first attempted by researchers at The University of Wyoming during the 90s by examining the effect of brine, crude oil, mineralogy and experimental procedure on rock surface-wettability. In the subsequent decade, the technology of low-salinity flooding was repeatedly implemented in the laboratory and in the field. Over the past decade low-salinity waterflooding has emerged as a viable enhanced oil recovery (EOR) method. Both laboratory tests and field trials have shown that injecting chemically modified water instead of seawater can lead to incremental oil recoveries. Although much research has been conducted, the governing physical and chemical mechanisms for this increase in recovery are not yet agreed upon, but are generally believed to involve some form of interaction between the rock, oil, and brine leading to changes in wettability, oil/water interfacial tension, or both.

Yildiz and Morrow (1996) conducted core floods using Berea sandstone, Moutray crude oil and either sodium-based brine composed of 4 % NaCl + 0.5 % $CaCl_2$ or calcium-based brine composed of 2 % $CaCl_2$. Recovery was higher with the calcium brine when the connate and injected brines were identical. The highest recovery was achieved by initially saturating the core with calcium brine, injecting Na brine until residual oil saturation was achieved, then injecting calcium brine. Tang and Morrow (1997) investigated the effects of connate and injection brine salinity, aging time and temperature on water flooding and imbibition with 3 different crude oils and 3 different brines. They concluded that in water floods with constant connate brine salinity and variable injected brine salinity, diluting injected brine 100 times produced ~5 % incremental oil recovery. In water floods with variable connate brine salinity and constant injected brine salinity, decreasing connate brine salinity dramatically improved recovery—about 40 % incremental oil recovery was achieved by diluted the connate brine 100 times.

Based on the above findings, researchers began to focus on the only variable that can be manipulated in a reservoir—the injection brine salinity. Researchers noticed that improved recovery by injection of low-salinity brine only occurred when crude oil and clay bearing sandstone mineralogy were present. Based on this observation, Tang and Morrow (1999a, b) offered the first theoretical interpretation of the mechanism responsible for the recovery improvement. They observed an increase in water flood and spontaneous imbibition recovery with a decrease in salinity in numerous cases. The authors used Berea cores, CS crude, and refined oil and 7 different brines ranging from 35,960 ppm TDS down to 151.5 ppm TDS. Oil Recovery improved significantly in the CS reservoir and Berea cores when low-salinity brine was injected instead of high salinity, but recovery improved only marginally in the more clay-free cores. Berea cores that were fired and acidized, to stabilize fines, were insensitive to brine salinity. Tang and Morrow (1999a, b) concluded that heavy polar components in the crude oil adsorb onto fine particles along the pore walls and that these mixed-wet fines are stripped by low-salinity brine, altering wettability and increase oil recovery.

Zhang and Morrow (2006) conducted water flood and spontaneous imbibition experiments using 4 different samples of Berea sandstone and three different crude oils. These authors observed improved recovery by injecting low-salinity brine in secondary and tertiary modes. The impact of low-salinity brine varied significantly between the different samples of Berea, suggesting that mineralogy was the most important variable affecting improved recovery. The lowest permeability block of Berea (k_{nitrogen} ~60–140 md) showed no sensitivity to salinity. The lack of response was attributed to the presence of chlorite. In several cases, cores responded to low-salinity brine in the secondary but not the tertiary mode. Low-salinity effects become more dramatic as the initial water saturation increased. In all cases, injection of low-salinity brine was accompanied by an increase of pressure followed by a gradual decrease. Effluent pH also increased.

Some publications indicated that there is no benefit of low-salinity waterflooding. Sharma and Filoco (2005) investigated the impact of connate and injection brine salinity and crude oil on oil recovery, residual saturations and wettability using Berea cores, 3 different oils and NaCl brine in various concentrations. In their imbibition experiments, decreasing connate water salinity increased recovery and significantly affected relative permeability. The salinity of the displacing brine had no significant impact. Drainage experiment's recovery and relative permeability were insensitive to salinity. During waterflooding of crude oil, oil recovery increased with decreasing connate brine salinity. However, during waterflooding of mineral oil, recovery was insensitive to connate brine salinity. In all cases, waterflood recovery was insensitive to the salinity of the injected brine. Sharma and Filoco suggested that low-salinity connate brine changes the wetting properties of the rock surface from water-wet to mixed-wet and thereby increase the recovery.

Webb et al. (2003) observed a reduction in residual oil saturation in the near wellbore region by injecting low-salinity brine. Three different brines were injected into a clastic formation from a producing well. Saturation was measured after each injection using a pulsed neutron capture log. A base line S_{or} was established with a synthetic native brine (250,000 ppm). Synthetic sea water (120,000 ppm), injected second, did not reduce oil saturation further. Low-salinity brine (3,000 ppm), injected last, reduced S_{or} significantly in two sand intervals and slightly in another.

Zhang et al. (2007) reported increased recovery in the tertiary mode by reducing reservoir brine salinity 20 times. Two consolidated reservoir sandstone cores were used. X-ray diffraction indicated that each of the cores were rich in chert and kaolinite. Two different crudes and a mineral oil were used. Almost 70 % incremental oil recovery was achieved in the secondary mode. Both the high and low-salinity secondary floods were conducted in the same core. Tertiary recovery was also quite large; 25 % incremental recovery in the best case. The recovery was achieved slowly, taking more than 10 injected pore volumes. In several cases, the pH fell upon injection of low-salinity brine; contrary to other researcher's observations. Pressure drop was closely tied to incremental recovery. In all cases, where significant incremental recovery was achieved pressure drop increased significantly then fell gradually.

Pu et al. (2008) observed low-salinity tertiary recovery from an almost clay-free core for the first time. They injected coalbed methane (CBM) water into 3 sandstone reservoir cores composed of quartz, feldspar, dolomite, and anhydrite cements but which had very little clay. The CBM water's salinity was about 1,316 ppm TDS. Cores were first waterflooded with high salinity formation brine (38,651 ppm). When oil production due to high salinity brine ceased CBM water was injected. In all cases CBM water liberated additional oil. In each core, the benefit of tertiary low-salinity flooding became less dramatic after each flood and restoration. A core was acidized to remove dolomite crystals and subsequently its recovery became insensitive to low-salinity flooding. Pu et al. proposed that dolomite crystals play an important role in the low-salinity recovery mechanism. Some of the dolomite crystals become mixed-wet as they contacted the oil phase during aging. During the low-salinity flood the dolomite crystals may detach from the pore walls releasing oil from the rock surface. The detached dolomite crystals will then reside at the crude oil/brine interface increasing resistance to flow of brine at the interface, delay snap-off at pore-throats and preventing the collapse of oil lamella.

Bagci et al. (2001) studied the effect of brine composition on oil recovery by water flooding using limestone cores. Ten different brine compositions were examined for injection through the study. The brines were NaCl, $CaCl_2$, KCl, and binary mixtures of them at two different concentrations (2 and 5 wt%). The highest oil recovery was 35.5 % of original oil in place (OOIP) for 2 wt% KCl. The authors concluded that any adjustment to the injected brine composition of a mature waterflood can offer a possible and economically feasible approach to increase oil production. Wettability alteration was mentioned as a reason for recovering more oil but without any further explanation. These authors used long core samples (20 inches) in the folding experiments and at a reservoir temperature of 122 °F. Low-salinity effluent brine samples showed higher pH and that was caused by ions exchange reaction.

Høgnesen et al. (2005) concluded that any modification to the injection water ions can impact rock wettability and that can result in additional oil to be recovered. They presented an imbibition study at high temperature condition using reservoir limestone, outcrop chalk cores, seawater and formation water. The results showed that increasing the sulfate ion concentration at high temperature can act as a wettability modifying agent in carbonates, and increased the oil recovery. Scale and souring problems will be enhanced as increasing the sulfate concentration in the injected water. Moreover, this strategy has limitations with regard to initial brine salinity and temperature. At low temperature condition, cationic surfactant was mixed with the aqueous solution and that increased the spontaneous imbibition through the cores.

Webb et al. (2005) presented a study that compared oil recovery from a North Sea carbonate core samples using sulfate-free formation simulated brine, with seawater, which contains sulfate. The imbibition capillary pressure experiments were performed at reservoir conditions using live crude oil and brine. The final results showed that the simulated seawater was able to modify the wettability of the carbonate system, changing the wettability of the rock to a more water-wet state. This conclusion was based on the saturation change noted in the spontaneous imbibition tests between simulated formation water and sea water.

Most of the low-salinity waterflood studies were conducted on limestone; seawater, also, was recommended as an injection fluid in chalk formations. Strand and Austad (2008) explained in preliminary experimental studies the chemical mechanism for the wettability alteration in fractured limestone after injecting seawater, sodium chloride brine, and formation water.

Synthetic seawater with and without sulfate ions was used to determine the sulfate ions effect on wettability. Spontaneous imbibition results at 248 °F showed 15 % increase in the oil recovery when limestone core was imbibed with seawater compared to seawater free of sulfate ions. Seawater has the lowest TDS compared to the other examined brines, but it did not include any brine test that has lower salinity than seawater. More details on reaction mechanism will be explained in the next section. Fjelde (2008) presented results on low-salinity water that increased oil recovery in limestone formation. Spontaneous imbibition experiments were conducted using formation water and low-salinity water. Low-salinity water showed similar oil recovery results to seawater experiments. No further details were mentioned in this work. Paul Vledder et al. (2010) work showed that the wettability can be changed at a reservoir scale, similar to more controlled experiments such as laboratory experiments and Log-Inject-Log tests. A large number of observations consistently prove this change in wettability that is shown to lead to an increased recovery factor. The incremental recovery due to the change in wettability on a field wide scale is 5–15 %. This range mostly overlaps with the range obtained from the experimental data (9–23 %) and therefore the final conclusion is that the incremental recovery due to low-salinity injection in Omar Field amounts to 10–15 % of the STOIIP. The main issues to be managed are water sourcing, water disposal and water mixing in the reservoir.

The objective of this work is to investigate the merits of using low-salinity flooding techniques to enhance oil recovery in Bu Hasa reservoir, identify the mechanism involved in the additional recovery of oil by low-salinity flooding, and evaluate the effect of different ions (SO_4^{2-} and Ca^{2+}) on the overall recovery process.

2. EXPERIMENTAL MATERIALS AND SETUP

2.1 Materials

Reservoir crude oil from Bu Hasa field was used in all experiments. The oil was filtered through a 5.0-μm filter paper (with a vacuum pump) to remove any possible solid particles. The oil is sweet oil and has no H_2S and about 2.6 mole% CO_2 which is very low. The oil density and viscosity are 0.825 g/cc and 3.08 cp measured at room temperature of 25 °C, respectively. Four types of water were used. The first type was Um-Eradhuma (UER) water which is the injection water that has been used in flooding the reservoir for many years.

A live sample of this water was filtered, degassed, and then used in the experiments. This water has a salinity of 197,357 ppm and density of 1.15 g/cc. The second type was Simsima (SIM) water which is formation water that has been used in water flooding the reservoir for many years. A live sample of this water was filtered, degassed, and then used in the experiments. Its salinity is 243,155 ppm and its density is 1.16 g/cc.

The third type is sea water (SW) which was collected from one of the Arabian Gulf beaches in Abu Dhabi. It was filtered, degassed, and then used in the experiments. Its salinity is about 40,980 ppm and its density is 1.029 g/cc. The fourth type was lab-prepared distilled water. Table 1 shows the analysis of these water samples.

Table 1. Analysis of the different types of water

Type	mg/L								TDS (ppm)
	Ca^{++}	K	Mg^{++}	Na^{+}	CO_3	HCO_3^{-}	Cl^{-}	SO_4^{--}	
SIM	20,808	–	3,047	68,214	–	119	150,617	350	243,155
UER	14,033	–	3,024	57.613	–	244	122,023	420	197.357
SW	600	–	1,560	13.900		200	24,300	420	40,980

Five core samples were selected from well number 589 in Bu Hasa field. The physical properties of these cores are listed in Table 2. Four of these core samples were used in the flooding experiments, namely, numbers 31, 39, 40, and 42. The fifth core sample (no. 46) was used for contact angle measurements. The mineral composition of these core samples and others are listed in Table 3.

Table 2. Properties of the selected core samples

Sample no.	Depth (m)	L (cm)	D (cm)	kw (md)	Dry weight (g)	Saturated weight (g)	Pore volume (cc)	ϕ Fraction
31	2650.206	7.090	3.843	8.7	164.080	186.700	20.423	0.248
39	2650.267	7.032	3.850	8.5	163.470	184.960	19.402	0.237
40	2650.937	7.247	3.853	8.7	170.210	192.370	20.007	0.237
42	2652.309	7.272	3.855	8.8	190.160	205.030	13.425	0.158
46	2652.400	7.160	3.860	7.7	190.800	204.130	12.035	0.144

Table 3. X-ray diffractograms mineral composition of core samples

Sample no.	Major mineral	Minor mineral
31	Calcite ($CaCO_3$)	Dolomite [$CaMg(CO_3)_2$]
39	Calcite ($CaCO_3$)	Dolomite [$CaMg(CO_3)_2$]
40	Calcite ($CaCO_3$)	Dolomite [$CaMg(CO_3)_2$]
42	Calcite ($CaCO_3$)	Dolomite [$CaMg(CO_3)_2$]
46	Calcite ($CaCO_3$)	Dolomite [$CaMg(CO_3)_2$]

2.2 Experimental setup

Oil floods were conducted at constant injection pressure of 689.476 kPa (100 psi) with overburden pressure of 5,515.808 kPa (800 psi) and at room temperature of 25 °C (77 °F). The cores were positioned in a vertical configuration and oil was injected at the top as illustrated in Figure 1. Effluent rate was regularly measured with a stop watch and pressure drop was monitored continuously. The core holder was designed for high pressure flood tests and to accommodate core samples of 3.86 cm (1.52 in.) in diameter and a maximum of 7.62 cm (3 in.) in length.

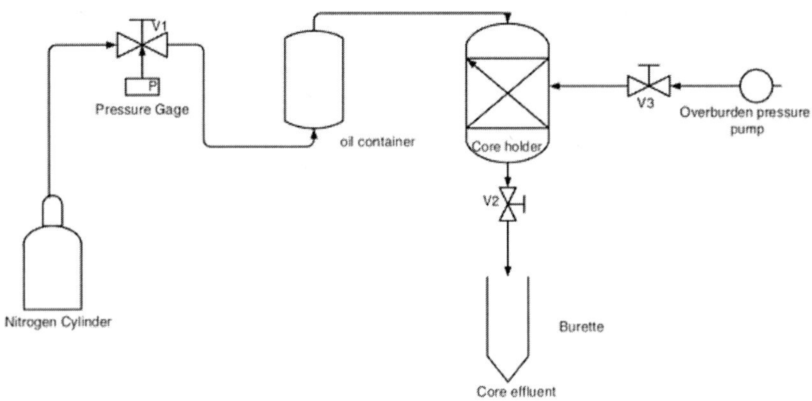

Figure 1. A schematic diagram of the core flooding setup

The core samples were first saturated with formation water and displaced by crude oil to S_{wc} using the standard Core Lab procedures. The analysis of this water is presented in Table 4. The crude oil-saturated cores were then aged in crude oil for 4 weeks and then flooded with various brines.

The volumes of produced brine and oil were recorded as a function of time at constant pressure drop. To avoid fingering of the displacing phase, these flooding processes were conducted at low rate between 1 and 3 cc/min. The injection of brine continued until oil ceased to show any production at the core outlet.

The measured stable pressure drop(s) and the corresponding flow rate(s) were used to calculate end-point water permeability to water $(k_{rw})_{Sor}$using Darcy's law. For example, the values of $(k_{rw})_{Sor}$ when flooding oil with 5,000 ppm UER water and 171,585 ppm UER water are 0.17 and 0.29, respectively. This observation is indicative of the system's wettability shift toward more water-wet as the end-point relative permeability to water decreases with decreasing of UER water salinity.

Table 4. Analysis of formation water

Element	Concentration, mg/L
Sodium	51,820
Calcium	15,992
Magnesium	1,282
Iron	277
Barium	0
Chloride	111,852
Bicarbonate	391
Sulfate	272
Carbonate	0
Hydroxide	0

The material balance was then applied to calculate the residual oil saturation (S_{or}). Figure 2illustrates a block diagram of all test runs performed in this work. Experiments were conducted with UER water, SIM water, SW and distilled water.

Experiments were then conducted using the dilutions of each type of water to determine the optimum water salinity. The effect of ionic concentration of sulfate and calcium ions on the flooding process was investigated using the optimum water salinity.

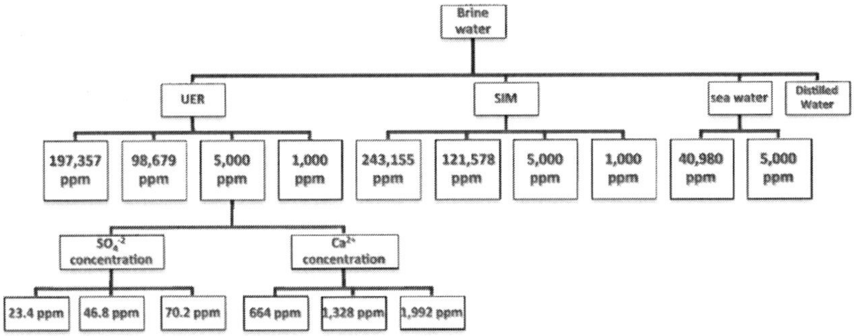

Figure 2. Block diagram showing the various experimental runs performed in this work

3. RESULTS AND DISCUSSION

3.1 Flooding with UER water

In order to evaluate the effect of different salinities on the recovery eleven core flooding tests were conducted. The UER brine was diluted to one-half of its original salinity, i.e., down to 98,679 ppm then to 5,000 ppm and finally to 1,000 ppm. The flooding results of this set of experiments are plotted as oil recovery percent of OOIP versus pore volumes injected of brine as shown in Figure 3.

Figure 3 shows that the highest oil recovery was obtained from the 5,000 ppm dilution which is about 84.5 % of the OOIP. The lowest oil recovery percent observed when using the distilled waters was about 60.5 % of the OOIP.

The 1,000 ppm dilution yielded a lower oil recovery than the 5,000 ppm solution and all the diluted UER waters resulted in a higher oil recovery than the distilled water. Therefore, the salinity of 5,000 ppm was considered as the optimum salinity for further evaluation.

3.2 Flooding with SIM water

Simsima brine with its original salinity of 243,155 ppm was first diluted by 50 % (salinity of 121,578 ppm) then down to 5,000 ppm and finally to 1,000 ppm. The results of this part are illustrated in Figure 4.

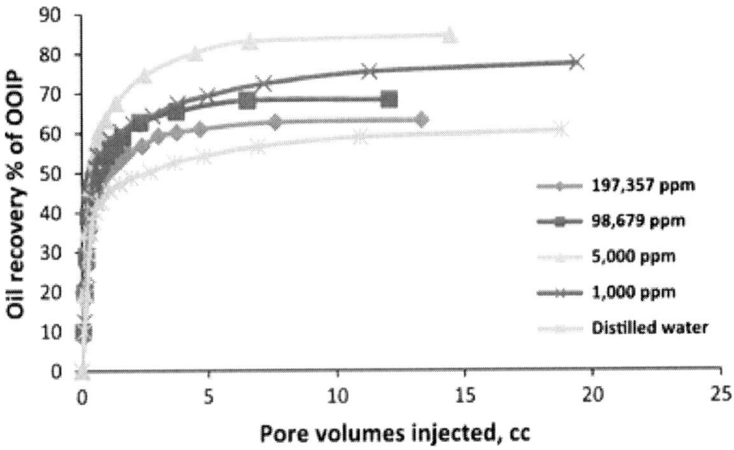

Figure 3. Oil recovery % of OOIP versus pore volumes injected of UER water with different salinity concentrations; *figures betweenbrackets* represent percent of dilution of original salinity concentration

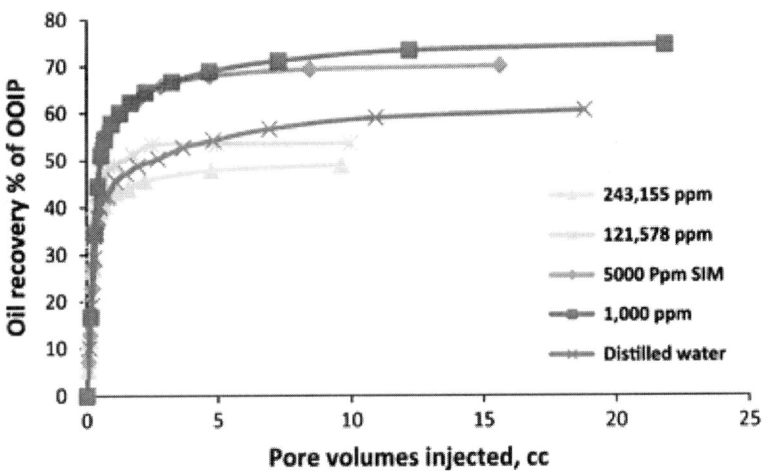

Figure 4. Oil recovery % of OOIP versus pore volumes injected of SIM water with different salinity concentrations; *figures betweenbrackets* represent percent of dilution of original salinity concentration

Figure 4 shows that the highest oil recovery of 74.4 % of the OOIP was obtained with the 1,000 ppm dilution. The lowest oil recovery of 48.9 % of the OOIP was observed with the original SIM water. The 50 % dilution resulted in a recovery percent of 53.5 % of thc OOIP then the 5,000 ppm dilution resulted in a recovery of 70.0 % of the OOIP.

3.3 Flooding with sea water

Sea water brine of original salinity of 40,980 ppm was diluted to 5,000 ppm. The results of this section are shown in Figure 5.

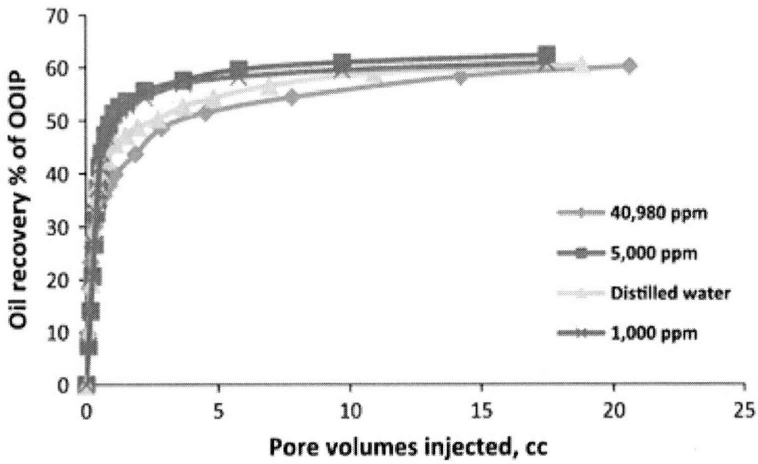

Figure 5. Oil recovery % of OOIP versus pore volumes injected of SW water with different salinity concentrations; *figure betweenbrackets* represent percent of dilution of original salinity concentration

Figure 5 shows that original SW resulted in a recovery of 60.2 % of the OOIP and the 5,000 ppm brine showed a recovery of 62.2 % of the OOIP and the distilled water resulted in a recovery of 60.5 % of the OOIP. The three waters resulted in comparable results as there is no significant increase in oil recovery by dilution.

The results presented in Figs. 3, 4, and 5 show that flooding core samples with the three different brines at their original salinities resulted in the lowest ultimate oil recoveries. Also, by diluting these brines to lower salinities the oil recovery increased significantly except for SW as its original salinity is not high as in UER and SIM brines. The highest oil recovery of 84 % of OOIP recovery was achieved by flooding with UER brine at 5,000 ppm. Tang and Morrow (1999a, b) observed similar trends and attributed these improvements to the presence of clay in their core samples. They concluded that the flow mechanism in low-salinity flooding is highly controlled by the clay. The core samples used in this work are clay-free and therefore another flow mechanism was responsible for the increased oil recovery by low-salinity flooding. Pu et al. (2008) observed increased oil recovery but in clay-free core samples. They proposed that dolomite crystals could play an important role in the low-salinity recovery mechanism.

3.4 Results of changing ionic composition of the brine

To evaluate the effect of Ca^{2+} and SO_4^{2-} concentrations on the performance of low-salinity flooding, two sets of core flooding runs were conducted. The 5,000-ppm dilution of UER brine was used as the base water for evaluating the effect of ionic composition because the highest oil recovery of 84 % of the OOIP was achieved by the 5,000-ppm dilution of UER brine. The first set of experiments involved flooding core samples with water UER at 5,000 ppm and sulfate concentrations of 11.7, 23.4, 46.8, and 70.2 ppm, respectively. Brine concentration was varied by changing the concentration of total dissolved solids of the synthetic brine in proportion to give sulfate content of 23.4, 46.8, and 70.2 ppm by adding Na_2SO_4. The ionic strength of these solutions was kept constant by adjusting the content of NaCl. The results of this set of experiments are shown in Figure 6.

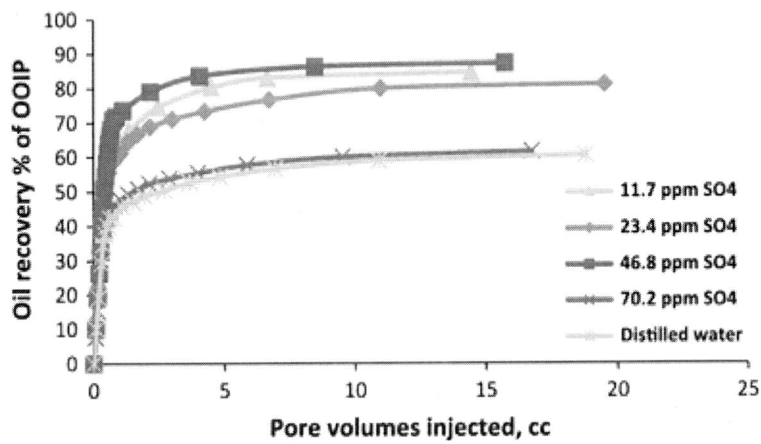

Figure 6. Oil recovery versus pore volumes injected brine at different SO_4^{2-} concentrations

Figure 6 shows that with sulfate concentration of 46.8 ppm (four times the original sulfate concentration in the UER water) the highest oil recovery of 87.2 % of the OOIP was achieved. The sulfate concentration of 70.2 ppm (six times the original sulfate concentration) resulted in the lowest oil recovery of 61.5 % of the OOIP which is comparable to the distilled water of 60.5 % of the OOIP. Figure 6 also shows that increasing the sulfate concentration in the brine solution could increase the oil recovery until a critical value of sulfate concentration is reached above which oil recovery decreased. Similar trends were observed by Webb et al. (2005). In their study, they compared oil recovery from a North Sea carbonate core samples using sulfate-free formation simulated brine with seawater, which contains sulfate. The final results showed that the simulated seawater was able to increase the recovery by 20 percent. There is no clear evidence on the critical sulfate concentration in the literature.

The second set of experiments in this section involved flooding core samples with UER water at 5,000 ppm containing calcium concentrations of 332, 664, 1,328, and 1,992 ppm, respectively. Brine concentration was varied by changing the concentration of total dissolved solids of the synthetic brine in proportion to give brine of 664, 1,328, and 1992 ppm of calcium content by adding $CaCl_2$. The ionic strength of these solutions was kept constant by adjusting the content of NaCl. The results of this set of runs are shown in Figure 7.

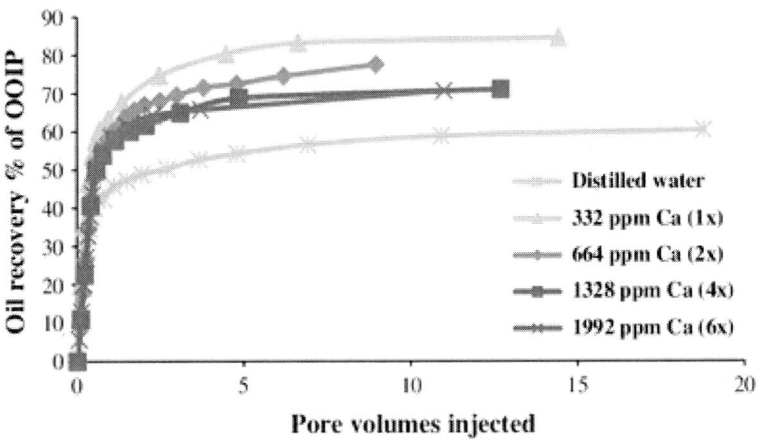

Figure 7. Oil recovery versus pore volumes injected brine at different Ca^{2+} concentrations

Figure 7 shows that calcium concentration has a negative effect on oil recovery. Flooding with the original calcium concentration of 332 ppm yielded the highest oil recovery and by increasing the calcium concentration the oil recovery decreased. The MIE (multicomponent ionic exchange) concept that was proposed by Lager et al. (2008) and the Double layer effect that was suggested by Ligthelm et al. (2009) indicates that calcium plays a key role in the oil recovery process in the presence of clay. In this work, however, it is shown that in clay-free core samples, flooding with calcium could have a negative effect on oil recovery.

3.5 Contact angle measurements

The results of the sessile drop method were analyzed with Sigma Scan Pro image analysis software to determine the change of contact angle measurements with time. The results of these measurements for the oil-UER brine-rock, oil-SIM brine-rock, and oil-SW-rock systems considered in this work are illustrated in Figs. 8, 9, and 10, respectively.

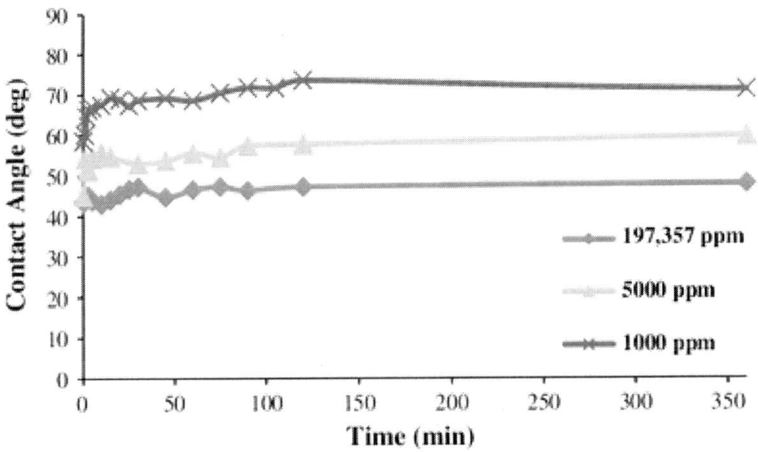

Figure 8. Results of contact angle measurements for UER water and its corresponding diluted solutions

The repeatability of the contact angle measurements was evaluated by calculating the percent deviation of a certain reading from the preceding value. The maximum deviation observed was approximately 1.37 % which could be partially attributed to machine error. The early time fluctuations were observed in all the experiments of contact angle measurements. These fluctuations are believed to be attributed to the continuous change in the shape of the oil drop observed during the early time of the experiments. Once the shape of the oil drop reached a stabilized state, the fluctuations disappeared and a stable value of contact angle was observed for the rest of the experiment. Although not shown in Figs. 8, 9, 10, 11, and 12, each contact angle measurement was allowed to continue for 650 min to make sure that stabilization has been reached. Thus, it can be deduced that the fluctuations did consume some 20 % of the testing time.

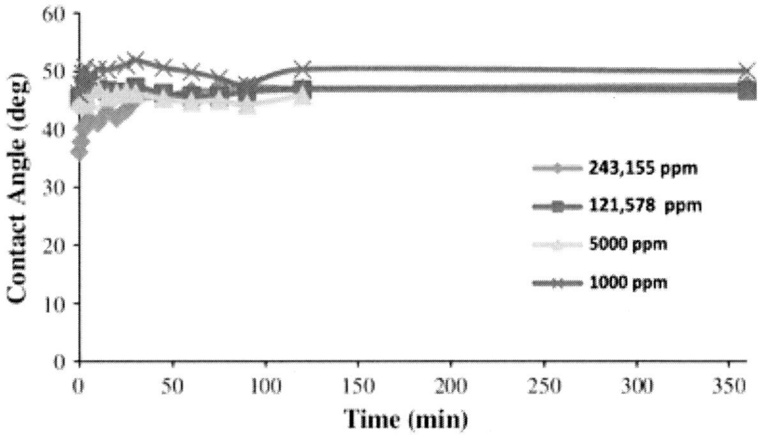

Figure 9. Results of contact angle measurements for SIM water and its corresponding diluted solutions

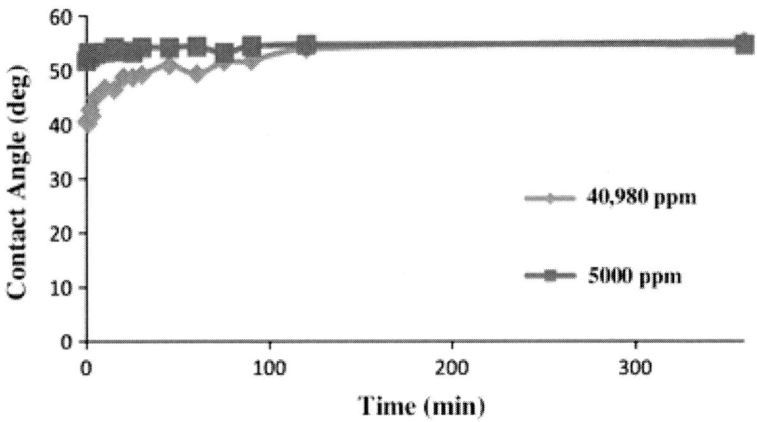

Figure 10. Results of contact angle measurements for SW water and its corresponding diluted solutions

The results of contact angle measurements presented in Figs. 9 and 10 and the results of core flooding tests presented in Figs. 4 and 5 indicate that oil recovery could increase if the contact angle is increased. In other words, the oil recovery could increase as the wettability is changed to more intermediate levels. This observation is in contradiction with Jadhunandan and Morrow (1995), Tang and Morrow (1999a, b), Zhang and Morrow (2006); they all concluded that water-wet wettability yields higher oil recovery. Agbalaka et al. (2008) conducted a review on the effect of reservoir rock wettability on oil recovery for secondary and tertiary oil recovery processes. Several field cases as well as laboratory studies were discussed. The fact that wettability affects oil recovery can affect oil recovery efficiency is widely acknowledged. However, the wetting phase that will result in optimal recovery of oil has been the subject of intense debate. Agbalaka et al. (2008) observed that the reason for this divergence in observed reports is attributable to a number of factors which includes (1) difficulty in wetting state reproducibility, (2) lack of unified standards and procedure for coring, core handling and core storage, and (3) the wetting state characterization method adopted. They concluded that strongly oil wet reservoirs give the least oil recovery and the best recovery appears to be the intermediate wet reservoirs. These findings are consistent with the results of the present work. Sharma and Filoco (2005) also suggested that low-salinity brine changes the wetting properties of the rock surface from water-wet to mixed-wet and thereby increase the recovery.

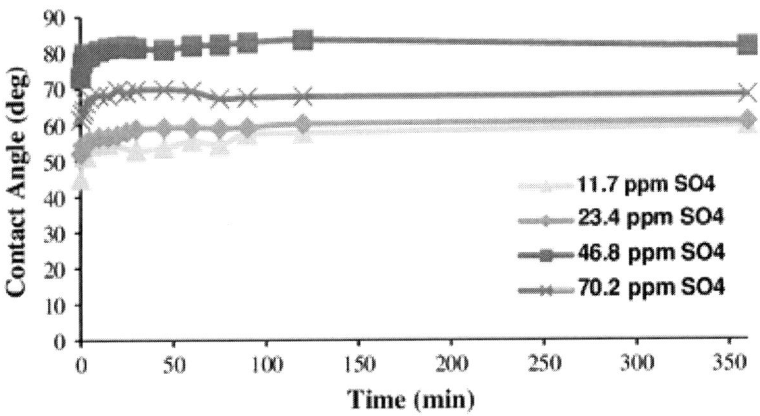

Figure 11. Contact angle measurements for different SO_4^{2-} concentrations

Figure 10 shows that there is hardly any change in the contact angle between the original SW brine and the 5,000-ppm solution of the SW brine. This observation may explain the results presented earlier in Figure 5 where flooding with low-salinity flooding SW has no effect on the oil recovery. Also shown in Figure 5 that the oil recovery performances in the three runs including distilled water are comparable.

In order to investigate the effect of changing the ionic composition on wettability, eight brines with different concentration of Ca^{2+} and SO_4^{2-} were used. The Ca^{2+} concentration ranges from 332 to 1992 ppm. The

SO_4^{2-} concentration ranges from 11.7 to 70.2 ppm. Figures 11 and 12show the results of the contact angle measurements for different SO_4^{2-} and Ca^{2+}concentrations, respectively.

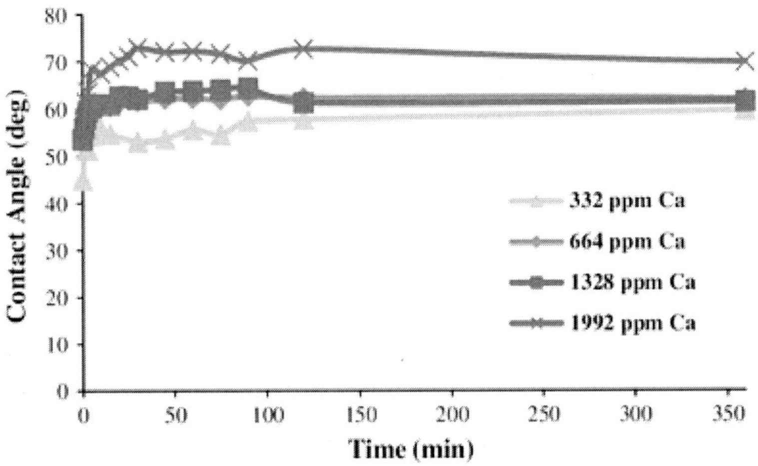

Figure 12. Contact angle measurements for different Ca^{2+} concentrations

Figure 11 shows that the highest angle of contact was observed at 46.8 ppm concentration of SO_4^{2-}. This observation could explain the results of Figure 6 as the 46.8-ppm concentration of SO_4^{2-} yields the highest recovery and confirms the results of Agbalaka et al. (2008). These authors concluded that the best recovery appears to be achieved in the intermediate wet reservoirs. Zekri et al. (2011) investigated the effect of EOR techniques on wettability and oil recovery of carbonate and sandstone formation.

They concluded that increasing the sulfate concentration in the injection brine changed the wettability of the chalky and microcrystalline limestone used in their experiments into more water-wet.

3.6 IFT measurements

Interfacial tension measurements (IFT) were conducted with the interfacial tensiometer using Bu Hasa crude oil and different brines to assess the effect of IFT on the mechanism of low-salinity flooding. All measurements were carried out at room temperature of 25 °C and the results of these tests are illustrated in Figs. 13, 14, 15, 16, 17, and 18.

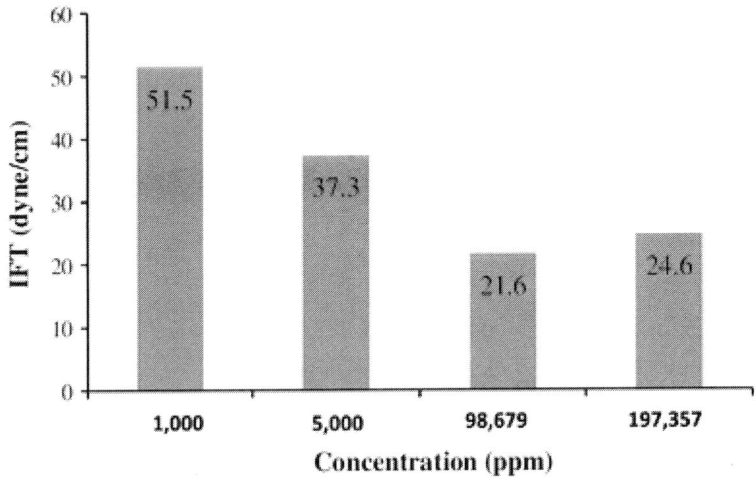

Figure 13. Results of IFT measurements for UER brines and its diluted samples

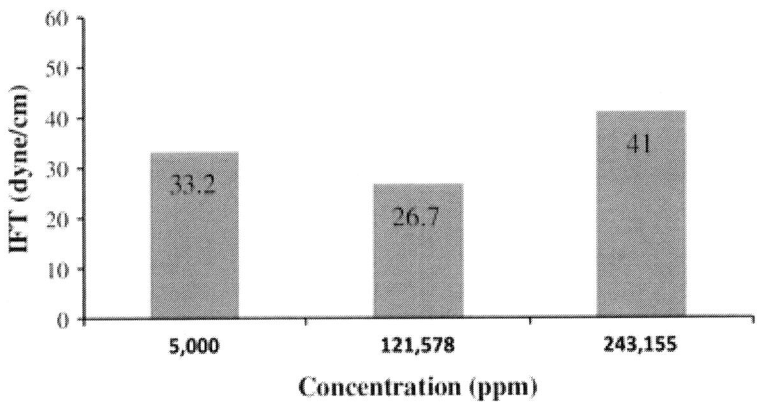

Figure 14. Results of IFT measurements for SIM brines and its diluted samples

No systematic trend on the effect of Ca^{2+} on IFT measurements was observed in this work and this may explain the high oil recovery with the lowest concentration of Ca^{2+} that exhibited lowest value of IFT. The results also indicate that there is an optimum salinity for different tested brines but the observed optimum salinity does not correlate with the optimum oil recovery by low-salinity flooding. Therefore, interfacial tension may not be responsible for the increase in oil recovery due to the injection of low-salinity water. This observation contradicts the results of

Okasha and Al-Shiwaish (2008) who investigated the effect of brine salinity on interfacial tension in Arab-D carbonate reservoir in Saudi Arabia. He concluded that the reduction of IFT with brine dilution reflects the potential implication of low-salinity flooding in improving oil recovery.

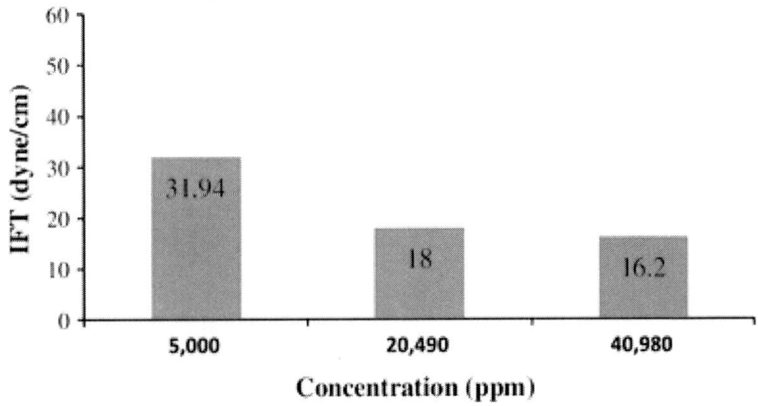

Figure 15. Results of IFT measurements for SW brines and its diluted samples

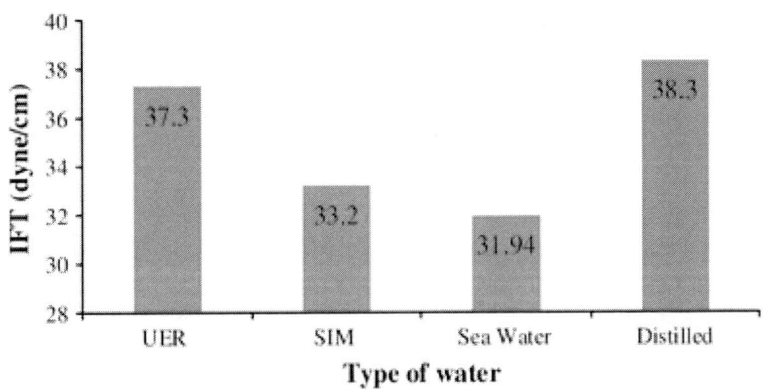

Figure 16. Results of IFT measurements for UER, SIM and SW brines at 5,000 ppm dilution

No fluctuations were observed in the IFT measurements and excellent repeatability of selected measurements was noted.

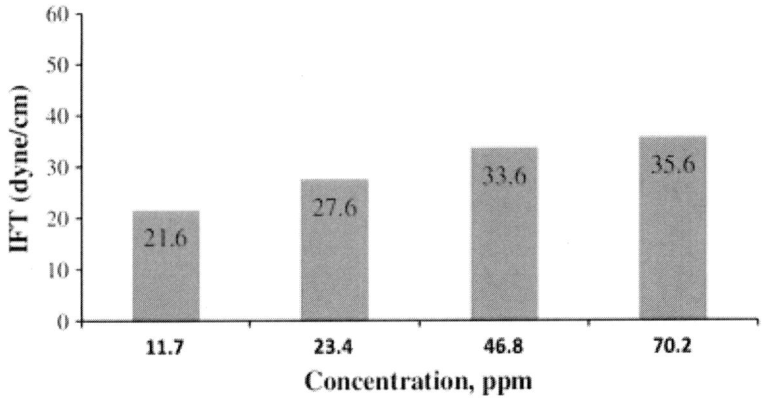

Figure 17. Results of IFT measurements for different SO_4^{2-} concentrations

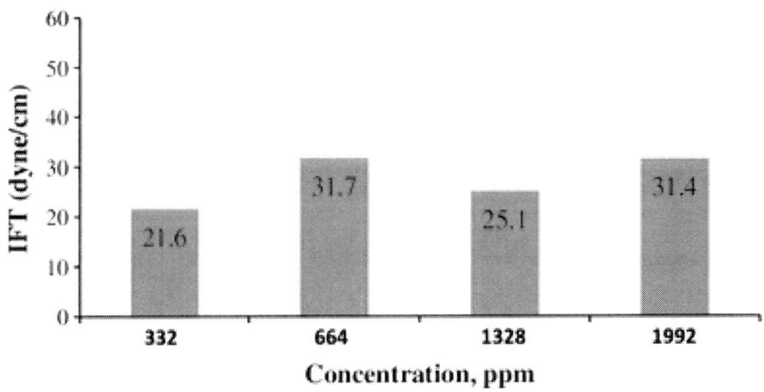

Figure 18. Results of IFT measurements for different Ca^{2+} concentrations

3.7 pH measurements

Some studies have shown a rise in pH during low-salinity flooding experiments. This rise in pH is due to two reactions: carbonate dissolution and cation exchange. The dissolution of carbonate (i.e., calcite and/or dolomite) results in an excess of OH− and cation exchange occurs between clay minerals and the invading water. The dissolution reactions are relatively slow and dependent on the amount of carbonate material present in the rock (Lager et al. 2006). Conflicting evidence throws doubt on this mechanism being the cause of the low-salinity flooding effect. The results obtained from the core flooding experiments and pH measurements of this study indicate no correlation between the pH variation and improved oil through low-salinity flooding as shown in Figure 19. Therefore, high pH may not be responsible for the increase in oil recovery with low-salinity waterflooding.

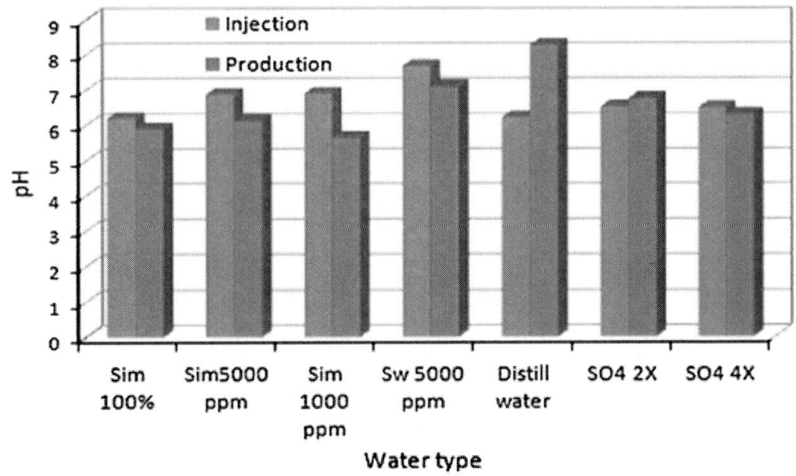

Figure 19. Results of pH measurements for different brines

4. CONCLUSIONS

Based on the results of the experimental work conducted in this study the following conclusions may be drawn:

1) Low-salinity flooding seems to have good potentials in Bu Hasa field.

2) Contact angle measurements indicate that lowering the solution salinity could shift the wettability of the system towards intermediate levels and this effect appears to be responsible for the improved oil recovery.

3) Increasing the Ca^{2+} concentration in the injected brine resulted in decreased ultimate oil recoveries.

4) Increasing the SO_4^{2-} concentration in the injected brine tends to change the wettability to more intermediate levels and resulted in improved ultimate oil recoveries. However, there is an optimum SO_4^{2-} concentration of 46.8 ppm in the 5,000 ppm UER which resulted in highest oil recovery. It is believed that this optimum concentration of

SO_4^{2-} in the flooding tests is responsible for shifting the system's wettability to intermediate water-wet.

5) The IFT of the oil-brine system may not have a direct effect on low-salinity flooding overall performance.

ACKNOWLEDGMENTS

The authors would like to express their appreciation for the financial support received from ADNOC and ADCO operating companies. Acknowledgment is extended to ADNOC R&D OSC EOGR Technical Committee for their technical support. Also we would like to acknowledge the faculty of Engineering and Research Affairs at UAEU for their continuous support and encouragements.

REFERENCES

1. Agbalaka C, Abhijit Y. Dandekar, Shirish L. Patil, Khataniar S, Hemsath JR (2008) The effect of wettability on oil recovery: a review, paper SPE 114496 presented at SPE Asia Pacific oil and gas conference and exhibition, Perth, Australia, 20–22 October 2008
2. Bagci S, Kok MV, Turksoy U (2001) Effect of brine composition on oil recovery by waterflooding. Pet Sci Technol 19(3–4):359–372
3. Fjelde I (2008) Low salinity water flooding experimental experience and challenges. Force RP work shop: low salinity water flooding, the importance of salt content in injection water, Stavanger, Norway, 15 May
4. Høgnesen EJ, Strand S, Austad T (2005) Waterflooding of preferential oil-wet carbonates: oil recovery related to reservoir temperature and brine composition. Paper SPE 94166 presented at the SPE Europec/EAGE Annual Conference, Madrid, Spain, 13–16 June
5. Jadhunandan PP, Morrow NR (1995) Effect of wettability on waterflooding recovery for crude-oil/brine/rock systems. SPE reservoir engineering, 40–46 February
6. Lager A, Webb KJ, Black CJJ, Singleton M, Sorbie KS (2006) Low salinity oil recovery—an experimental investigation. Paper SCA2006-36 presented at the International Symposium of the Society of Core Analysts held in Trondheim, Norway, 12–16 September
7. Lager A, Webb KJ, Collins IR, Richmond DM (2008) LoSal™ enhanced oil recovery: evidence of enhanced oil recovery at the reservoir scale. Paper SPE 113976 presented at the 2008 SPE/DOE improved oil recovery symposium held in Tulsa, Oklahoma, USA, 19–23 April
8. Ligthelm DJ et al (2009) Novel waterflooding strategy by manipulation of injection brine composition, EUROPEC/EAGE conference and exhibition. Society of Petroleum Engineers, Amsterdam, The Netherlands
9. Okasha TM, Al-Shiwaish A-JA (2008) Effect of Brine Salinity on Interfacial Tension in Arab-D Carbonate Reservoir, Saudi Arabia, paper SPE presented at SPE Middle East Oil and Gas Show and Conference, Bahrain, Bahrain 15–18 March 2009
10. Pu H, Xie X, Yin P, Morrow NR (2008) Application of coalbed methane water to oil recovery from tensleep sandstone by low salinity waterflooding, SPE/DOE symposium on improved oil recovery. Society of Petroleum Engineers, Tulsa, Oklahoma, USA
11. Sharma MM, Filoco PR (2005) Effect of brine salinity and crude-oil properties on oil recovery and residual saturations, SPE 65402: September 2000, 5(3):293–300

12. Strand S, Austad T (2008) Effect of temperature on enhanced oil recovery from mixed-wet chalk cores by spontaneous imbibition and forced displacement using seawater. Energy Fuels 22(5):3222–3225
13. Tang GQ, Morrow NR (1997) Salinity, temperature, oil composition and oil recovery by waterflooding SPE 36680. 12(4):269–276
14. Tang G, Morrow NR (1999a) Oil recovery by waterflooding and imbibition—invading brine cation valency and salinity. Paper SCA9911
15. Tang GQ, Morrow NR (1999b) Influence of brine composition and fines migration on crude oil/brine/rock interactions and oil recovery. J Pet Sci Eng 24:99–111
16. Vledder P, Carrera Fonseca J, Gonzalez I, Ligthelm D (2010) Low salinity water flooding: proof of wettability alteration on a field wide scale. Paper SPE 129564 presented at SPE improved oil recovery symposium, Tulsa, Oklahoma, USA, 24–28 April 2010
17. Webb KJ, Black CJJ, Al-Ajell H (2003) Low salinity oil recovery -log-inject-log, SPE 81460 presented at middle east oil show, Bahrain, 9–12 June 2003
18. Webb KJ, Black CJJ, Tjetland G (2005) A laboratory study investigating methods for improving oil recovery in carbonates. Paper IPTC 10506 presented at the International Petroleum Technology Conference, Doha, Qatar, 21–23 November
19. Yildiz HO, Morrow NR (1996) Effect of brine composition on recovery of Moutray crude oil by waterflooding. J Petrol Sci Eng 14:159–168
20. Zekri AY, Nasr M, Zaid A (2011) Effect of *LoSal™* on the wettability and oil recovery of carbonate and sandstone formation. Paper SPE 14131 presented at the international petroleum technology conference, Bangkok, Thailand, 15–17 November 2011
21. Zhang Y, Morrow NR (2006) Comparison of secondary and tertiary recovery with change in injection brine composition for crude oil/sandstone combinations, SPE 99757
22. Zhang Y, Xie X, Morrow NR (2007) Waterflood performance by injection of brine with different salinity for reservoir cores. Paper SPE 109849 presented at the 2007 SPE annual technical conference and exhibition in Anaheim, California, USA, 11–14 November

CHAPTER 4

Sensitivity of Coastal Flood Risk Assessments to Digital Elevation Models

Bas van de Sande [1*], Joost Lansen [2] and Claartje Hoyng [2]

[1] *Royal HaskoningDHV, Barbarossastraat 35 6522 DK Nijmegen, The Netherlands*

[2] *Royal HaskoningDHV, George Hintzenweg 85 3068 AX Rotterdam, The Netherlands*

ABSTRACT

Most coastal flood risk studies make use of a Digital Elevation Model (DEM) in addition to a projected flood water level in order to estimate the flood inundation and associated damages to property and livelihoods. The resolution and accuracy of a DEM are critical in a flood risk assessment, as land elevation largely determines whether a location will be flooded or will remain dry during a flood event. Especially in low lying deltaic areas, the land elevation variation is usually in the order of only a few decimeters, and an offset of various decimeters in the elevation data has a significant impact on the accuracy of the risk assessment. Publicly available DEMs are often used in studies for coastal flood risk assessments. The accuracy of these datasets is relatively low, in the order of meters, and is especially low in comparison to the level of accuracy required for a flood risk assessment in a deltaic area. For a coastal zone area in Nigeria (Lagos State) an accurate LiDAR DEM dataset was adopted as ground truth concerning terrain elevation. In the case study, the LiDAR DEM was compared to various publicly available DEMs. The coastal flood risk assessment using various publicly available DEMs was compared to a flood risk assessment using LiDAR DEMs. It can be concluded that the publicly available DEMs do not meet the accuracy requirement of coastal flood risk assessments, especially in coastal and deltaic areas. For this particular case study, the publically available DEMs highly overestimated the land elevation Z-values and thereby underestimated the coastal flood risk for the Lagos State area. The findings are of interest when selecting data sets for coastal flood risk assessments in low-lying deltaic areas.

Keywords: coastal flood risk; LiDAR; GIS; DEM; data accuracy

1. INTRODUCTION

Coasts are experiencing the adverse consequences of hazards related to climate change and sea level rise and will be exposed to increasing risks of coastal flooding [1,2,3]. Coastal zones and deltas already contain a larger share of the world's population and will become more heavily populated. Ten percent of the world's population lives in the 10 m Low Elevation Coastal Zone. This share covers two percent of the world's land area [4]. This emphasizes the need for coastal flood risk assessments to which GIS (Geographic Information System) components and mapping can be incorporated.

There are many (online) tools to estimate the impact of sea level rise on land area. One can visualize flood extents by setting an elevation relative to sea level in the chosen application (e.g., Globalfloodmap [5] and the Wikipedia Sea Level Rise Explorer [6]). These tools are mainly used to indicate impacts on a global scale. As of yet it is still unclear and not easy to estimate to what extent these tools and its underlying data can be used in other (smaller) scaled spatial analyses (Figure 1).

1.1. GIS Approach

Coastal flood risk mapping is used to estimate flood risk as shown schematically in Figure 2 [8]. The combination of the occurrence probability of a flood event and the associated (socio-economic) consequences of the flood events determines the total flood risk of a region. In case of GIS-based flood risk assessments, three types of geo data are generated in coastal flood risk models (see Figure 2) representing the flood event (inundation depth maps defined as the difference of a water level and land elevation), the potential consequences of the flood, and a flood risk map.

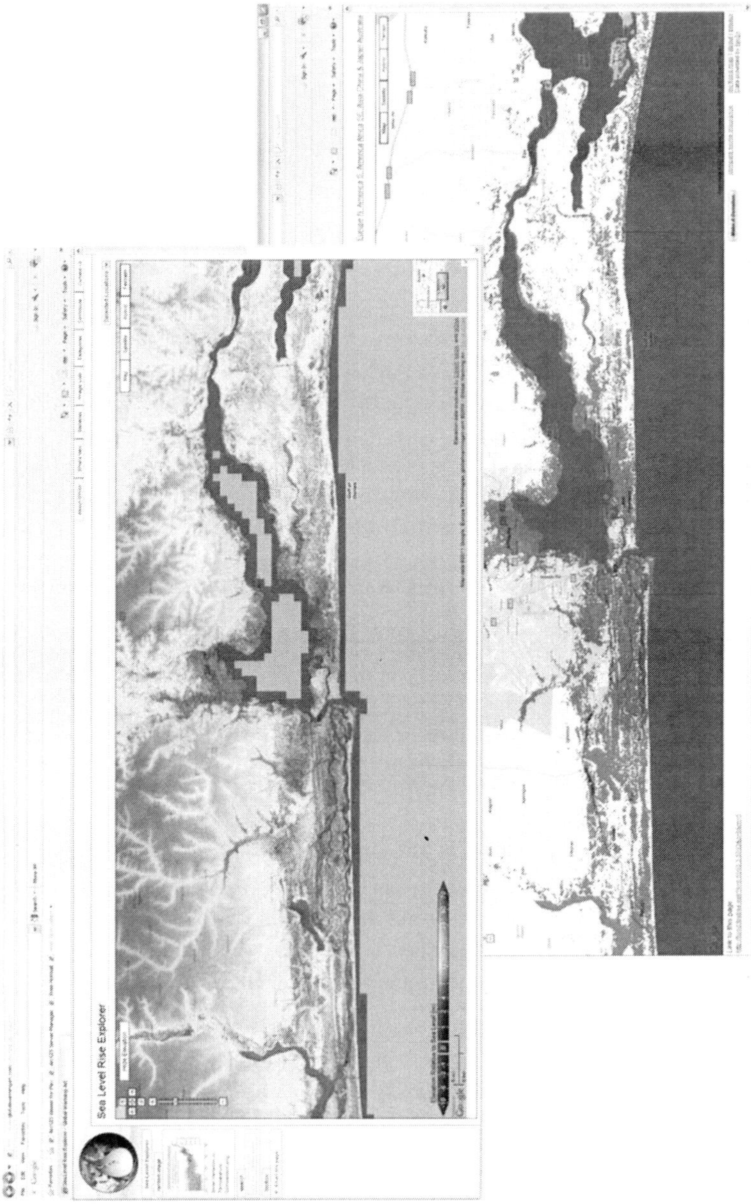

Figure 1. Online tools are available to estimate the impact of sea-level-rise on land area; tools which are based on public Digital Elevation Models (DEMs). Views were zoomed to Lagos State area. Source of upper left image: Wikipedia Sea Level Rise Explorer [6]; source of lower right image: Flood Maps [7].

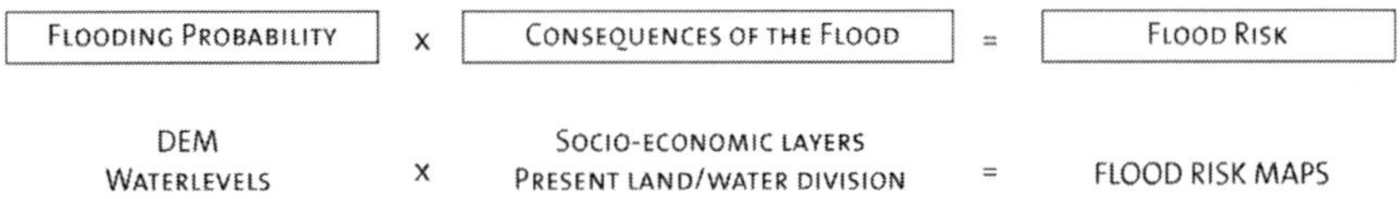

Figure 2. Conceptual risk equation.

1.2. DEM as Critical Factor

Most coastal flood risk assessments are based on a conceptual risk approach where Digital Elevation Models (DEM) is used in order to estimate the flood hazard for coastal zones according to projected water levels. The resolution and accuracy of DEM data are critical regarding the results of coastal flood risk assessments, because these data determine whether a location will be flooded by means of a flood simulation. Topographic data are the most important factor in determining the extent of flooding and therefore the accuracy of flood maps [9]. Especially in low-lying coastal zones, the simulation of flood extents was found to be very sensitive to the terrain representation [10].

Yet in general mainly these areas are most vulnerable to coastal floods and therefore demand adequate risk assessments. Commonly publicly available DEMs like ASTER GDEM (Advanced Spaceborne Thermal Emission and Reflection Radiometer) 30 m. horizontal resolution data set [11] and SRTM DEM (Shuttle Radar Topography Mission) 90 m. horizontal resolution data set [12] are used for coastal flood risk assessments [13], although the resolution and accuracy of these data sets are relatively low. In some areas of the world DEMs generated from LiDAR data are publicly available, but this is rarely the case. Many users end up using an SRTM DEM and ASTER GDEM dataset mainly because it is freely available and ready to use for modeling [14,15,16].

Also many online coastal flooding applications are based on these publicly available DEM data sets (see Figure 1), as are the earlier mentioned applications of Globalfloodmap and the Wikipedia Sea Level Rise Explorer.

However, the sensitivity of coastal flood risk assessments to the publicly available DEMs and the associated effects on coastal flood hazard and flood risk estimations based on these data sets are currently overlooked.

2. MATERIALS AND METHODS

2.1. Comparison Analyses

A case study for Lagos State (Nigeria) was conducted to show the effects on flood hazard and flood risk estimations by comparing flood risk estimations based on two publicly available DEM datasets (ASTER GDEM and SRTM DEM) to flood risk estimations based on LiDAR (Light Detection and Ranging) DEM estimations.

DEMs (Digital Elevation Models) can be split up in Digital Terrain Models (DTM) and Digital Surface Models (DSM). DTMs represent the terrain's elevation, whereas DSMs represent the surface elevation of objects in the landscape (e.g., roofs of buildings and canopies of trees) [17]. In fact ASTER GDEM and STRM DEM datasets are Digital Surface Models instead of Digital Terrain Models. Both sensors apply space borne measure techniques for capturing surface elevations. Unlike the publicly available DEM datasets the LiDAR DEM is a DTM. Although this essential difference is known, a comparison analysis is worth performing since STRM DEM and ASTER GDEM are widely used and ubiquitous as terrain or topography models.

The LiDAR survey was flown with a plane sensor platform in May 2009 by the PBH Group (based in Canada). The nominal point density of the bare earth point data is 2 points per m^2, and the provided accuracy assessment reported a root mean square error (RMSE) of less than 1 cm. These accuracy and resolution characteristics amply comply with the guidelines for accuracy and resolution of terrain data as input for flood risk assessments [18]. The LiDAR bare earth point measurements were adopted as ground truth, since there was no reference elevation dataset available. The bare earth point data was processed to a grid with a horizontal resolution of 5 m. by natural neighbor interpolation techniques. This processed grid was used for the flooding simulations. With a Root Mean Square Error of 0.10 m. and a Mean Error of 0.00 m. of this grid compared to the bare earth points, the deviations of the 5 m. resolution LiDAR DEM were considered acceptable.

In this case study coastal flooding is approached from a "bathtub" approach. In order to model hydrological connectivity, you can simulate inundation by selecting the values of your DEM which have a lower value than the projected water level [19]. The simulation can be executed in two manners; with and without surface connectivity (see Figure 3).

With surface connectivity only the grid cells which are lower than the projected water level and which are connected to an adjacent flooded grid cell or open water indicate inundation. Simulation with surface connectivity is more appropriate because a coastal flood only has direct influence on the adjacent land zone.

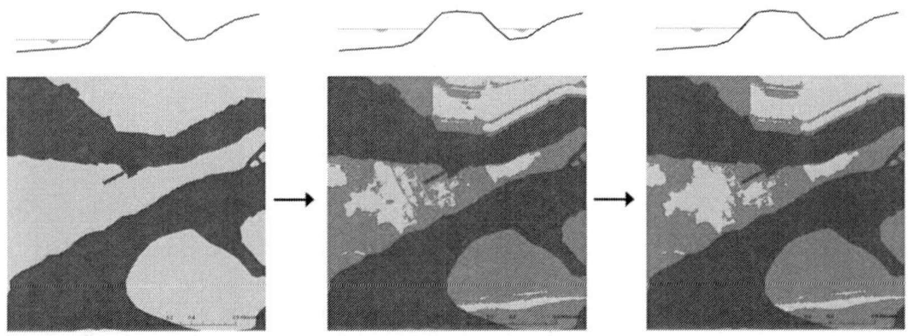

Figure 3. Bathtub principle (red indicates simulated inundated areas).

Coastal flood risk assessments can be performed at different horizontal and vertical scale levels (respectively the geospatial scale and the elevation scale). Assessments differ from estimations of the share of the global population living in low-lying deltas to estimations of the probability of a flood in a small harbor. Two case study scales were included in order to reveal the effects of usage of the public DEMs on coastal flood risk assessments and to give an indication of the suitability for use of these public datasets. Lagos State scale and Lagos City scale represent the geospatial scales with accompanying elevation scales; respectively 5 meter Low Elevation Coastal Zone (LECZ) and storm surge water levels (see Figure 4). These two case study scales were chosen as they have been used relatively often in past executed coastal flood risk assessments based on both publicly available DEM datasets.

2.2. Lagos State Scale and Low Elevation Coastal Zone

McGranahan *et al.* [4] conducted a study on the estimation of the world's population living in LECZ. An estimation of the population of the world living in a 5 m. or 10 m. LECZ was made for many countries based on SRTM data [12] and GRUMP population data [20]. The concept of the Low Elevation Coastal Zone represents the scale level comparable with the global sea level rise tools which are available online. These tools support a quick and indicative analysis.

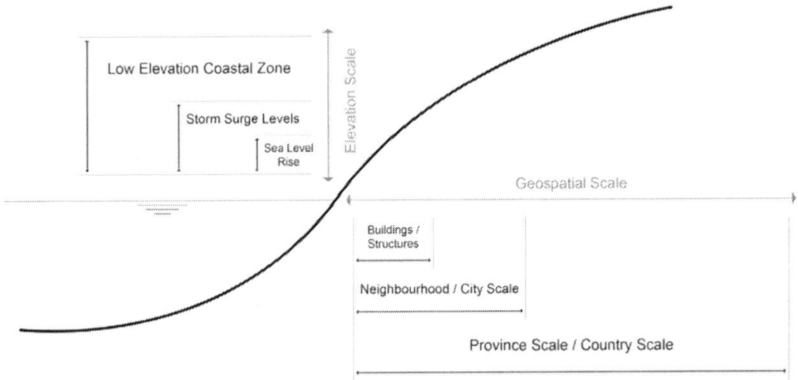

Figure 4. Different elevation and geospatial scales can be applied on coastal flood risk assessments.

In this focusing scale with Lagos State area as geospatial scale level, and the 5 m. LECZ as elevation scale level, the goal was to assess the land cover of the 5 m. LECZ for Lagos State area by the public DEMs and the LiDAR DEM. Because the entire bare earth point dataset of the LiDAR DEM is too extensive to be used for the entire area, a subset was used. In the subset strips of tiles were chosen to handle calculation time and reduce memory space.

The strips are oriented north-south because most elevation differences occur in north-south direction (perpendicular to the coastline and the major lagoons) and have an equal distance of 20 km from each other (see Figure 5).

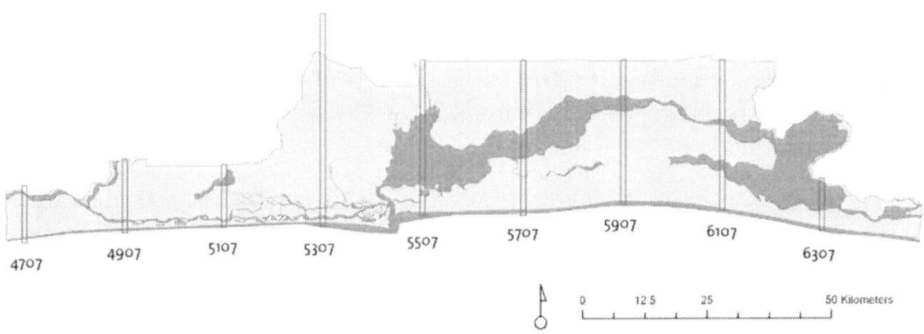

Figure 5. Strips of data tiles were selected because the full dataset for the Lagos State area was too extensive.

2.3. Lagos City Scale and Storm Surge Water Levels

For this focusing scale, both geospatial and elevation scale levels are smaller. In this case probable flood hazard areas of the city of Lagos were estimated by simulating a storm surge hitting the city of Lagos. Storm surges can differ in intensity and have a relation to the return interval. Storm surge height is directly related to the intensity of the storm. The storm intensity is also related to the frequency of storm surges in terms of return periods. For Lagos State the return periods of storm surges were investigated as well (see Table 1). Flooded areas due to storm surges can be simulated by exposing these heights to the public DEMs and the LiDAR DEM.

Table 1. Water levels used for the case study of Lagos City.

Return period		Storm surge height in m. relative to geopotential model EGM96
Present	1/100 yr	1.7 m
(without Sea Level Rise)	1/200 yr	2.0 m
Future	1/100 yr	2.7 m
(with Sea Level Rise)	1/200 yr	3.0 m

2.4. Statistical Analysis of Elevation Differences

The statistical elevation differences among the DEM datasets can be revealed by error statistics, because the bare earth points of the LiDAR DEM dataset were adopted as ground truth. The differences in estimated flood extents can be supported by error measurements. These error measurements can be used as a means to quantify the vertical accuracy of a DEM. The errors are calculated for each point by taking the difference of elevation of the GRID and the true elevation of the point. The error statistics were calculated for the 9 strips of the Lagos State case study scale.

The error statistics of the global public DEM datasets are assessed by three functions; the root mean square error of elevation measurements (see Equation 1), the mean error of elevation measurements (see Equation 2) and the standard deviation of elevation measurements (see Equation 3).

$$RMSE = \frac{\sum (Z_p - Z_{cell})^2}{n} \quad (1)$$

$$ME = \frac{\sum(Z_p - Z_{cell})}{n} \quad (2)$$

$$SD = \sqrt{\frac{\sum((Z_p - Z_{cell}) - ME)^2}{n-1}} \quad (3)$$

where Z_p is elevation of ground truth; Z_{cell} is elevation of model; n: sample size.

The RMSE is a widely used measure of conformity between a set of estimates and the actual values and has become a standard measure of map accuracy [21]. The mean error (ME) is an important component for elevation accuracy assessments because it provides an indication of whether the dataset is under- or overestimated in relation to ground truth. The standard deviation provides information on the spread of the errors, which indicates to what extent there is a variety in the magnitude of the errors in the dataset.

3. RESULTS

3.1. Results Lagos State Scale

The amount of LiDAR data was the inhibiting factor which meant the case study could not be assessed for the entire extent of Lagos State. In order to compare the LECZ areas based on the SRTM DEM, the ASTER GDEM, and the LiDAR DEM, only nine strips were used for comparison. A result for one of the strips is shown in Figure 6. The cumulative results are shown in Table 2.

Table 2. Numbers and percentages of the estimated LECZ areas.

DEM Datasets	LECZ area in m^2	% of LECZ area compared to LiDAR DEM
LiDAR DEM	78,095,118	-
ASTER GDEM	11,855,514	15
SRTM DEM	19,690,838	25

The results show that the estimated LECZ area estimated by the SRTM DEM is four times smaller than the LECZ area estimated by the LiDAR DEM. The results generated by the ASTER GDEM provide an area that is almost seven times smaller than LiDAR DEM estimations.

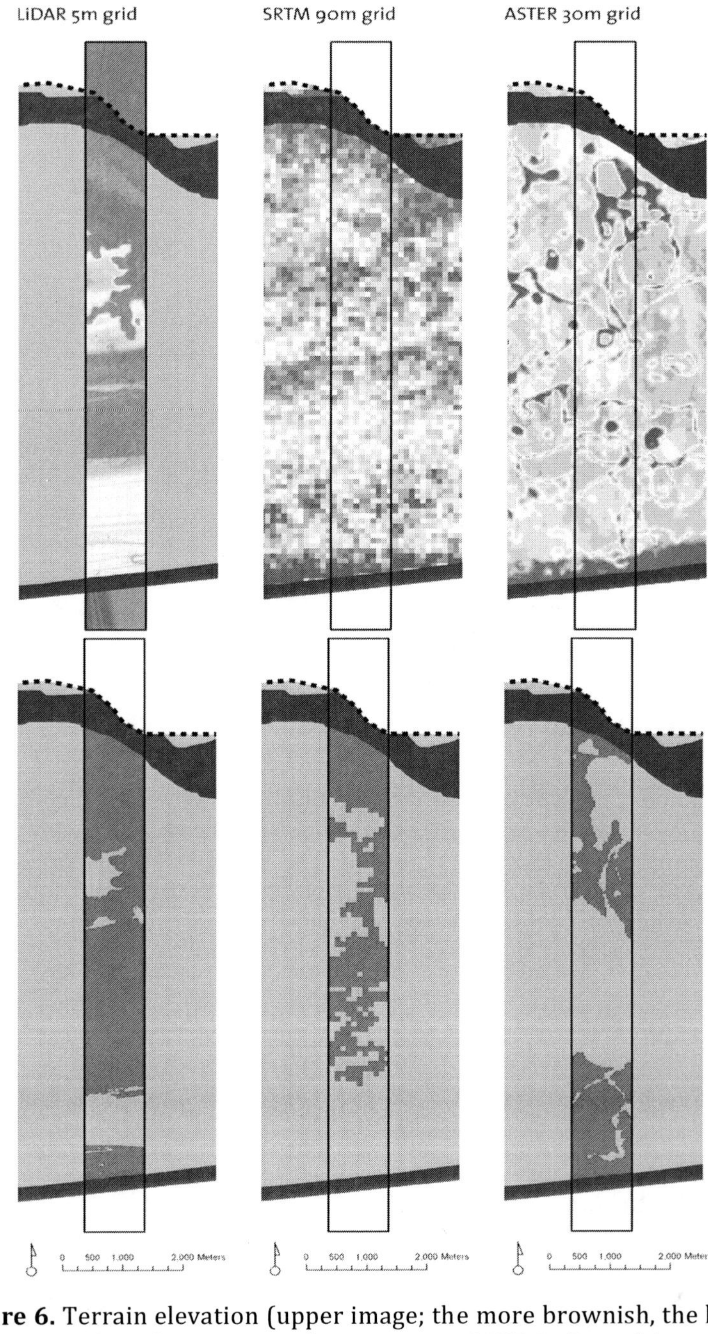

Figure 6. Terrain elevation (upper image; the more brownish, the higher the elevation) and LECZ area (lower image, LECZ indicated in red) for one of the strips.

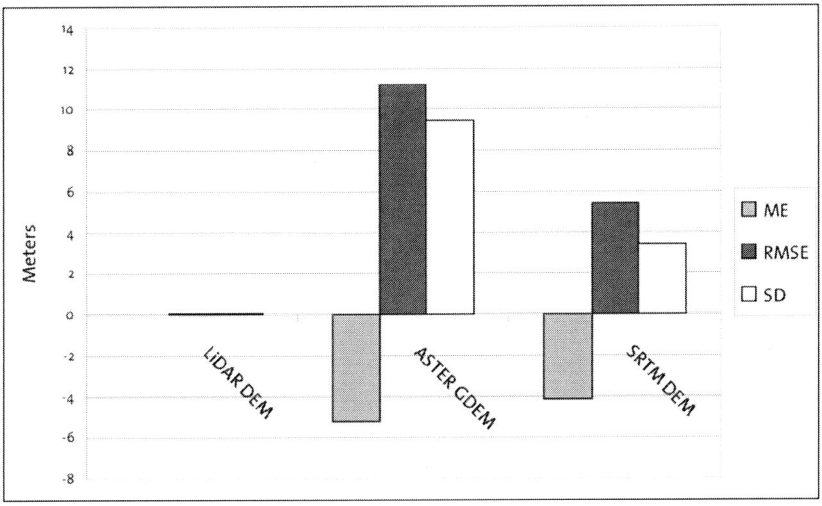

Figure 7. Error statistics.

For the error statistics you see larger magnitudes in all accuracy components for ASTER GDEM than for SRTM DEM (see Figure 7). SRTM DEM shows an ME of −4.2 m, whereas ASTER GDEM shows an ME of −5.25 m.

This means that both public DEM datasets overestimate the elevation for Lagos State area and consequently underestimate the risk of flooding. From the perspective of the elevation focusing scale of this case study, you see that the RMSE of both public DEMs (respectively 11.3 m and 5.3 m.) is even larger than the defined range of the LECZ (0–5 m above mean sea level).

3.2. Results Lagos City Scale

For the Lagos City study area it was possible to make use of the full data coverage of the study extent.

The focusing scale applied in this case study is more precise and is meant as a case study where the precision of estimations approximates detail levels which are tangible, such as affected neighborhoods or affected houses which are inundated in flood areas during a plausible event. The results are shown in Figure 8 and Table 3.

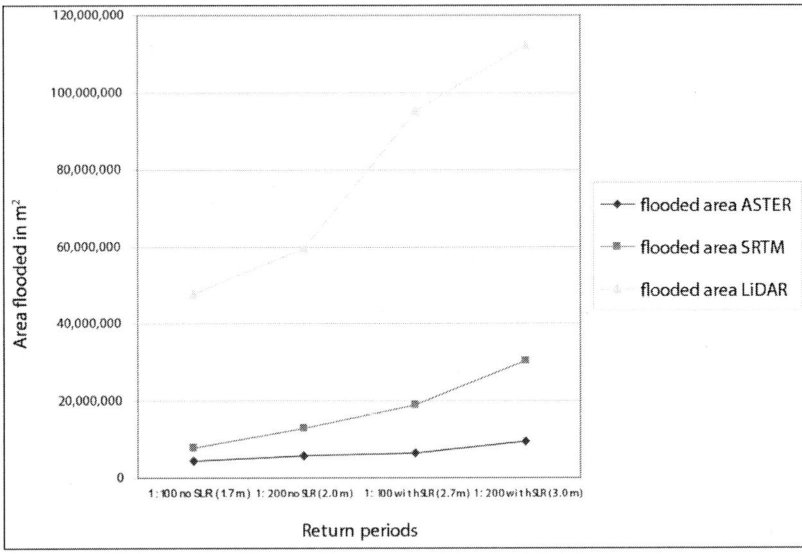

Figure 8. Flooded area estimated by ASTER, SRTM and LiDAR for Lagos City.

Table 3. Percentages of ASTER GDEM and SRTM DEM of flooded areas compared to LiDAR DEM.

Return period	Waterlevel in [m] + MSL	Sea level rise	% Flooded area by ASTER compared to LiDAR DEM	% Flooded area by SRTM compared to LiDAR DEM
1/100 yr	1.7	Without	9	16
1/200 yr	2.0		9	21
1/100 yr	2.7	With	7	20
1/200 yr	3.0		8	27

Again the flood areas (in Lagos State case, the LECZ areas) estimated by the ASTER GDEM and the SRTM DEM are significantly smaller than the estimations by LiDAR DEM. SRTM DEM estimates results that are closer to LiDAR DEM than ASTER GDEM.

It could be observed that in the Lagos City case scale, the deviations of both public DEMs in relation to the LiDAR DEM estimations were larger than the deviations in the Lagos State case scale. Especially ASTER GDEM (with percentages not higher than 10%) provides large hazard indication deviations.

4. CONCLUSIONS AND DISCUSSION

This study shows that attention should be paid in the application of publically available DEMs to support coastal flood risk assessments at different geospatial and elevation scales. In this case study, a comparison was made between public DEMs and very accurate vertical elevation measurements (LiDAR DEM). It can be concluded that the resulting flood risk can deviate significantly in such magnitudes that ASTER GDEM and SRTM DEM do not meet the accuracy requirements for the coastal flood risk assessment applied on Lagos State. Both ASTER GDEM and SRTM DEM overestimate the land elevation and, therefore, underestimate the flood hazard and associated risk. The overestimation is mainly due to the sensor techniques used in both publicly available datasets. The DSMs include elevations of vegetation and buildings. More research is needed to reveal the deviation magnitudes of elevations for different land-uses like forests, urban areas, and open areas. Although ASTER GDEM and SRTM DEM are Digital Surface Models instead of Digital Terrain Models, these datasets are widely used across the world for coastal flood risk assessments. Therefore these datasets should be compared and analyzed in original state to elevation data with a higher resolution and accuracy to reveal flood risk outcome differences. Underestimating hazards and risks in coastal flood risk assessments based on SRTM DEM or ASTER GDEM data is of concern, especially when investment decisions are made or policy procedures are adopted.

The flood inundation maps using SRTM DEM or ASTER GDEM show an inundated area at least four times smaller than a LiDAR DEM based inundation map. Although the horizontal resolution of ASTER GDEM is higher than SRTM DEM, the risk estimate of ASTER GDEM deviates significantly more from the LiDAR DEM risk estimate than the SRTM DEM does. A comparison of a coastal flood risk assessment at two geospatial scales (Lagos State and Lagos City) using the ASTER GDEM land elevation shows that the bias increases when the risk assessment becomes more geographically detailed, for instance when zooming in on a map. The same conclusion can be made for flood risk assessments using the STRM DEM land elevation, although the bias is smaller.

Despite the fact that this case study has been conducted for one location only (Lagos State area) the conclusions are relevant for other delta areas in the world as well. SRTM DEM and ASTER GDEM are extensively used across the globe in order to assess coastal flood risks. The findings are important to take into account when selecting publically available data sets for coastal flood risk assessments for low-lying deltas. In addition, conclusions based on a coastal flood risk assessment performed using these publically available data sets should be reevaluated taking into consideration this accuracy level. In addition to this, the absolute terrain elevation (relative to mean sea level, chart datum, *etc.*) is critical information for any GIS-based coastal flood risk assessment. If the elevation data (SRTM DEM or ASTER GDEM) shows an overestimation throughout its entire global dataset, this influences the results of any flood risk assessment with highly plausible underestimations of the flood risk. However, there is a need for further research to conclude whether there is a trend that SRTM DEM and ASTER GDEM show an overestimation of the elevation in low-lying delta areas across the globe with resulting underestimation of flood hazards for the studied areas.

REFERENCES

1. Intergovernmental Panel on Climate Change. Summary for Policymakers, The Physical Science Basis, Cambridge University Press, New York, NY, USA, 2007.
2. Nicholls, R.J.; Wong, P.P.; Burkett, V.R.; Codignotto, J.; Hay, J.; McLean, R.; Ragoonaden, S.; Woodroffe, C.D. Coastal Systems and Low-Lying Areas; Cambridge University Press: New York, NY, USA, 2007.
3. Nicholls, R.J.; Wong, P.P.; Burkett, V.R.; Woodroffe, C.D.; Hay, J. Climate change and coastal vulnerability assessment: Scenarios for integrated assessment. Sustain. Sci. 2008, 3, 89–102.
4. McGranahan, G.; Balk, D.; Anderson, B. The rising tide: Assessing the risks of climate change and human settlements in low elevation coastal zones. Environ. Urban. 2007, 19, 17–37.
5. Globalfloodmap Home Page. Available online: http://www.globalfloodmap.org (accessed on 26 July 2012).
6. Globalwarmingart. Available online: http://www.globalwarmingart.com/wiki/ Special: SeaLevel (accessed on 26 July 2012).
7. Flood Maps Home Page. Available online: http://flood.firetree.net (accessed on 26 July 2012).
8. EXCIMAP (European exchange circle on flood mapping), Handbook on Good Practices for Flood Mapping in Europe; Martini, F., Loat, R., Eds.; Drukkerij Feiko Stevens: Emmeloord, The Netherlands, 2007.
9. National Research Council of the National Academies. Mapping the Zone, Improving Flood Map Accuracy; The National Academies Press: Washington, DC, USA, 2009.
10. Colby, J.D.; Dobson, J.G. Flood modelling in the coastal plains and mountains: Analysis of terrain resolution. Nat. Hazards Rev. 2010, 11, 19–28.
11. METI & NASA Home Page. Available online: http://www.gdem.aster.ersdac.or.jp/ (accessed on 16 July 2012).
12. CGIAR: Consortium for Spatial Information (CGIAR-CSI) Home Page. Available online: http://srtm.csi.cgiar.org/(accessed on 16 July 2012).
13. Federal Emergency Management Agency. Airborne light detection and ranging systems. LiDAR Specifications for Flood Hazard Mapping; FEMA: Washington, DC, USA, 2002.
14. Demirkesen, A.C.; Evrendilik, F.; Berberoglu, S.; Kilic, S. Coastal flood risk analysis using Landsat-7 ETM+ imagery and SRTM DEM: A case study of Izmir, Turkey. Environ. Monit. Assess. 2007, 131, 293–300.
15. De Roo, A.; Barredo, J.; Lavalle, C.; Bodis, K.; Bonk, R. Potential flood hazard and risk mapping at Pan-European scale. In Digital Terrain Modelling; Springer: Berlin, Germany, 2007; pp. 183–202.

16. Lugeri, N.; Kundzewicz, Z.W.; Genovese, E.; Hochrainer, S.; Radziejewski, M. River flood risk and adaptation in Europe-Assessment of the present status. Mitig. Adapt. Strateg. Glob. Chang. 2010, 15, 621–639.
17. Li, Z.; Zhu, Q.; Gold, C. Digital Terrain Modeling: Principles and Methodology; CRC Press: Boca Raton, FI, USA, 2005.
18. Holmes, K.W.; Chadwick, O.A.; Kyriakidis, P.C. Error in a USGS 30-meter digital elevation model and its impact on terrain modeling. J. Hydrol. 2000, 233, 154–173.
19. Poulter, B.; Halpin, P.N. Raster modelling of coastal flooding from sea-level rise. Int. J. Geogr. Inf. Sci. 2008, 22, 167–182.
20. GRUMP dataset developed by Center for International Earth Science Information Network Home Page. Available online: http://sedac.ciesin.columbia.edu/data/set/grump-v1-urban-extents (accessed on 16 July 2012).
21. Fisher, P.; Tate, T. Causes and consequences of error in digital elevation models. Prog. Phys. Geogr. 2006, 30, 467–489.

CHAPTER 5

Technologies to Support Community Flood Disaster Risk Reduction

Ian McCallum1, Wei Liu1, Linda See1, Reinhard Mechler1, Adriana Keating1, Stefan Hochrainer-Stigler1, Junko Mochizuki1, Steffen Fritz1, Sumit Dugar2, Miguel Arestegui3, Michael Szoenyi4, Juan-Carlos Laso Bayas1, Peter Burek1, Adam French1, Inian Moorthy

1 International Institute for Applied Systems Analysis (IIASA), 2361 Laxenburg, Lower Austria, Austria

2 Practical Action Consulting South Asia, Kathmandu, Nepal

3 Practical Action Latin America, Lima, Peru

4 Zurich Insurance Group, 8022 Zurich, Switzerland

ABSTRACT

Floods affect more people globally than any other type of natural hazard. Great potential exists for new technologies to support flood disaster risk reduction. In addition to existing expert-based data collection and analysis, direct input from communities and citizens across the globe may also be used to monitor, validate, and reduce flood risk. New technologies have already been proven to effectively aid in humanitarian response and recovery. However, while ex-ante technologies are increasingly utilized to collect information on exposure, efforts directed towards assessing and monitoring hazards and vulnerability remain limited. Hazard model validation and social vulnerability assessment deserve particular attention.

New technologies offer great potential for engaging people and facilitating the coproduction of knowledge.

Keywords: Crowdsourcing, Disaster, risk reduction, Flood resilience, Social media, Volunteered geographic information (VGI)

1 INTRODUCTION

Climate-related hazards such as flooding, droughts, and windstorms are collectively responsible for the majority of recent disaster losses globally (IPCC 2012; UNISDR 2015) with floods affecting more people globally than any other type of natural disaster. This is partially a result of the rising trend in urbanization, which is altering land use and increasing the number of people living in areas that are highly exposed to the effects of natural hazards, such as communities located in river basins and coastal areas (Mileti 1999; Hallegatte 2011; GFDRR 2015).

Understanding and addressing each component of risk, and their interactions, is essential for effective flood disaster risk reduction (DRR). Technologies such as crowdsourcing [literally defined as outsourcing tasks to the crowd (Howe 2006)] is increasingly playing a role in DRR. In addition, social media services such as Twitter and Facebook are frequently used to monitor public reaction to floods, earthquakes, and fires (Abel et al. 2012; Earle et al. 2012; Kongthon et al. 2012), and volunteered geographic information (VGI) tools are now frequently used to complement official channels of humanitarian relief operations, as seen in the 2010 Haiti Earthquake (Norheim-Hagtun and Meier 2010; Zook et al. 2010), Hurricane Sandy in New York, 2012 (Schnebele et al. 2014), and the 2015 Nepal Earthquake (Clark 2015). The decentralized nature of disaster response and recovery makes new technologies a valuable tool for local monitoring and coordination and this is especially relevant in the area of ex-ante risk reduction. In this article, we outline recent technological applications in DRR and identify important areas for further engagement of technologies in reducing flood risk.

2 TECHNOLOGIES AND FLOOD DISASTER RISK REDUCTION

Flood risk may be defined as a function of hazard, exposure, and vulnerability (Cardona et al. 2012) and new technologies are revolutionizing the way in which information is collected and analyzed for each of these components (Jung 2011; Gómez et al. 2013; Poblet et al. 2014). Flood risk reduction takes place at different spatial scales, from local to global, with the scale of assessment often driving data requirements. Global mapping efforts such as Aqueduct and near-real-time early warning such as the Global Flood Awareness System (GloFAS) exist, but scales are often too coarse to be applied locally and empirical validation remains a challenge. In general, community level information is notoriously difficult to obtain, yet this is typically the focal scale of flood impacts and interventions.

Traditionally, community level information has been collected at the community level by nongovernmental organizations (NGOs), governments, and others in flood risk zones using participatory methods such as Participatory Vulnerability and Capacity Assessment (PVCA) resulting in a variety of different products, for example, maps of hazard, flood extent, exposure, and risk (Figure 1a). While this information is the necessary catalyst, it quickly becomes outdated and is difficult to merge with other information or to share with donors or other stakeholders. Efforts to digitize this information are under way (Figure 1b), achieving their greatest impact if the results are placed into shared open databases online, for example, OpenStreetMap (OSM) (Figure 2) (Kienberger 2014; de Andrade and Szlafsztein 2015; Chingombe et al. 2015; Thaler and Levin-Keitel 2016).

While extensive community-scale data are available in industrial countries and the more developed areas of the developing world, data deficiency is still a major obstacle to effective flood risk management across much of the developing world. The governments of many least developed countries (for example, Haiti and Nepal) usually have low human and technological capacity in disaster risk monitoring, but the emergence and popularization of mobile technology usage across the developing world is changing this. As a result, new technologies are now widely employed in flood response and recovery. The potential for these technologies gained international attention in the context of the catastrophic 2010 Haiti Earthquake where humanitarian responders had little or no access to detailed maps. Voluntary contributors rapidly began mapping road networks using satellite images and OSM—an effort later known as the Humanitarian OpenStreetMap Team (HOT). They went on to include baseline (transportation, education, health, water and sanitation facilities), humanitarian (hurricane disaster shelters and cholera-response structures), and community mapping as well as capacity building programs. In recent years, global initiatives such as the Standby Task Force and Digital Humanitarian Network are also harnessing the power of near-real-time mapping to support humanitarian response efforts. However, the application of technologies in ex-ante risk reduction is still limited, in particular in assessing critical risk factors—hazards and vulnerability. The following sections review the state of technologies used and identify important areas for further application in community flood risk management.

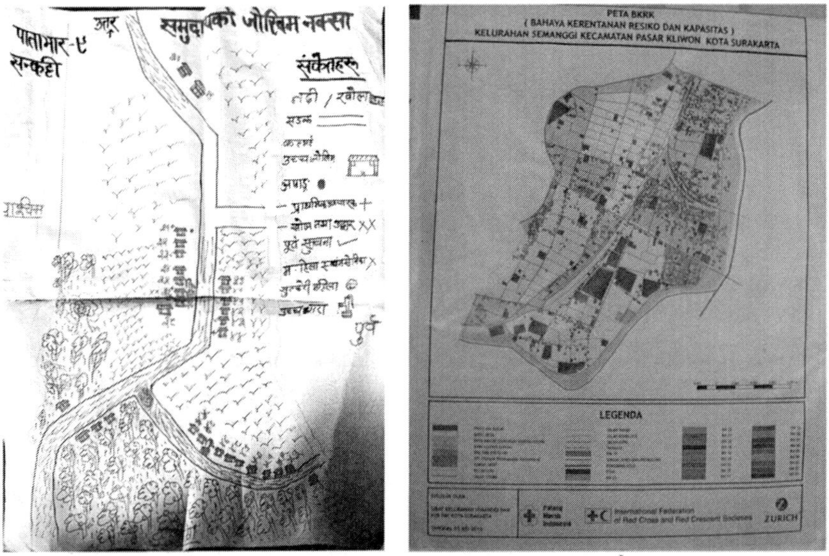

a b

Figure 1. **a** A Community flood risk map created by residents in the Sankatti community in the Karnali River basin, Nepal as part of an NGO-initiated, community-based vulnerability assessment conducted by Practical Action; **b** GIS-based risk map made by the Indonesian Red Cross Society, Palang Merah Indonesia (PMI) in the Bengawan River basin, Central Java, Indonesia

2.1 Hazard

To supplement traditional community-level participatory approaches such as PVCA, mobile phones are increasingly being used for the collection of data on flood events, for example, asking communities and/or NGOs to document the height that the flood level has reached and where the flooding has occurred or is occurring in real time.

This type of information is revolutionizing the speed and ways in which the outputs of a hydrodynamic model may be validated. For example, the PetaJakarta twitter application in Jakarta, Indonesia collected 150,000 tweets within 24 h of the onset of flooding on 5 February 2014, dynamically mapping where floods occurred across the city (Holderness and Turpin 2015). Patterns of mobile use can also give similar clues as to which regions are impacted most severely by flooding (Pastor-Escuredo et al. 2014). In these studies, mobile phone usage was seen to increase in areas hardest hit by flooding. These technologies may be used to complement, validate, or even replace flood mapping undertaken by other, more traditional means.

For example, crowdsourcing strategies costing as little as USD 60 may yield information equivalent to a conventional gauging station costing as much as USD 15,000 (Lowry and Fienen 2013).

Figure 2. The result of digitizing social, risk, and capacity maps into OSM for the Chakkhapur village, Rajapur Municipality, Bardiya District in the Karnali basin, Nepal. In addition to housing units, the local first aid and emergency shelter are indicated

2.2 Exposure

Mapping of exposed elements from the community up to the global level is possible through efforts such as OSM (for example, HOT and Missing Maps) and this is where a large portion of efforts have been targeted globally. Participatory and large-scale community mapping of exposed assets such as schools, hospitals, and other public and private buildings now exist in many areas across the globe, including developing countries such as the Philippines, Indonesia, Nepal, and South Sudan (GFDRR 2014).

Successful engagement of local human resources—including universities, NGOs and residents—is key to effective collection of information.

A variety of mobile applications exist that allow for the collection of in situ information related to infrastructure (for example, GPSLogger for Android devices and Pushpin OSM for Apple devices) with subsequent direct import into OSM. For effective flood risk reduction, it is desirable to create a detailed map in which the spatial distribution of the elements at risk are delineated, although a thorough assessment is often difficult and expensive due to the sheer number of the elements that should be considered (Hochrainer 2006).

2.3 Vulnerability

Vulnerability (divided into physical and socioeconomic aspects) is crucial for flood risk assessment, but is particularly poorly understood and quantified (Mechler and Bouwer 2014). Many vulnerability assessments undertaken in low-income, at-risk communities are focused on raising risk awareness and developing organizational capacity, and only a few local studies and assessments have used systematic techniques for recording, generating, and analyzing data (Government Office for Science 2012). Physical vulnerability is strongly linked to exposure and hence easier to determine.

For example, building taxonomies are needed that describe the characteristics of individual buildings or a class of building with similar characteristics. A recent effort such as the Community Based Early Warning System at the Búzi River in Central Mozambique is conducting community-level physical vulnerability mapping (Kienberger 2014).

Efforts to quantify socioeconomic vulnerability are extremely limited, and information of this kind is rarely integrated into risk assessments (GFDRR 2014). Studies such as Helgeson et al. (2013) and Enenkel et al. (2015) are exceptions, having recorded socioeconomic vulnerability such as the incidence of malnutrition ex-ante within affected communities through geospatial surveys using, for example, Open Data Kit (ODK) or Geographical ODK (GeoODK).

Given the cost-effectiveness and ease of implementation offered by many of these mobile technologies, vulnerability surveys may be easily repeated over time, greatly facilitating the tracking of long-term changes in socioeconomic vulnerability, as well as the detection of adaptation behaviors that become increasingly important in the context of altering patterns of flood risks and climate change adaptation.

3 TOWARD USER-CENTRIC FLOOD RISK ASSESSMENT

Community knowledge and feedback is essential in improving flood risk assessment and risk reduction interventions. Increasingly, development and humanitarian organizations working in the field of disaster risk reduction and climate change adaptation are using new technologies to facilitate data collection and monitoring (GFDRR 2014). In this context, the idea of "group-sourcing" or "expert-sourcing," that is, information collected by trained individuals by means of mobile technologies (Enenkel et al. 2015) may complement these community interventions.

Besides mapping critical spatial information through PVCA, for example, open-source tools like ODK are extensively used for collecting and sharing information on socioeconomic vulnerability of communities and the disaster risks they face. For example, ODK was used in Indonesia by the community-based disaster preparedness team (SIBAT) and volunteers of PMI to conduct baseline surveys in thousands of households from 21 communities in three river basins. Information collected includes the local people's living, health, environment, and disaster history and was used to assist in flood risk mitigation decision-making processes. Unlike a conventional paper-based survey, this mobile-based survey system has been quicker, more accurate, cost-effective, and easier to analyze. ODK has also been linked with InaSAFE (software that produces realistic natural hazard impact scenarios for better planning, preparedness, and response activities) by the same groups of SIBAT members and volunteers to map actual information about river area, housing, livelihood, vulnerable areas, and evacuation routes at community and regional scales to make spatially explicit risk maps. These preliminary but effective efforts have the potential to be replicated throughout the country to establish a nationwide database on disaster risk and community vulnerabilities.

Furthermore, the global Zurich Flood Resilience Alliance has established the Risk Geo-Wiki effort, a branch of the Geo-Wiki application (See et al. 2016), allowing for the sharing of data and information gathered by communities, NGOs, the private sector, and research institutions. A key gap that is being tackled is to produce actionable preevent information on risk, in particular on physical and socioeconomic vulnerability. For example, the approach and process can be used to provide information on physical vulnerability (level of flooding for a certain flood intensity), which is essential for developing local stage-damage curves (Moel et al. 2015). Additionally, the approach can be used to survey and validate latent resilience properties, such as socioeconomic characteristics of households and communities. In fact, a large focus of the global Zurich Flood Resilience Alliance involves measuring resilience and implementing a scorecard approach, which will be subjected to in situ validation via a crowdsourcing approach.

4 CONCLUSION

With a growing number of technologies available to aid flood disaster risk reduction, we have considered a selection of these across the disaster management cycle (Figure 3).

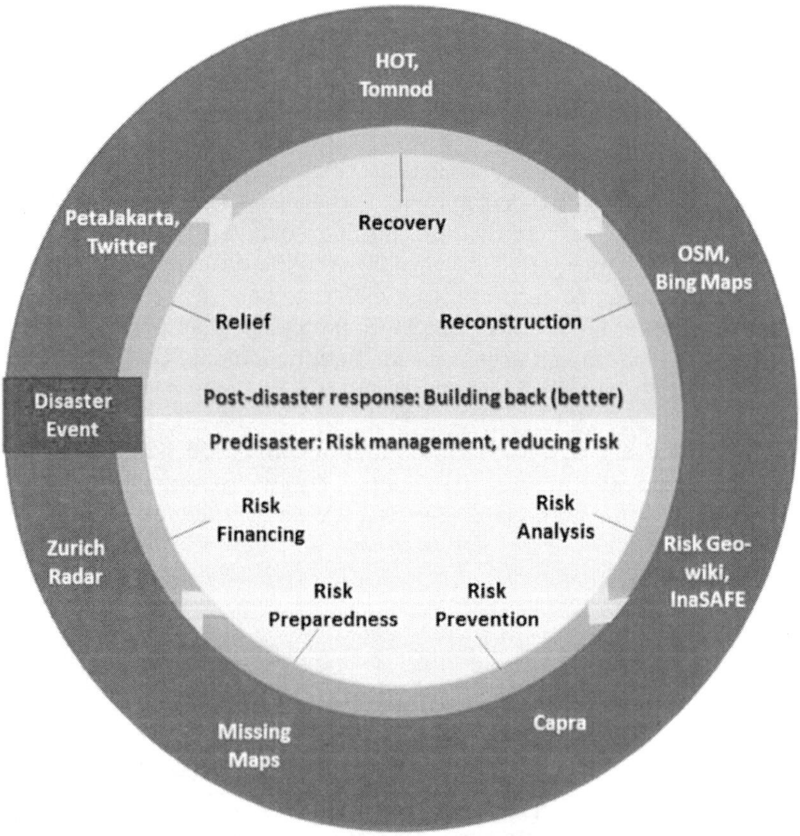

Figure 3. Examples (outer ring) of technologies and tools in the disaster management space. Relief typically relies on social media tools, along with crowdsourcing and VGI. Recovery and reconstruction tools are often reactive, with predisaster tools typically proactive

Post-disaster response begins with relief efforts that typically utilize social media (for example, Twitter) and various forms of crowdsourcing and VGI. Shifting to recovery efforts, tools such as HOT and Tomnod (a high-resolution satellite imagery-based crowdsourcing tool) are followed by reconstruction needs that often rely on more reactive tools (for example, OSM with Bing Maps).

Predisaster efforts tend to involve increasingly proactive technologies (for example, Risk Geo-Wiki, InaSAFE, CAPRA (the Central American Probabilistic Risk Assessment), Missing Maps (mapping the most vulnerable places in the developing world, in order that international and local NGOs and individuals can use the maps and data to better respond to crises), and Zurich Radar (the Zurich Radar for Natural Hazards allows for conducting a solid site and property analysis for real estate in Switzerland, based on hazard index maps).

While increasing use is being made of new technologies ex-post in disaster response and recovery, large potential lies in the application of these technologies ex-ante in disaster-prone regions. In particular, efforts that capture the newly acquired information in open databases (for example, OSM) offer multiple benefits beyond just the natural hazard community. While the demand is growing for disaster preparedness solutions, numerous barriers remain, including a lack of technical capacity and perhaps most importantly a lack of awareness of existing technologies.

ACKNOWLEDGMENTS

Funding from the global Zurich Flood Resilience Alliance is gratefully acknowledged. We thank Andreas Tanadi and Arfik Triwahyudi from Zurich Insurance Indonesia for providing detailed information on the implementation of the Community Flood Resilience Program in Indonesia.

REFERENCES

1. Abel, F., C. Hauff, G.-J. Houben, R. Stronkman, and K. Tao. 2012. Twitcident: Fighting fire with information from social web streams. In *Proceedings of the 21st International Conference Companion on World Wide Web*, 305–308. WWW'12 Companion. Lyon, France: Association for Computing Machinery. doi:10.1145/2187980.2188035.
2. Cardona, O.D., M.K. van Aalst, J. Birkmann, M. Fordham, G. McGregor, R. Perez, R.S. Pulwarty, E.L.F. Schipper, and B.T. Sinh. 2012. Determinants of risk: Exposure and vulnerability. In *Managing the risks of extreme events and disasters to advance climate change adaption. A special report of the Working Groups I and II of the Intergovernmental*

Panel on Climate Change, ed. C.B. Field, V. Barros, T.F. Stocker, D. Qin, D.J. Dokken, K.L. Ebi, M.D. Mastrandrea, et al., 65–108. Cambridge and New York: Cambridge University Press.

3. Chingombe, W., E. Pedzisai, D. Manatsa, G. Mukwada, and P. Taru. 2015. A participatory approach in GIS data collection for flood risk management, Muzarabani district, Zimbabwe. *Arabian Journal of Geosciences* 8(2): 1029–1040.
4. Clark, L. 2015. How Nepal's earthquake was mapped in 48 hours. *Wired UK*. http://www.wired.co.uk/news/archive/2015-04/28/mapping-nepal-after-the-earthquake. Accessed 23 May 2016.
5. de Andrade, M.M.N., and C.F. Szlafsztein. 2015. Community participation in flood mapping in the Amazon through interdisciplinary methods. *Natural Hazards* 78(3): 1491–1500.
6. Earle, P.S., D.C. Bowden, and M. Guy. 2012. Twitter earthquake detection: Earthquake monitoring in a social world. *Annals of Geophysics* 54(6): 708–715.
7. Enenkel, M., L. See, M. Karner, M.D. Alvarez, E. Rogenhofer, C. Baraldes, C. Lanusse, and N. Salse. 2015. Food security monitoring via mobile data collection and remote sensing: Results from the Central African Republic. *PLoS One* 10(11): e0142030.
8. GFDRR (Global Facility for Disaster Reduction and Recovery). 2014. *Understanding risk: The evolution of disaster risk assessment*. http://www.drrinacp.org/sites/ drrinacp.org/files/publication/_Understanding_Risk-Web_Version-rev_1.7.3.pdf. Accessed 23 May 2016.
9. GFDRR (Global Facility for Disaster Reduction and Recovery). 2015. *Bringing resilience to scale, GFDRR Annual Report'14*. Washington, DC: GFDRR. https://www.gfdrr.org/sites/gfdrr/files/publication/GFDRR%20ANNUAL%20REPORT%202014.pdf. Accessed 23 May 2016.
10. Gómez, D., A.M. Bernardos, J.I. Portillo, P. Tarrío, and J.R. Casar. 2013. A review on mobile applications for citizen emergency management. In *Highlights on practical applications of agents and multi-agent systems*, ed. J.M. Corchado, J. Bajo, J. Kozlak, P. Pawlewski, J. M. Molina, V. Julian, R.A. Silveira, R. Unland, and S. Giroux, 190–201. Berlin: Springer.
11. Government Office for Science. 2012. Foresight reducing risks of future disasters: Priorities for decision makers. Executive Summary. London: The Government Office for Science. https://www.gov.uk/government/uploads/system/uploads/attachment_ data/file/286480/12-1322-reducing-risks-of-future-disasters-summary.pdf. Accessed 23 May 2016.
12. Hallegatte, S. 2011. *How economic growth and rational decisions can make disaster losses grow faster than wealth*. Policy research working paper. Washington, DC: World Bank.
13. Helgeson, J.F., S. Dietz, and S. Hochrainer-Stigler. 2013. Vulnerability to weather disasters: The choice of coping strategies in rural Uganda. *Ecology and Society* 18(2): Article no. 2.
14. Hochrainer, S. 2006. *Macroeconomic risk management against natural disasters*. Wiesbaden: German University Publisher.
15. Holderness, T., and E. Turpin. 2015. PetaJakarta.org: Assessing the role of social media for civic co-management during monsoon flooding in Jakarta, Indonesia. White paper. SMART Infrastructure Facility, University of Wollongong.
16. Howe, J. 2006. The rise of crowdsourcing. *Wired Magazine* 14(6): 1–4.
17. IPCC (Intergovernmental Panel on Climate Change). 2012. *Managing the risks of extreme events and disasters to advance climate change adaptation. A special report of Working Groups I and II of the Intergovernmental Panel on Climate Change*. ed. C.B. Field, V. Barros, T.F. Stocker, D. Qin, D.J. Dokken, K.L. Ebi, M.D. Mastrandrea, et al. Cambridge and New York: Cambridge University Press.
18. Jung, C. 2011. Mobile data collection systems: A review of the current state of the field. Centre National d'Etudes Spatiales. http://www.parkdatabase.org/files/documents/nomad_mdc_research.pdf. Accessed 23 May 2016.
19. Kienberger, S. 2014. Participatory mapping of flood hazard risk in Munamicua, District of Búzi, Mozambique. *Journal of Maps* 10(2): 269–275.
20. Kongthon, A., C. Haruechaiyasak, J. Pailai, and S. Kongyoung. 2012. The role of Twitter during a natural disaster: Case study of 2011 Thai flood. In *2012 Proceedings of PICMET'12: Technology Management for Emerging Technologies*, 29 July–2 August 2012, Vancouver, BC, 2227–2232.

21. Lowry, C.S., and M.N. Fienen. 2013. CrowdHydrology: Crowdsourcing hydrologic data and engaging citizen scientists. *Ground Water* 51(1): 151–156.
22. Mechler, R., and L.M. Bouwer. 2014. Understanding trends and projections of disaster losses and climate change: Is vulnerability the missing link? *Climatic Change* 133(1): 23–35.
23. Mileti, D. 1999. *Disasters by design: A reassessment of natural hazards in the United States.* Washington, DC: Joseph Henry Press.
24. Moel, H., B. Jongman, H. Kreibich, B. Merz, E. Penning-Rowsell, and P.J. Ward. 2015. Flood risk assessments at different spatial scales. *Mitigation and Adaptation Strategies for Global Change* 20(6): 865–890.
25. Norheim-Hagtun, I., and P. Meier. 2010. Crowdsourcing for crisis mapping in Haiti. *Innovations: Technology, Governance, Globalization* 5(4): 81–89
26. Pastor-Escuredo, D., A. Morales-Guzman, Y. Torres-Fernandez, J.-M. Bauer, A. Wadhwa, C. Castro-Correa, L. Romanoff, J.G. Lee, A. Rutherford, V. Frias-Martinez, N. Oliver, E. Frias-Martinez, and M. Luengo-Oroz. 2014. Flooding through the lens of mobile phone activity. In *Proceedings of the IEEE Global Humanitarian Technology Conference*, 10–13 October 2014, San Jose, CA, 279–286.
27. Poblet, M., E. García-Cuesta, and P. Casanovas. 2014. Crowdsourcing tools for disaster management: A review of platforms and methods. In *AI Approaches to the Complexity of Legal Systems,* ed. P. Casanovas, U. Pagallo, M. Palmirani, and G. Sartor, 261–274. Berlin: Springer.
28. Schnebele, E., G. Cervone, and N. Waters. 2014. Road assessment after flood events using non-authoritative data. *Natural Hazards and Earth System Science* 14: 1007–1015.
29. See, L., P. Mooney, G. Foody, L. Bastin, A. Comber, J. Estima, S. Fritz, et al. 2016. Crowdsourcing, citizen science or volunteered geographic information? The current state of crowdsourced geographic information. *ISPRS International Journal of Geo-Information* 5(5): Article no. 55.
30. Thaler, T., and M. Levin-Keitel. 2016. Multi-level stakeholder engagement in flood risk management—A question of roles and power: Lessons from England. *Environmental Science & Policy* 55: 292–301.
31. UNISDR (United Nations International Strategy for Disaster Reduction). 2015. *Making development sustainable: The future of disaster risk management. Global assessment report on disaster risk reduction.* Geneva: UNISDR.
32. Zook, M., M. Graham, T. Shelton, and S. Gorman. 2010. Volunteered geographic information and crowdsourcing disaster relief: A case study of the Haitian earthquake. *World Medical & Health Policy* 2(2): 7–33.

CHAPTER 6

Assessment of Prolonged Unsafe Conditions through Tidal Flood and Rain water Flood at Lower Bengal Basin, India

Barman NK[1*], Chatterjee S[2], Paul AK[1] and Pradhan MK[1]

[1] *Department of Geography and Environment Management, Vidyasagar University, Midnapore-721102, West Bengal, India*

[2] *Department of Geography, Presidency University, Kolkata, India*

ABSTRACT

Khejuri police station is located at the lower Bengal Basin near the Hugli river mouth under Midnapore (east). It is considered as a littoral tract region having monsoon climatic character. Naturally this area is known for its agricultural potential. Rice, Betel leaves, Lemons and various types of vegetables like potato, Brinzal, Chilli, Sugarcane stick etc. are the main crops produced which are sent to local and national markets which is important in the context of national trade. The present paper seeks to analyze the problems of land degradation due to recurring water logging that has been instrumental in losing the earlier glory of the area as an agriculturally potential tract.

Keywords: Water logging; Bengal basin; Littoral tract; Agricultural potential

1. INTRODUCTION

Khejuri a part of Kanthi coastal tract (regionally extended from Rupnarayan River to Subarnarekha delta around Hugli estuary) is associated with several beach ridges and sand dunes and intervening tidal flats at the south west of this area [1]. The area is also formed under deltaic floodplain environment of Hugli estuary. Later the area was modified by tidal process. This coastal low land tract lies at a height about 2-3 m above mean sea level. The beach ridges and sand dunes of this area are partly obliterated, segmented and cut by tidal waves, **ANTHROPOGENIC** activities and weathered in sub aerial environment [2]. These coastal features suggest that the area was once lying under older coastal control and estuarine environment. Belong a low land area, Khejuri is also prone to water logging, flood risk and storm hazards due to estuarine location [3]. The entire area is protected by earthen embankment to avoid water logging during storm and high tide episodes [4,5].

Water logging is one of the major problems of land degradation in India. Unscientific management of soil, water and crops in irrigated lands and obstruction of natural drainage systems by various developmental activities are the main factors responsible for disrupting the balance of inflow and outflow of water, leading to water logging [6]. While irrigation has increased by leaps and bounds, its attendant problem of water logging is now plaguing substantial area of agricultural lands.

1.1 The Study Area

Khejuri Police Station is formed of two blocks namely, Khejuri- I and Khejuri- II, extending between 21°47'4" N to 22°4' N latitudes and 87°45'4" E to 88°18' E longitudes covering an area of 267.97 Sq.km. (**TABLE 1**).

Table 1: The physical properties of the study area according to block.

Block	Area (Km^2)	% of Area to total district area	Physical Boundary
Khejuri-I	130.51	3.04	Talpati channel in the North. Rasulpur in the South. Khejuri – II in East. Thakurnagar Khal and Bhagwanpur-II Block in West.
Khejuri II	137.46	3.20	Talpati channel in the North. Hugli river in the East. Rasulpur in the South-West. Khejuri – I in the West.

Khejuri is a part of coastal plain of west Bengal by the western bank of giant funnel shaped Hugli estuary, lying just to north of Bay of Bengal (**FIGURE 1**). Hence this area is frequently attracted by cyclones that take of course along the Hugli estuary.

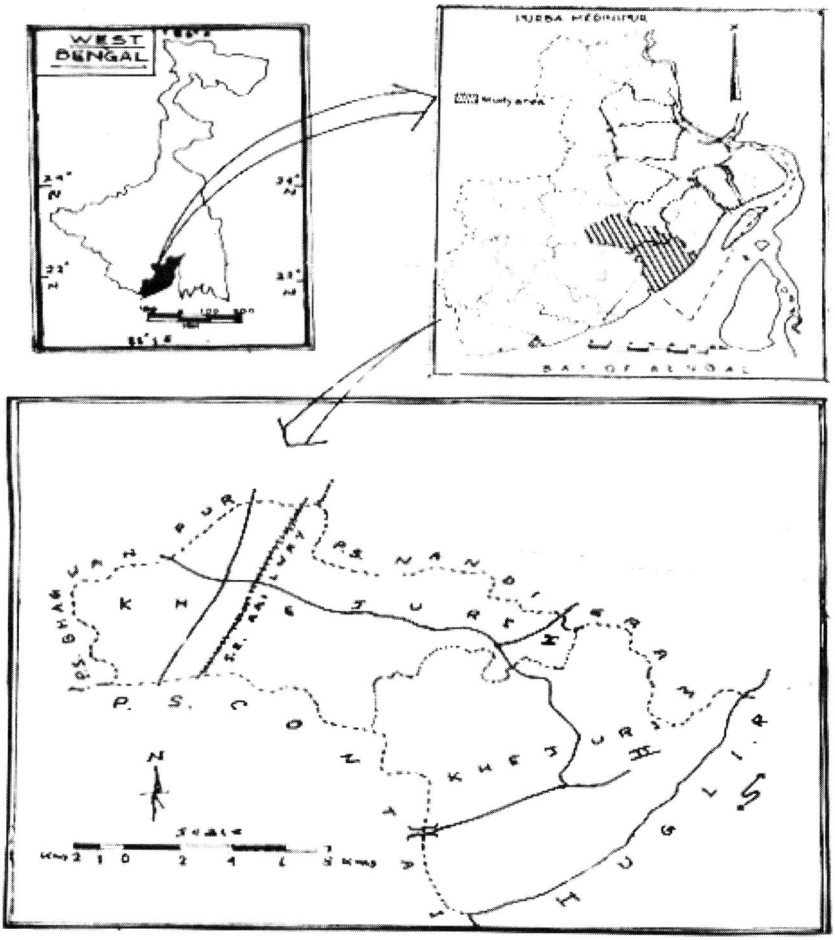

Figure 1: Location Map of the Study Area extending between 21°47'4" N to 22°4' N latitudes and 87°45'4" E to 88°18' E longitudes.

At the event of a cyclone amplification of storm surges causes inland flooding of vast area [7]. Some devastating cyclones like Calcutta cyclone (1737) Midnapore cyclone (1864, 1942) Sundarban cyclone (1988 and 1989) many other cyclones occurred during October and November and even 'Aila' of 2009 struck Khejuri due to its estuarine location [8]. The south-west monsoon (June- September) appears in the lower Bengal with a series of depressions developed at the head of the Bay Bengal and also recedes from west Bengal (Oct. Nov.) by forming depression over Bay of Bengal [9].

Among these depressions at least 80% (**TABLE 5**) turn into storms and lash on the coastal area of Bengal Basin. Usually, high monsoon rainfall over Bengal coastal plain, downstream discharge of Bengal Rivers and the seasonal high seas of the Bay of Bengal modify tide and surges during cyclonic episodes.

2. IDENTIFICATION OF WATER LOGGING CONDITIONS

An irrigated area is said to be waterlogged when the surplus water stagnates due to poor drainage or when the shallow water table rises to an extent that soil pores in the root zone of a crop become saturated, resulting in restriction to the normal circulation of the air, decline in the level of oxygen and increase in the level of carbon dioxide [7]. The actual depth of water table, when it starts affecting the yield of the crops adversely, may vary over a wide range from zero for rice to about 1.5 meters for other crops.

A Working Group constituted in 1991 by the Ministry of Water Resources to identify the problem areas affected by water logging/ salinity/ alkalinity in existing irrigation projects in the country and to suggest suitable remedial measures for their reclamation adopted the following norms for identification of waterlogged areas (**TABLE 2**):

Some States, however, have adopted different norms for defining waterlogged areas according to their own conditions and perceptions.

Table 2: The norms for identification of waterlogged areas according to their own conditions and perceptions.

Water logged Areas **(Due to rise in water Table)**	**Water table within 2 meters of the land surface**
Potential Areas for water logging	Water table between 2 to 3 meters below land surface
Safe Areas	Water table below 3 meters of land surface

3. MAJOR CAUSES OF WATER LOGGING CONDITION OVER THE STUDY AREA

The prime causes of water logging condition of this geomorphic part have been identified by the present researchers as well as the Administrative Department of East Medinipore. These causes are very frequent in nature in this part of lower coastal littoral tract. The causes are as follows:

a) The earthen embankment of the river Rasulpur and Hugli are in very poor condition.

b) The excess water of the river Subarnarekha from Orissa over flown and becomes stagnant during the rainy season which inundates the coastal part of Khejuri.

c) Sea coast saving embankment is in weak condition. Some time it is broken by the lashing of wave water and allows the tidal water to enter into the low lands of the area.

d) In some parts, the old embankments are very so weak that after heavy rain that are unable to withstand the water pressure and thence undergo wreckage.

e) Over flow of the river Rasulpur, Hugli and the water surges of Bay of Bengal frequently inundate the three sides (Excluding western) of Khejuri even in pre and post- monsoon period.

District administration have also declared Khejuri- I, as light flood zone and Khejuri- II, block as the cyclonic and flood prone zone (Disaster Management Plan of Purba Medinipur, 2002, Office of district Magistrate). In course of time the tidal canals of the study area have fallen victim of decay and causes water logging.

3.1 Man-made causes

1) The entire southern part of Khejuri belt adjoining the Hugli and Rasulpur River is known as "JALPAI" which means an area from where huge fuel-wood was collected for making salt from brine [10]. Once one a time Khejuri was noted for making this type of salt. So it paved the way for massive deforestation causing enormous soil erosion in this area. Naturally this erosion helped silting in the river bed.

2) The density of population here demands varieties of trees and plants in place of mangrove swamps seemingly less valuable in the eyes of the people of this place. But its saline environment, only suitable for

mangrove swamps, cannot stand the indiscriminate felling of these trees and so it causes quick erosion of soil and silting.

3) The natural flow of the tidal channels here is mostly impeded for fishing purpose by way of casting nets and making petty cross earthen embankments in these channels. For example Hijili Tidal Canal (H.T.C.), once an important waterway for trade and commerce, is now mostly obstructed for fishing purposes.

4) As per need of human habitation here many 'khals' (narrow tidal channels linked with the big tidal channels) have been occupied forcibly by way of filling mud into the beds of these. Secondly, culverts and foot-bridges have been constructed for communication and transport. All these cause obstruction in the natural flow of water in these channels.

5) In the past, up to mid-19th century brine was allowed to enter into the channels for making salt. But now salt making is totally stopped and brine is not allowed to enter as was previously done. So there is dearth of easy flow of water.

6) More over brine is not helpful for agriculture. So it is not allowed to enter through channels. It can be referred that, in 1832 an earthen embankment was constructed by the then British government around the river side for the purpose of restraining brine for agriculture.

3.2 Natural causes

1) This area is recognized by National Atlas and Thematic Mapping Organization, Govt. of India as recent alluvial zone. Consequently, it facilitated the quick erosion of soil.

2) Actually Khejuri is situated to the north part of the Digha-Kanthi coastal sand dune belt. Once there were also many sand dunes over the area. Due to repeated cyclones, storm surges and some man-made activities, these sand dunes were destroyed prompting future siltation in channel beds.

3) For its geographical location by the side of Hugli mouth, sea wind and waves are strong enough to cause erosion and silting simultaneously.

4. WATER LOGGING SCENARIO IN THE STUDY AREA

Water logging may be a result of both natural and man-made factors. Natural factors may include poor natural drainage as a consequence of unfavorable sub-soil geology like existence of hardpan at shallow depths; spilling of rivers resulting in submergence of agricultural lands; heavy storm rainfall event etc. [11]. Water logging is, however, caused mainly because of man-made factors like deforestation. Developmental activities such as construction of roads, bridges and dwelling houses lead to the choking of natural drainage.

Even a rain having duration of few hours causes water-logging for few days. All the Gram Panchyayets (G.P.) of Khejuri-I block except the Kamarda and Kalagachia G.P of the study area water logging is very common. Inadequacy of outlets for drainage of rain water; entry of high tide water along the channels hindering passage of excess water of rain and flooding due to embankment breaching are the causes behind water logging in this area. Not only in 1942, had most of the area under study remained under the water of height about 5 feet by only one day cyclonic rain of 5 hours duration on 25th May, 2009 kept the entire of Khejuri block especially the river side low lands, under water of considerable depth. On the year 2005, the one day rain of 3 hours duration kept the water standing up to two feet height in the central part of Khejuri II block along the river side low lands.

The following **TABLES 3** and **4** gives an idea of severity of the problem in the area by representing the effects of a 7 days rainfall event in 2004. So far as the month wise distribution of storms and depressions originating in Bay of Bengal is considered. There is a remarkable concentration of such events in October and November. During this post **MONSOON** phase of hydrologically saturated regime, occurrence of cyclones with high rainfall leads the area to be affected severity by flooding.

The tropical storms and depressions for the period 1971-1990, except for a small number that formed over land during the south west monsoon months, June to September, all the remaining disturbances were of sea origin. The monthly frequencies of occurrence of a period for the last 20 years are given below (**TABLE 5**).

In effect of the Cyclone 'Aila' a large section of backward and marginal people of the area became homeless and their suffering knew no bounds. Mainly sea facing villages like Pachuria under Khejuri –II block submerged after bursting of sea dyke [9-11]. 6 rescue camps at Khejuri-II were set by the Government to provide immediate relief and support to about 20,000 affected people.

Table 3: The Prime Causes of Water Logging Condition During the October 3rd to 7th in 2004 at Khejuri-II Block.

Khejuri-II Block	
Causes of calamity	Continuous Rain fall
Occurrence period of rain	3rd to 7th October, 2004
Affected area	116 sq.km. out of 133.96 sq.km
No. of mouzas affected	95 out of 99
No. of people affected	1,13,940 out of 1,17,409
No.of house affected	368
Damage of crops in hectares	2417 out of 4616
Pan boroj (Beetle leaf Garden) affected in hectare	122
Relief measures	Tarpaulins sheets, cloths, ice, medicines, water pouches distributed from different sectors.
Source: Block Development Office, Khejuri-II D.R.Cell,Memo No. 1165 /14.10.2004.	

Table 4: Statement Showing the Damage Due to Water-Logging by Heavy Rain Fall on 2004, Khejuri Police Station, West Bengal.

Sl	Names of the G.P.	Types of the Crops						
		Beetle vine yards in hectare			**Summer vegetables in hec.**		**Orchard in hec.**	
		Area Covered	Affected /Damaged	No. of Boroj	Area Covered	Affected/ Damaged	Area Covered	Affected / Damaged
1	Heria	3	2	30	23	10	12	4
2	Lakshi	4	2.4	65	25	15	10	5
3	Tikashi	5	5	50	15	10	10	5
4	Birbandar	5	5	60	10	5	5	3
5	Kamarda	10.5	5	150	5	3.5	4	3
6	Kalagachia	7.5	5	120	5	3.5	5	3
	Total	35	24.4	475	83	47	46	23

Source: Agricultural Development Office, Khejuri –I Block, (30/5/2009).

Table 5: Monthly Frequencies of Storms and Depressions in Bay of Bengal Region (1971-1990).

Type of Calamities	Jan	Feb	Mar	Apr	May	Jan	July	Aug	Sep	Oct	Nov	Dec	Total
Storms	1	0	0	2	11	3	3	4	7	16	27	8	82
Depressions	2	0	0	0	1	13	11	29	10	15	7	8	96
Total	3	0	0	2	12	16	14	33	17	31	34	16	178

Source: An addendum to the storm track atlas 1877-1970. Indian Meteorological dept.-1996, Mousom Bhaban New Delhi, Page i.

Table 6: Damage Impacts of 'Aila', 2009 in Purba Medinipur District in Terms of House and Population.

Name of District	Name of Sub-Division	Name of Block	No. of Affected Village	No. of House Fully Destroyed	No. of House Partly Destroyed	Population Affected	Sector of Population Most affected
Purba Medinipur	Contai	Contai-I	50	60	1000	25000	Fishermen and Marginalized Families, Children, and women
		Contai-II	40	75	500	20000	
		Bhagwanpur-II	25	7	400	15000	
		Ramnagar-I	75	15	750	30000	
		Ramnagar-II	65	10	500	25000	
		Khejuri-I	20	10	500	20000	
		Khejuri-II	30	215	350	12000	

Source: Cyclone Aila report of W.B.Govt. 2009.

Due to breaching of embankment along Rupnarayan River area in its western part of Khejuri –I Block got inundated. Following table (**TABLE 6**) has given the detail of damage impacts of 'Aila' 2009 in Purba Medinipore District.

A few low lying pockets in the area faces water logging during kharif season due to poor drainage system resulting from deposition of silt along the drainage channels. The total drainage length is sufficient for the area but the problem may be solved properly. Most of the channels are virtually connected to the two rivers-Hugli and Rasulpur river. Hence, there happens ingress of saline tidal water dominantly during summer months, which increases the salinity of the area. This may be prevented by constructing sluice gates on selected drainage points.

The inherent salinity being at considerably lower level is not a problem for the area. Some areas which are very frequently flooded by saline tidal water during high tide become highly saline. This problem gets minimized on repeated washing by the rain water during Kharif season. In some Entisol area sub-soil salinity increases during Rabi season if faulty irrigation practice is followed. The growing of salt tolerant crops with a limited irrigation is suggested. In the pre-monsoon summer months, the unsaturated coastal soil of the area may be encrusted with salt by saline water when large cyclone passes over the area with high tidal waves [12].

Logging of saline water has not only worsened the agricultural practices but also it prompted the spread of some tropical disease like Malaria, Blood dysentery, Small Pox and Cholera. The saline ocean water is stored in the ditch like low lands for the salt preparation in this coastal region. Salt is manufactured during the months from January to May. After that, in the rainy season rain water mixes with the saline water disabling the process of making salt. This stored water creates an unhygienic environment offers an ideal place for mosquito breeding and Malaria spreads. So the people of this area are used to be attacked by a native proverb goes as -'It is one thing to go to Hijli but quite another to come back alive'. When Job Charnok took shelter over here in 1687, Khejuri was then attacked with a severe Malaria episode [13].

Characteristically Khejuri is an alluvial tract, similar to most districts of the Gangetic plain in lower Bengal. So the environment had always been conducive for living. Just 500 years ago people from Western side of this tract came over to here for living and were mostly engaged in salt manufacturing, agriculture and collection of sea fish. The anthropogenic modifications of the landscape upon increasing population pressure have added much fuel to the water logging problem (**TABLE 7**) [14].

Table 7: Temporal Change in Number of Mouzas of the Study Area under Different Population Density Classes (1951-2001).

Population density Classes	1951	1961	1971	1981	1991	2001	Average growth rate/ year 1951-1961	Average growth rate/ year 1961-1971	Average growth rate/ year 1971-1981	Average growth rate/ year 1981-1991	Average growth rate/ year 1991-2001
Below 401	79	55	34	21	14	7	-3.04	-3.82	-3.82	-3.33	-5
401-800	44	62	74	77	54	45	4.09	1.94	0.41	-2.99	-1.67
801-1200	7	14	18	24	44	50	10	2.86	3.33	8.33	1.36
1201-1600	3	1	8	8	17	18	-6.67	70	0	11.25	0.59
Above 1600	0	2	1	4	8	17	0	-5	30	10	11.25
Depopulation	7	7	6	7	4	4	0	-1.43	1.67	-4.29	0
Total	140	141	141	141	141	141					

Source: District Census Hand Book, Govt. of India, 1951, 1961, 1971, 1981, 1991, 2001. N.B.: After 1951 AMouza Named Kamdebnagar (J.L. No. 142) was formed.

5. LONG TERM PROBLEMS PRODUCED BY WATER LOGGING CONDITIONS

The study area, a part of the **GANGETIC** alluvial plain in lower Bengal, receives medium-high rainfall (average 175.26 cm/y) and the soils have developed from sand and clay brought down by the river Ganga from the country above. Moreover, the area is situated near the sea coast and low-lying areas are traversed by tidal river and creeks. All these factors collectively play and increased the soil salinity and set negative impacts on agricultural crops [8-11] (**TABLE 8**). Other rivers being incapable of holding excess water during monsoon are apt to overflow and cause serious damage to standing crops. In the events the period of inundation, large section of the area covered by deposits of sands. To keep the area out of these floods, many embankments have been constructed around flood prone areas, called 'circuits'. As such much of the alluvial tract is consequently crisscrossed with a network of embankments. Transverse embankments are constructed across the channels which prohibit ingress of tidal saline water but accelerate the silting process causing decline in the capacity of channels to hold water.

In the coastal area, embankments of extensive length are required for protecting the coast line from the invasion of river, for preventing the inrush of salt water along tidal creeks and for avoiding the submersion of low lands by the rivers overflowing their banks. In most cases, the river have ceased to be a drainage channel and water holding capacity diminishes as its approaches towards mouth, while its bed rises in elevation.

Embankments are thus urgently required in the tidal areas for **PROTECTION** of the lands in the interior, which slope away from the rivers and form saucer-shaped hollows between them. It is said in Hijili, that a single overflow of salt water is fatal to crops for the next three years [15]. Poorly managed embankment of river system and utterly neglected drainage systems are instrumental for water logging in the area which not only causes serious agricultural damages but also various infectious diseases to spread. Cholera took an epidemic form in 1901, 1902, 1906 and 1907. A death density of 25.8 people per square mile and 3.8 people per square mile were recorded in 1906 and 1907 respectively due to blood dysentery. A special investigation about the disease was made in 1906-07 by captain W.E.H. Forster, I.M.S. by whom a method of vaccine therapy was introduced. After that the death rate was checked. About 17841 lives were lost by smallpox in 1902, and in the coastal area the death density was 6.39 people per square mile [15].

Table 8: Soil Problems in Terms of Salinity, Alkalinity and Acidity in Khejuri Block.

Name of Block	Saline			Alkaline			Acidic			Micro Nutrient Deficiency			Soil Erosion			Water Logged			Total
	Mild	Severe	Area (Hec.)	Mild	Severe	Area (Hec.)	Mild	Severe	Area (Hec.)	Mild	Severe	Area (Hec.)	Mild	Severe	Area (Hec.)	Mild	Severe	Area (Hec.)	
Khejuri - I	725	0	725	0	0	0	0	0	0	300	0	300	0	0	0	300	200	0	1525
Khejuri - II	545	0	545	0	0	0	0	0	0	0	0	0	0	0	0	250	150	0	945
District Total	8640	0	8640	425	0	425	2420	0	2420	5585	0	5585	2660	75	2585	10415	5635	2175	37405

Source: S.R.E.Plan PurbaMedinipur, 2006, page-III/2.

6. CONCLUSIONS

Storm, tidal waves, flood, water-logging conditions etc. play jointly as natural causes of the degradation of agricultural lands. This has significantly reduced the agricultural field of the blocks under study. In addition to the natural factors, the anthropogenic factors have contributed much to the situation. Deforestation has emerged as a serious problem in the area that has naturally worsened the water logging problem more serious by weakening and exposing the embankments to the hazards. Hence there is a need for increasing safety of the area against water logging.

REFERENCES

1. Chakroborty A (1991) History of Bengal. Buardwan University.
2. Das NN (1956) History of Midnapore 1: 1760-1942.
3. (2002) Disaster Management Plan of PurbaMedinipurpp: 90-124.
4. Huntter WW (1868) The Annals of Rural Bengal. W.B.D.Z pp: 30-53.
5. Huntter WW (1868) A Statistical Account of Bengal. W.B.D.Z pp:66-89, 220-247.
6. Karan MN (2002) Khejuri Bandar. K.E.S.P, Khejuri, Midnapore.
7. Mukharjee RK (2008) The Changing face of Bengal - A Study in Riverine Economy. CU Cal pp:98-117.
8. Paul AK (2002) Coastal Geomorphology and Environment. ACB Pub, Cal pp: 355-380.
9. Pradhan MK, Chatterjee S, Paul AK, (2008) Decline of a historical port in thechanging face of Hugli mouth estuary. I.G.I.Benaras Hindu University 20: 113.
10. Pradhan MK, Chatterjee S, Paul AK (2008) Methods of the identification of shore line changes of the historical port- A study at Khejuri of Hugli downstream section near Bay of Bengal. I.G.I. Tripura University 21: 82.
11. Pradhan MK, Chatterjee S, Paul AK (2009) Identification of paleo shorelines basis on excavated mangroves tree trunks and Existing geomorphological features at Khejuri sector of Bay of Bengal. I.G.I. Allahabad University 22: 134.
12. Pradhan P (2003) Hijilinama. ContaiHOA,Contai.
13. (1999) Stevenson-Moore Committee Report-Reverse of Bengal. W.B.D.G 2.
14. Willson CR (1996) The Early Annals of English in Bengal. Asiatic Socity Cal.
15. O,Malley LSS (1905) Bengal District Gazetteer. Govt. of W Bpp: 91-109,124-131.

CHAPTER 7

Saving Lives: Timely Flash Flood Warnings in the UK

Colin Clark

CHRS Shute Lane, Bruton, Somerset, UK

ABSTRACT

Flash floods are a major cause of death and destruction to property on a worldwide scale. In the UK sudden flooding has been the cause of the loss of over 60 lives during the last century. Forecasting these events to give enough warning is a major concern: after the 2004 flood at Boscastle, Cornwall UK the Environment Agency (2004) stated that it was not possible to provide a warning in such a fast reacting and small catchment. This is untrue since the Agency had already implemented a real time non-linear flow model as part of a flood warning system on the upper Brue in Somerset UK. This model is described in this paper as it has been applied to the Lynmouth flood of 1952, and briefly for the Boscastle catchment, both of which have an area of about 20 km^2. The model uses locally measured SMD and saturated hydraulic conductivity data. With the addition of further parameters the model has been successfully used nationwide.

Keywords: Flash Floods, Flood Warning, Non-Linear Flow, Model Historic Floods

1. INTRODUCTION

Flash flooding in the UK has been a major cause of water related deaths during the 20th century. The worst of these events was the Lynmouth flood disaster in which 34 people were drowned in 1952. This is followed by the Louth flood of 1920 in which 23 souls were lost. More recently people have been drowned during the floods of 1958 (Boscastle), 1968 (Mendip), 1998 (Midlands), 2007, 2009 and 2012. For those who survive such ordeals the psychological effects can be dramatic [1]. Having suffered such losses in the past the authorities have been woefully slow to provide good quality flood warnings. Sene [2] provides an overview of the methods used to warn people of serious flooding. The requirements for warnings include rainfall and radar data [3], soil moisture deficit data [4] [5], and for some probabilistic forecast methods [6] [7] [8] suggest that improvements in data collection, assimilation and modelling are needed, as are the human responses to flood warnings. Real time flood warning systems are not yet the norm in the UK . There is too much reliance on radar data which is either not detailed enough for small catchments or is simply inaccurate [3]. Following recent flooding the UK Government produced a review of flood resilience [9]. This tended to focus on the provision of rainfall simulations for probabilistic forecasting, the effects of climate change, but missed out small scale effects such as neglected drainage ditches and road levels becoming higher than household thresholds over a long period of successive road resurfacing. While there has been a call to replace the 1 in 100 year style of describing a flood [10] since it is said to be "confusing" to the public, the general level of understanding of flood frequency and especially flash flooding is so low that a much more practical approach is needed. This is especially true for surface water flooding which in the main, is caused by changes in surface hydraulic properties and inadequate road drainage, especially raised road levels. For example, in the Upper Brue at Bruton the main road has been raised by 0.5 m in the past 150 years, and at Combe in Oxfordshire the road had been raised about 0.3 m which led to floodwaters entering nearby houses. While flash floods caused by surface water is unlikely to cause death from drowning a much more serious risk is from river flooding in small rural catchments. The Land Drainage Act [11] states that where a flood warning system can be produced and is requested, the authorities have a duty to act accordingly. However, the level of understanding following the Boscastle flood of 2004 in Cornwall UK can be summed up:

"It's not possible to accurately forecast flooding in some areas such as parts of north Cornwall, where steep valleys mean that rivers can rise so rapidly after heavy rain that, with current technology, there's not enough time to issue warnings" [12].

However, a real time warning system had already been provided for the upper Brue [13] and which was tested against the smaller Valency river at Boscastle where the catchment area is 20 km^2, to give 1 hour warning on 16th August 2004 [14]. The non-linear flow model has now been extended over the whole of England and Wales [15] for design flood estimation, and with suitable soil hydraulic conductivity data it can be tested against actual storm events such as the most severe flood that has ever taken place in recent history, namely the Lynmouth flood on the East and West Lyn . whose catchment areas are 77 km^2 and 22 km^2 respectively. To date, this is the most severe test of the method. While Ciervo et al. [16] have used post event surveys to simulate a flash flood the present example is the first which uses both real time rainfall measurements and soil data gathered in the field. In addition where there is a significant soil moisture deficit at the start of the storm the model uses field based measurements of SMD which have been found to be much more reliable than calculated values [5]. It is contended that only with locally based field measurements can reliable flood warnings for fast reacting catchments be made both in the UK and elsewhere. This approach is in contrast to WMO [17] which states: "The difficulty of in situ monitoring of catchment wetness by conductivity or lysimeter measurements makes the use of these techniques very uncommon in flood forecasting operations".

The remainder of this paper will give a description of the evolution of the flow model, how it has been tested in over 600 catchments in England and Wales , and how the measurement of in situ SMD has been carried out using low cost weighing lysimeters. Unlike some flow models which are computationally demanding, the current model can be used on an ordinary PC. With the SMD data posted daily on a website it can be accessed and input to the model. The model can also be used to produce flood frequency estimates. These are made more robust when field surveys of bankfull discharge are made, which in many rivers worldwide has a frequency of about 1.1 - 2.0 years. While the biggest catchment that the model has been tested is around 650 km^2 it is on small flashy catchments that its value for providing timely warnings may be appreciated.

2. THE NONLINEAR UH FLOW MODEL

The Unit hydrograph (UH) method has a long history being based on the work of Sherman [18]. Young and Bevan [19] have used a non-linear response since it is widely recognised that many hydrological processes are non-linear. The need to model non-linear responses in hydrology has also been promoted by Sivakumar [20]. The unit hydrograph method needs the following data and analysis in order to produce an estimate of the flood before the flood arrives:

1) Real time measurement of storm rainfall intensities
2) The antecedent soil moisture deficit (SMD) based on weighing lysimeter observations
3) An estimate of slope runoff when the SMD is zero
4) The percentage runoff at each stage of the storm event based on measured soil hydraulic conductivity
5) Conversion of the runoff depth into quickflow discharge
6) Conversion of the remaining rainfall into delayed flow discharge
7) Summation of quickflow and delayed flow to give total discharge

The first version of the flow model was developed as a flood warning tool for the Environment Agency to be used for flood warning at Bruton in East Somerset which is situated below a flood detention dam [13]. The approach to calibration of the model is data based mechanistic in that the autographic rainfall and estimates of peak discharge and timing of the flood events of 1968, 1974, 1979, and 1982 were used to produce the algorithms. The non-linear behaviour of the upper Brue catchment with an area of 31 km^2 suggested that the change in discharge rate was of a logistic borrowed from biology which mimics autocatalytic monomolecular reactions Street & Opik, [21]:

$$\mathrm{Log}x / a - x = k(t - t_1) \tag{1}$$

where t and t_1 are time intervals, k = constant, a = maximum size, x = a given value.

This equation can be rewritten and optimised for the Brue dam site for the rising stage of the hydrograph:

$$Y = [\mathrm{INVLOG2}(t - 0.7Tp)/1 + \mathrm{INVLOG2}(t - 0.7Tp)]Qp \tag{2}$$

where Tp = time to peak, t = time since start of storm, Qp = peak runoff rate per mm net rainfall and is related to Tp as follows:

$$Qp = (330/Tp)\ A/1000 \tag{3}$$

where A = catchment area (km^2).

For the falling limb:

$$Y = [\mathrm{INVLOG}(t_1 - 0.85(TB - Tp))/1 + \mathrm{INVLOG}(t_1 - 0.85(TB - Tp))]Qp \quad (4)$$

where TB = time base and is related to Tp as follows:

$$TB = 2.52\ Tp \quad (5)$$

The time interval t_1 is then related to TB as follows:

$$t_1 = TB - t \quad (6)$$

The delayed flow was calculated using a maximum value of 0.3896 $m^3 \cdot s^{-1} \cdot mm^{-1}$. There is a time lag of 2 hours before delayed flow starts and a further 2 hour delay for peak delayed flow to occur. Thereafter a linear decay function was used. The model was calibrated using the storm events of 1968, 1974, 1979, and 1982. For these events the time to peak decreased with increasing rainfall intensity (13), and by implication storm rarity, an observation that has been noted elsewhere [22]. The relationship between maximum rainfall intensity (R) and time to peak (Tp):

$$Tp = (2.4073R + 10.1005)/R \quad (7)$$

There are no measured flow records for the upper Brue but autographic measurements of rainfall have been made at North Brewham since 1966. As a further check, the estimated rainfall hyetograph of the 1917 storm [23] was also used. The results are shown in Table 1.

These were based on field observations of the timing of over-bank and peak discharge and estimates of the peak discharge based on hydraulic calculations [24].

The result for the 1917 storm gave a peak discharge similar to that estimated from wrack marks and hydraulic calculations.

The flow model was adopted by the Environment Agency during 2004 and has been in use for over 10 years. On 1 December 2005 it was able to predict the water level behind the flood detention dam upstream of Bruton to within 0.1 m. This result is considered to be accurate enough to issue a flood warning with a lead time of about 2 hours if the predicted water level is expected to get within 0.2 m of the level at which flow down the spillway takes place.

Table 1. Comparison of estimated peak discharge (cumecs) in the field and using the runoff model.

Date	Slope area method	Flow model
11/7/1968	52	45
27/9/1974	25	27
30/5/1979	58	54
12/7/1982	78	77
29/6/1917	178	180

3. CALIBRATION AND APPLICATION TO UNGAUGED CATCHMENTS

There are many hydrological models which can be calibrated for one catchment. It is quite another to produce a method that can be applied when there is no riverflow data and even more demanding to provide a sensible flood warning. Hydrological reasoning suggested that the time to peak, Tp, not only depends on the rainfall intensity but also on the permeability of the soil. Thus the equation for the upper Brue was generalised:

$$Tp = c(R^{-0.17})\ \text{MSL catch/MSL Brue} \tag{8}$$

MSL = mainstream channel length (km) for the study catchment and Brue.

$$c = 7[\text{INVLOG}(0.06633(50 - \%R))]/1 + \text{INVLOG}(0.06633\ (50 - \%R))] \tag{9}$$

where %R = % runoff at rainfall equal to 10 mm $\cdot$ hr^{-1}, the %R being based on the Ksat survey data for values 1% - 50%. For %R > 50% at rainfall 10 mm $\cdot$ hr^{-1}, the equation for c becomes:

$$c = 3.5\ (0.98\ \exp\ (\%R - 50)) \tag{10}$$

Adjustments for catchment slope and the mainstream catchment slope were also made. These changes were based on the Dudwell at Burwash, Ancholme at Toft Newton, and Brue at Bruton calibration catchments. The catchment and mainstream slope have correction factors. For catchment slope the correction factor (S_{cf}) which best fit the data from 8 calibration catchments:

$$Scf = 1.7\ [INVLOG(0.35(S - 3.30))/1 + INVLOG(0.35(S - 3.30)] \tag{11}$$

where S = average catchment slope in degrees. This was assessed by counting the number of 10m changes in elevation on 1:50000 OS maps in each grid square or part grid square over a distance of 1km. The results are summed and then applied:

$$Tan_{slope} = 0.01[Sx/A] \tag{12}$$

where Sx = sum of changes in land height, A = catchment area.

The lowest value of Scf is 0.292 as found for the Ancholme at Toft Newton. Its general form is that of a logistic in that there is a minimum and maximum value which is attained at an ever decreasing rate at both low and high slopes. The constant 3.30 is the average slope for both the Brue and Lud catchments. The constant 0.35 influences the rate at which the value of Scf changes with slope.

The correction for mainstream channel slope is based on the gradient of the last 10 m fall in the mainstream (m/km). Where the gradient is lower further upstream up to a point which includes 50% or more of total catchment area, this gradient should be used. Again, considerable experiment yielded the correction factor for mainstream slope (MSS_{cf}):

$$MSS_{cf} = (2/L)^{F} - (L^{1.001} - L) \tag{13}$$

where L = distance between two 10 m contours on 1:50,000 OS maps, measured along the mainstream channel closest to the outlet.

$$F\ 0.7[(INVLOG(Log220 - LogA)/1 + INVLOG(Log220 - LogA) \tag{14}$$

with a minimum value of 0.3. Where the point of interest is close enough to a confluence so that both tributaries are included in the 10 m height drop, the area weighted value of L is used. If L = or < 2.0 km then MSScf = 1.0.

Further adjustments for the lag time and time to peak of the delayed flow were made:

$$\text{Lag/Tp} = \text{MSL catchment/MSL Brue} \tag{15}$$

where MSL = mainstream channel length as shown on 1:50,000 OS maps. Delayed flow increases linearly upwards until it reaches the peak. Thereafter it decreases exponentially according to the decay constant k, (Equation (24)), which varies with catchment area (A) via:

$$k = 0.0247\text{LogA} + 0.909 \tag{16}$$

This is applied to the delayed flow via:

$$\text{Del T} + 1 = \text{Del T exp k} \tag{17}$$

The full description of how the model was calibrated for permeable catchments and the effects of urban areas and abstraction-mainly of importance for common events, is described in Clark [15].

For the estimation of SMD a weighing lysimeter is used in the upper Brue catchment. The design of this has already been described [4] and a comparison of the results with both MORECS [5] and MOSES [25] have been given. Big differences between the measured and calculated values makes its effect on provision of a flood warning considerable.

4. FLOOD WARNING FOR THE LYNMOUTH STORM AUGUST 16 1952 USING THE FLOW MODEL

This flood resulted from over 9 inches (229 mm) taking place in about 9 hours on a catchment which was already at field capacity. The effects of the resulting flood have been described by Delderfield [26], Prosser [27]. Figure 1 shows the daily rainfall which has recently been revised according to an analysis of the effect of elevation on rainfall depth during the storm.

There were no rainfall recorders in the local area but 22 km to the west at Chivenor and 15 km to the east at Wootton Courtenay the combined continuous records enabled an estimated rainfall profile to be constructed. Figure 2 shows the results.

The percentage runoff at different rainfall intensities was estimated by measuring the saturated hydraulic conductivity of the soils in the field using the core method [28] Arellano & Clark, [29]. For each soil Association in the catchment at least 30 tests were made. The results were classified into 5 mm bands viz 0 - 5, 5.1 - 10 and so on as shown in Table 2.

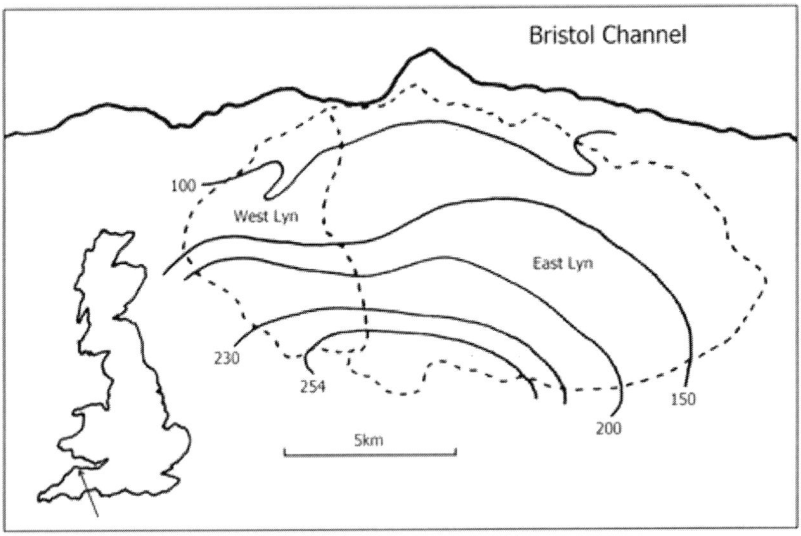

Figure 1. Daily rainfall for 16 August 1952.

Percentage runoff calculated from the midpoint in each class of conductivity: Eg. 20 mm · hr^{-1}percentage runoff = (20 – 2.5) ´ 0.31) ´ (20 – 7.5) ´ 0.23 + (20 – 12.5) ´ 0.20 + (20 – 17.5) ´ 0.11 = 50%. The percentage cover of each Soil Association was then measured from the Reconnaissance Map [30], and the catchment average % runoff calculated based on this and the conductivity values of each soil. Figure 3 shows the results.

The flow model is non-linear because of the response of soils to increasing rainfall intensity and the UH ordinates or rates of runoff per mm net rainfall being non linear with time.

Variables of the unit hydrograph method:

Time to peak (hours) = c MSL catchment/ MSL Brue (9) ($R^{-0.17}$), where R = rainfall intensity mm · hr^{-1}.

MSL = mainstream channel length (longest tributary) East Lyn = 17.27 km.
c = 7[INVLOG 0.06633 (50 – %R)/1 + INVLOG (50 – %R)].

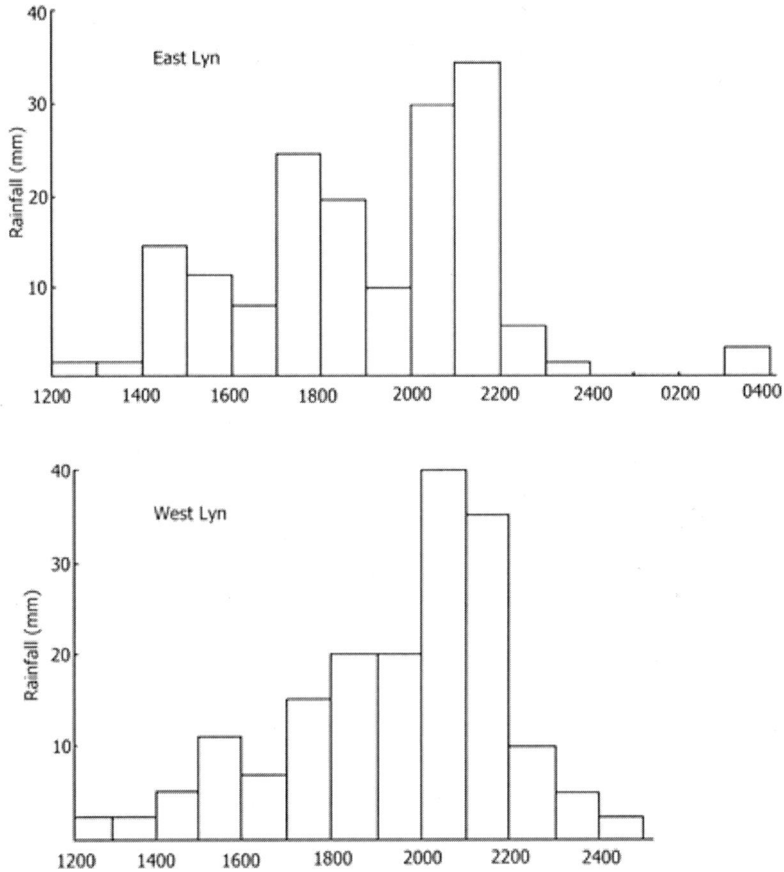

Figure 2. Estimated rainfall profiles for the East and West Lyn catchments.

%R = percentage runoff at 10 mm · hr^{-1} rainfall, based on soil hydraulic conductivity.

For the East Lyn %R = 36. Thus c = 6.2619.

Time to peak = 12.015 ($R^{-0.17}$).

Time base (TB) = 2.52Tp.

Peak quickflow ordinate (Qp) = 330/Tp x catchment area/1000.

Ordinates (Y) of the rising limb of the hydrograph:

Y = [INVLOG2(t – 0.7Tp]/1 + [INVLOG2(t – 0.7Tp)]Qp.

For the falling limb: Y = [INVLOG(t_1 – 0.85(TB – Tp))/1 + INVLOG(t_1 – 0.85(TB – Tp))]Qp. Where t_1 = TB – T where T = time since start of storm.

Table 2. Saturated hydraulic conductivity of the Lydcott Association (654b).

Conductivity class mm·hr^{-1}	Number of tests	%	% runoff
0 - 5	11	31	16
5.1 - 10	8	23	29
10.1 - 15	7	20	41
15.1 - 20	4	11	50
20.1 - 25	3	9	58
25.1 - 30	2	6	64

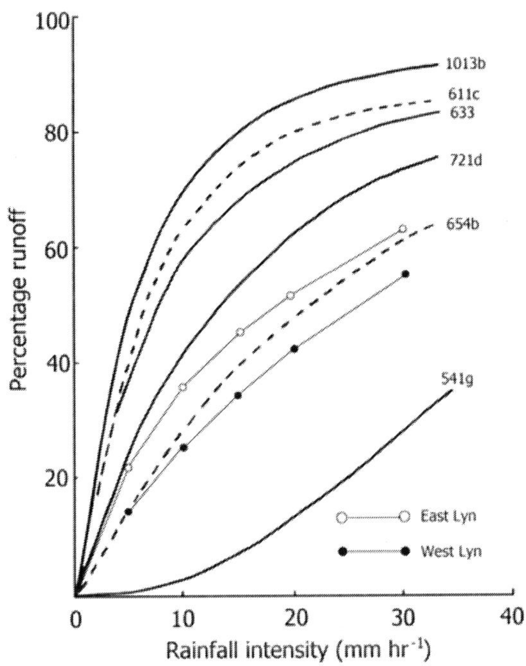

Figure 3. Saturated hydraulic conductivity of soils in the East and West Lyn.

The model is unique in that Tp, TB, and Qp will change at each stage of the storm according to the rainfall intensity. Slope runoff = R [sine average catchment slope (degrees)] = R [0.01 (N/A)] where N = sum of contour crossings in each grid square or part thereof on a 1:50,000 OS map over a straight or curved distance of 1km. A = catchment area. For East Lyn sine slope = 0.142.

Slope correction factor (Scf) = 1.7 [INVLOG(0.35(S – 3.30))/1 + INVLOG(0.35(S – 3.30))] where S = average catchment slope (degrees). For the East Lyn Scf = 1.667

Mainstream channel slope correction factor = 1.0 since distance between river channel contours < 2.0 km.

Hourly storm intensity, East Lyn : 2, 2, 15, 12, 8, 25, 20, 10, 30, 35, 6, 2

Extract of results:

Rainfall intensity ($mm \cdot hr^{-1}$)	Tp	TB	0.85 (TB – Tp)	Qp
2	10.679	26.912	13.798	2.384
8	8.437	21.261	10.901	3.017
10	8.123	20.470	10.495	3.133
15	7.582	19.106	9.796	3.357
12	7.875	19.485	10.174	3.232
20	7.220	18.194	9.328	3.526

Runoff:	Hourly rainfall intensity (mm)								
SR	2	2	15	12	8	25	20	10	30
K	0.28	0.28	2.13	1.79	1.70	3.55	2.84	1.42	4.26
T	0.13	0.13	5.29	3.81	1.87	11.30	8.16	2.74	14.90
SR	0.41	0.41	7.42	5.51	3.00	14.85	11.00	4.16	19.16
Del	1.59	1.59	7.58	6.49	4.00	10.15	9.00	5.84	10.84

where SR = slope runoff; K = runoff from the soil (hydraulic conductivity) T = SR + K. Del = Rainfall – T.

	Clock Hour									
T	1800	1900	2000	2100	2200	2300	2400	0100	0200	0300
0.41	0.002	0.240	2.188	2.383	2.379	2.365	2.214	1.347	0.274	0.030
0.41	-	0.002	0.240	2.188	2.383	2.379	2.365	2.214	1.347	0.274
7.42	0.008	0.655	3.224	3.356	3.200	2.253	0.567	0.067	0.006	-
5.51	-	0.003	0.278	2.922	3.228	3.081	2.171	0.549	0.064	0.006
4.00	-	-	-	0.043	1.829	2.997	3.016	2.890	2.100	0.562
14.85	-	-	-	0.133	2.969	3.589	3.586	2.467	0.598	0.069
11.00	-	-	-	-	0.027	1.544	3.481	3.526	3.003	1.493
4.16	-	-	-	-	-	0.001	0.127	2.555	3.125	3.132
19.16	-	-	-	-	-	-	0.133	2.969	3.589	3.596
23.90	-	-	-	-	-	-	0.004	0.335	3.358	3.872
2.00	-	-	-	-	-	-	-	-	-	0.011
0.41	-	-	-	-	-	-	-	-	-	-
	0.06	4.97	26.4	45	95	118	125	165	214	198

Extract from the unit hydrograph calculations is shown below. Body of the Table are ordinates ($m^3 \cdot s^{-1} \cdot mm^{-1}$ runoff. T = total runoff for each rainfall intensity. Total discharge is sum of the products at each time: eg. at 7 pm = 0.41 ´ 0.240 + 0.41 ´ 0.002 + 7.42 ´ 0.655 + 5.51 ´ 0.003 = 4.97. The discharges are cumecs ($m^3 \cdot s^{-1}$). The ordinates for delayed flow are 0.15 $m^3 \cdot s^{-1}$ 31 km^{-2} which is for rock Group D2 and D1 [14]. The time to peak and lag time of delayed flow = MSL catch/MSL Brue. Linear increase in the ordinates up to peak and exponential decay rate thereafter viz: decay constant = 0.0247LogA + 0.909.

Total discharge = Scf ´ quickflow + delayed flow: (1.667 ´ 214) + 23.5 = 380 $m^3 \cdot s^{-1}$. Baseflow is added via Q = CA (INVLOG[0.0005372 SAAR – 2.3114]) = 3 cumecs therefore Q = 383 $m^3 \cdot s^{-1}$.

Figure 4 shows the complete hydrographs for the East and West Lyn . The main events of the flood have been simulated. This gives a measure of the robustness of the method in view of the extreme nature of the storm with more than one rainfall peak. The falling limb should have been about 30 minutes earlier. This would have allowed PC Derek Harper to get across the parapet of Countisbury Bridge by 0400 h [27]. The peak discharges of 384 and 162 cumecs for the East and West Lyn respectively are 8% and 36% lower than the results of Dobbie and Wolf [31].

Taking the West Lyn first the heavy rainfall up to 1800 hrs gives no cause for concern. By 1900 hrs considerable overflow in Lynmouth is predicted since the discharge has become about double the channel capacity. People warned might simply move upstairs perhaps taking special belongings. By 2000 h the situation would look not much worse but by 2030 h there was a clear sign that people should evacuate at once. Recall now the events 2 km upstream at Barbrook [26]. At some time after 2130 h Tom Floyd looked out to see the raging river, but by then it was too late. Fortunately Tom, his daughter, and dog Tim survived, but the rest of his family were drowned [27]. We can now say that at least 1 hour's warning could have been given to the people at Barbrook, where four of the cottages were destroyed. On the East Lyn the longer travel distance gave more time to evacuate since by 1800 h the prediction of a life threatening flood is clear (Figure 4). Comparing the timing of this knowledge with events on the ground gives us hope for the future provision of flood warnings in the UK . At Bevan's Cottages on the East Lyn the doors burst open at about 2150 h. Thus over three hours warning could have been given. Such is the sad loss of life for want of a timely flood warning!

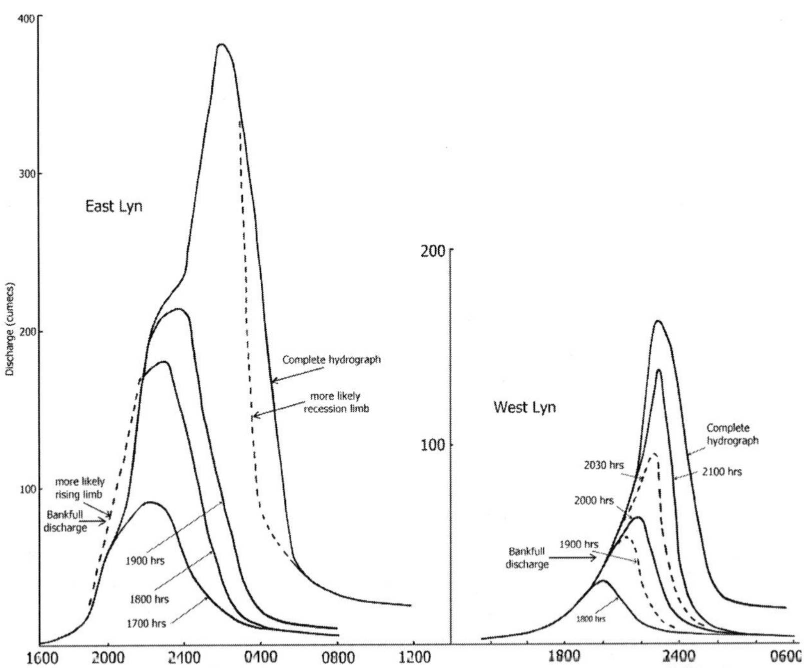

Figure 4. Predicted hydrographs for the East and West Lyn. The time of each forecast is shown.

5. FLOOD WARNING FOR BOSCASTLE

This event is included in this paper because the flow model was able to predict the flood with just over one hour warning. What has been described as the miracle of Boscastle is the fact that no one was drowned. However, several people, including visitors in the Wellington Hotel, had an escaped with a matter of a few minutes to spare, while a child sat on the bonnet of a car was grabbed seconds before the car was swept away. That the event happened in the day may have meant fewer people were present as visitors but the situation could have been a lot worse like at Lynmouth where people were tending to have a quiet evening indoors.

The storm was caused by a complex area of low pressure to the west with unstable air developing during the morning [32]. During a four hour period over 180mm were recorded at Lesnewth which is close to the centre of the Valency catchment which discharges into the Atlantic Ocean at Boscastle. Figure 5 shows the storm rainfall at a resolution of 0.25 hr.

The flow model used locally gathered soil hydraulic conductivity data. Full details of the application are given in Clark [14]. The river Valency started to overflow by 1515 hr. Three simulations were made and that for 1415 hr showed that a substantial flood was about to begin: a serious flood warning would then have been issued. Figure 6 shows the results.

The success of the model in giving a realistic estimate of the complete flood is shown in Figure 7. Calculations of the likely SMD showed that it was about 12 mm [33]. In contrast to this the MOSES PDM [34] suggested a value in excess of 100 mm which was far too high based on the rainfall regime earlier in the year and that at CHRS the weighing lysimeter on the same day was about 30 mm SMD but with a lower rainfall in the previous three months.

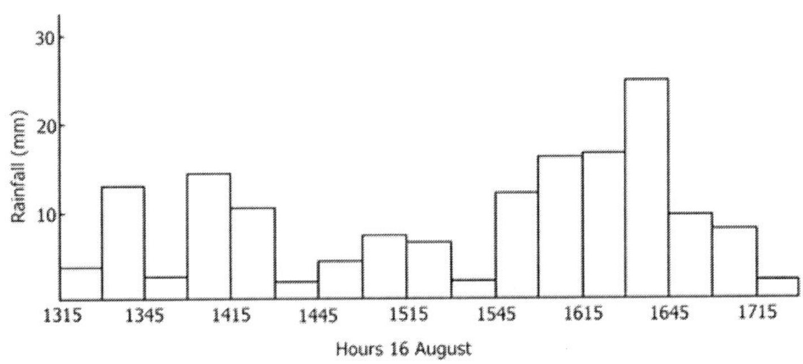

Figure 5. Storm rainfall at Lesnewth 16 August 2004.

The lysimeter is 235 mm diameter and the soil about 330 mm deep, and costs about £25 to construct.

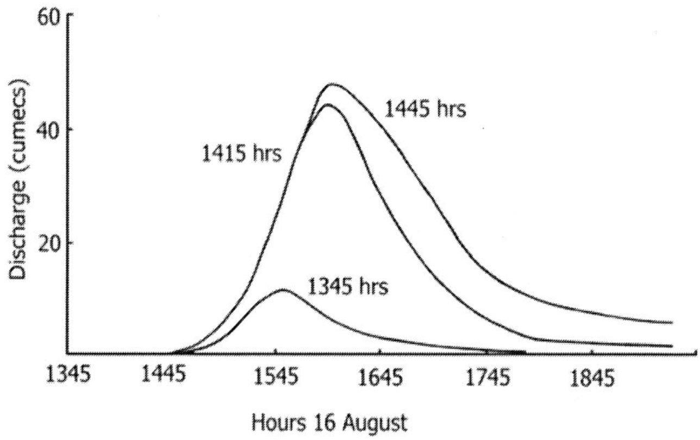

Figure 6. Estimated hydrographs for 16 August. The time of each forecast is shown.

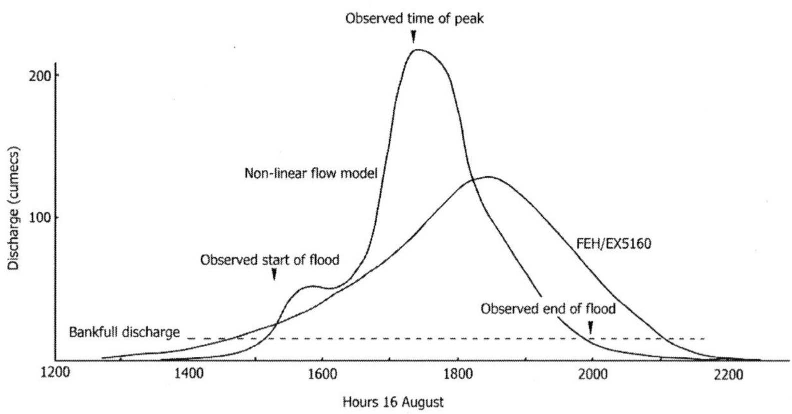

Figure 7. Estimates of the discharge for 16 August 2004 at Boscastle [35].

6. DISCUSSION

Timely flood warnings are essential during a serious flash flood if lives are to be saved. The method described here uses local data for rainfall and soil hydraulic conductivity. The use of a low cost lysimeter in order to get realistic estimates of SMD has been briefly mentioned. This approach is much more realistic than other methods which fail to produce timely warnings. In a country which already has some places protected up to about the 1 in 100 year design standard the need for flood warning is even more important in the event of the scheme being overwhelmed by a very rare flood. There are also other places without any protection or even a flood warning scheme. For catchments below about 15km^2 the lead time for a warning becomes less than 1 hour. Giving a false warning will undermine the confidence of future warnings so it is essential that accurate rainfall and catchment data are used. It is very much hoped that the techniques described here can be adopted in other countries.

REFERENCES

1. Penning-Rowsell, E., Tapsell, S. and Wilson, T. (2005) Key Policy Implications of the Health Effects of Floods. Extreme Weather Events and Public Health Responses, WHO, Springer, 207-223.
2. Sene, K. (2012) Flash Floods: Forecasting and Warning. Springer, 385 p.
3. Collier, C.G. (2007) Flash Flood Forecasting: What Are the Limits to Predictability? Q.J.R. Meteorol. Soc., 133, 3-23.
4. Clark, C. (2013) Measurements of Actual and Pan Evaporation in the Upper Brue Catchment UK: The First 25 Years. Weather, 68, 200-208.
5. Clark, C. (2002) Measured and Estimated Evaporation and Soil Moisture Deficit for Growers and the Water Industry. Meteorl. Appl., 9, 85-93.
6. Smith, P. and Bevan, K. (2014) When to Issue a Flood Warning: Towards a Risk- Based Approach Based on Real Time Probabilistic Forecasts. Proc. 2nd International Conference and Analysis and Risk Management. ASCE Conference, 1395-1404.
7. Kimura, N., Hsu, M.-H., Tsai, M.-Y., Tsao, M.-C., Lu, S.-L. and Tai, A. (2016) A River Flash Flood Forecasting Model Coupled with Ensemble Kalman Filter. J. Flood Risk Mangt, 9, 178-192.
8. Parker, D.J. (2007) Advances and Challenges in Flash Flood Warnings. Env. Hazards. , 7, 173-178.
9. HMG (2016) National Flood Resilience Review. 141 p.
10. DEFRA (2016) Future Flood Prevention. Environment Food and Rural Affairs. 53 p. House of Commons.
11. Land Drainage Act (1976) HMSO. London.
12. Environment Agency (2005) The Floods in Boscastle and North Cornwall 16 August 2004. 29 p.
13. Clark, C. (2004) A Nonlinear Unit Hydrograph Model for Real Time Flood Forecasting below a Dam. Internat. Water Power & Dam Constr, 56, 20-26.
14. Clark, C. (2006) A Real Time Flood Warning System for Boscastle, North Cornwall. Report to Environment Agency, Exeter, 18 p.
15. Clark, C. (2012) Spillway Design Flood and Flood Volume Estimated Using the New Guide to Flood Estimation. In: Wong, T.S.W. Ed., Flood Risk and Flood Management, Nova Science, New York, 219-255.
16. Ciervo, F., Papa, M.N., Medina, V. and Bateman, A. (2015) Simulation of Flash Floods in Ungauged Basins Using Post Event Surveys and Numerical Modelling. J. Flood Risk. Mangt, 8, 343-355.
17. WMO (2011) Manual on Flood Forecasting and Warning. WMO No. 1072.

18. Sherman, L.K. (1932) Streamflow from Rainfall by Unit-Graph Method. Eng. News Record, 108, 501-505.
19. Young, P.C. and Bevan, K.J. (1994) Data Based Mechanistic Modelling and the Rainfall-Flow Nonlinearity. Environometrics, 5, 335-363.
20. Sivakumar, B. (2008) The More Things Change, the More They Stay the Same: The State of Hydrologic Modelling. Hydrol. Process, 22, 4333-4337.
21. Street, H.E. and Opik, H. (1970) The Physiology of Flowering Plants. Edward Arnold, London, 263 p.
22. Ashfaq, A. and Webster, P. (2000) The Timing of Runoff Response in Design Flood Analysis. Hydrol. Process, 14, 1217-1233.
23. Clark, C. and Pike, W.S. (2007) The Bruton Storm and Flood after 90 Years. Wea- ther, 62, 300-305.
24. Clark, C. (1996) Floods at Bruton past Present and Future. 2nd Ed. Charldon Publications, Bruton, 86 p.
25. Clark, C. (2009) The Use of Local Data in Pre-dicting and Estimating Extreme Floods. 635-646 in Water, Environment, Energy, and Society. (WEES 2009). Vol II Statistical and Systems Analysis Techniques. New Delhi 12-16 January 2009. Allied Publishers, Mumbai. [Citation Time(s):1]
26. Delderfield, E. (1981) The Lynmouth Flood Disaster. E.R.D. Publications Ltd. Exmouth. 10th Ed.
27. Prosser, T. (2001) The Lynmouth Flood Disaster. Lyndale Photographic Pub. 99 p. Lynmouth.
28. Hollis, J.M. and Woods, S.M. (1989) The Measurement and Estimation of Saturated Soil Hydraulic Conductivity. Soil Survey & land res. Centre, Silsoe, 19 p.
29. Arellano, A.L. and Clark, C. (2004) The Louth Storm and Flood after 80 Years. Weather, 59, 71-76.
30. Soil Survey England and Wales (1984) Legend for the 1:250,000 soil map of England and Wales. 21 p.
31. Dobbie, C.H. and Wolf, P.O. (1953) The Lynmouth Flood of August 1952. Pro. Inst. Civ. Engrs., 2, 522-588.
32. Golding, B., Clark, P. and May, B. (2005) The Boscastle Flood: Meteorological Analysis of the Conditions Leading to Flooding on 16 August 2004. Weather, 60, 230-235. https://doi.org/10.1256/wea.71.05
33. Clark, C. (2005) Catastrophic Floods: Magnitude and Frequency Investigations. Internat. Water Power & Dam Constr., 57, 14-21.
34. HR Wallingford (2005) Flooding in Boscastle and North Cornwall, August 2004. EX5160. 169 p.
35. Clark, C. (2006) Planning for Floods: Will We Ever Learn? Internat. Water Power & Dam Constr, 58, 20-27.

CHAPTER 8

The 2011 Brisbane Floods: Causes, Impacts and Implications

Robin C. van den Honert * and John McAneney

Risk Frontiers, Macquarie University, NSW 2109, Australia

ABSTRACT

On 13th January 2011 major flooding occurred throughout most of the Brisbane River catchment, most severely in Toowoomba and the Lockyer Creek catchment (where 23 people drowned), the Bremer River catchment and in Brisbane, the state capital of Queensland. Some 56,200 claims have been received by insurers with payouts totalling $2.55 billion. This paper backgrounds weather and climatic factors implicated in the flooding and the historical flood experience of Brisbane. We examine the time history of water releases from the Wivenhoe dam, which have been accused of aggravating damage downstream. The dam was built in response to even worse flooding in 1974 and now serves as Brisbane's main water supply. In our analysis, the dam operators made sub-optimal decisions by neglecting forecasts of further rainfall and assuming a 'no rainfall' scenario. Questions have also been raised about the availability of insurance cover for riverine flood, and the Queensland government's decision not to insure its infrastructure.

These and other questions have led to Federal and State government inquiries. We argue that insurance is a form of risk transfer for the residual risk following risk management efforts and cannot in itself be a solution for poor land-use planning. With this in mind, we discuss the need for risk-related insurance premiums to encourage flood risk mitigating behaviours by all actors, and for transparency in the availability of flood maps. Examples of good flood risk management to arise from this flood are described.

Keywords: flood; Brisbane River; January 2011; water release strategy; flood risk management; insurance; land use planning

1. INTRODUCTION

On Thursday 13th January 2011 Brisbane, the state capital of Queensland, Australia (27°30' S, 153°1' E) (Figure 1), experienced its second highest flood since the beginning of the 20th Century. Major flooding occurred throughout most of the Brisbane River catchment, most severely in the catchments of the Lockyer Creek and Bremer River (major tributaries of the Brisbane River) where numerous record flood heights were experienced. The flooding caused the loss of 23 lives in the Lockyer Valley and one in Brisbane, and an estimated 18,000 properties were inundated in metropolitan Brisbane, Ipswich and elsewhere in the Brisbane River Valley.

This paper collates information on the flood from a number of public sources, notably the Bureau of Meteorology (BoM) and submissions to the Queensland Floods Commission of Inquiry [1], and offers some commentary on the operational management of the Wivenhoe Dam, and insurance and land use planning issues.

The paper is organised as follows. Section 2 provides a description of the Brisbane River system and catchment, while Section 3 reports on the history of flooding in Brisbane. Section 4 details the rainfall recorded in eastern Australia over the period December 2010 to January 2011. Section 5 describes and comments on operational strategies for water releases from Brisbane's flood mitigation and storage dams, while Section 6 examines the impacts of the flooding on the urban environment, and includes an aerial photographic comparison of the extent of flooding with the largest previous flood in Brisbane in January 1974.

Subsequent sections report on the Interim Report of the Queensland Floods Commission of Inquiry, and consider insurance and land-use planning issues and their implications for reducing future flooding risks, concerns that, at the time of writing, have yet to be fully examined by the Commission.

Figure 1. Location of Brisbane.

2. THE BRISBANE RIVER CATCHMENT

2.1. The Brisbane River

The Brisbane River is the longest river (309 km) in south east Queensland. Its source is located in the Brisbane Range some 120 km north west of the city of Brisbane. From there it makes its way south before meeting the Stanley River, just downstream of Somerset Dam, to run into Lake Wivenhoe, the main water supply for Brisbane. Lake Wivenhoe was created by the Wivenhoe Dam completed in 1984 in response to severe flooding in 1974 and with the principal aim of protecting Brisbane from future floods. Since its construction many considered that this dam would eliminate the flood risk to Brisbane, notwithstanding warnings against this way of thinking [2]. Downstream from the Wivenhoe dam the river flows eastwards, meeting the Bremer River near Ipswich before making its way through Brisbane's western suburbs towards the Pacific Ocean.

2.2. The Brisbane River Catchment

The Brisbane River's catchment is some 13,570 km^2 in area, bounded to the west by the Great Dividing Range and to the southeast and north by a number of smaller coastal ranges. Most of the catchment is rural—forestry and grazing land—but also includes the major metropolitan areas of Brisbane and Ipswich, as well as a number of smaller townships (Figure 2).

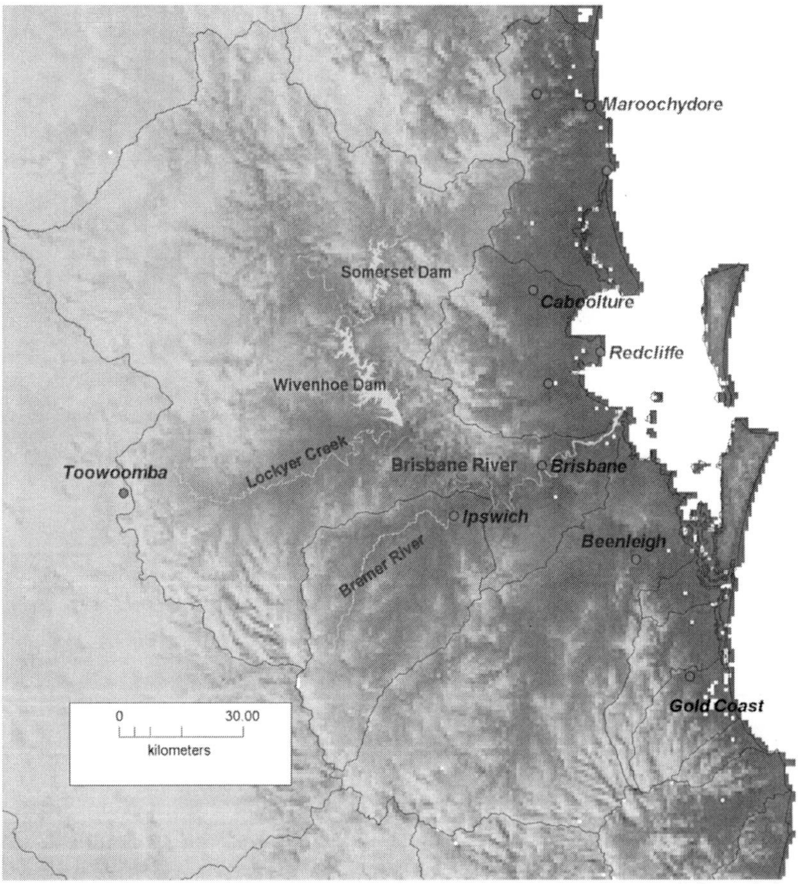

Figure 2. A topographical map of the Brisbane River catchment.

The physical characteristics of the catchment have changed over the last 100 years due to progressive settlement and development along the banks of the river [3]. This development has included the construction of the Wivenhoe and Somerset Dams. At full supply level (FSL) Wivenhoe and Somerset Dams have water supply capacities of 1.15 million mL and 0.37 million mL respectively. Total available storage capacities for flood mitigation are 1.45 million mL and 0.52 million mL respectively [3].

Just *where* and *how much* rain falls within the catchment will critically determine the dams' effectiveness as flood mitigation measures. Situated upstream from where the Lockyer Creek and the Bremer River join the Brisbane River, the dams can only have a limited influence on the Bremer River by reducing the downstream level of the Bremer River as it enters the Brisbane River. In extreme scenariosin small reservoirs, incoming waters will have to be released at at least the same rate in order to avoid destruction of the dam itself; larger reservoirs (like Wivenhoe and Somerset Dams) can provide significant flood attenuation until the flood compartments are filled. In such scenarios, the dam will have no moderating influence on downstream flooding. We will return to the role of the dam in moderating or aggravating the recent flooding in later discussion.

3. HISTORICAL FLOOD EVENTS ON THE BRISBANE RIVER

The city of Brisbane is built on a flood plain [4]. Flood records extend back as far as the 1840's, only a few years after European settlement of the area in 1824 (Figure 3).

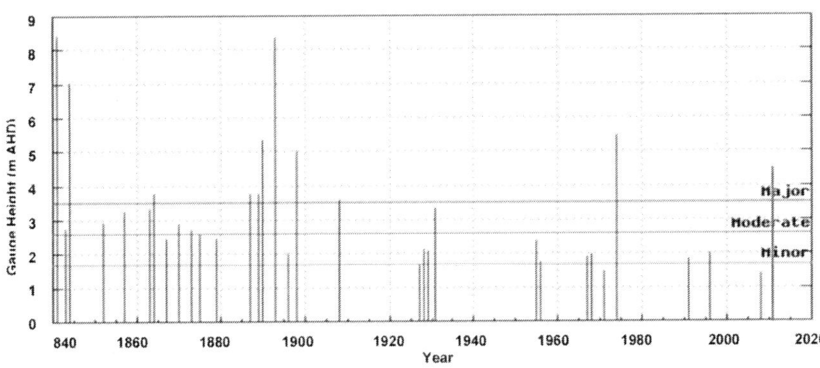

Figure 3. Maximum recorded gauge height on the Brisbane River at the Brisbane City gauge for significant flood events (Adapted from [2]) (Brisbane River at Brisbane City Gauge Highest Annual Flood Peaks).

The largest recorded gauge heights were in 1841 and 1893 (over 8.35 m AHD (Australian Height Datum)), which represents a depth of approximately 6.5 m above the highest tide level. The largest flood of the 20th Century was in January 1974, when a gauge height of 5.45 m was recorded at the Brisbane City gauge. As the Brisbane River flooded, it backed up the Bremer River, resulting in 4 to 5 days of record water heights in Ipswich [2].

The floods of 1893 (actually a series of three flood peaks over a period of three weeks) and 1974 were both caused by rainfall from decaying tropical cyclones.

In January 1974 Tropical Cyclone *Wanda* crossed the coast about 150 km north of Brisbane, and then followed a southwesterly track. In addition to providing rain that saturated the catchment, it forced a monsoonal trough southwards to Brisbane, which persisted for four days and resulted in several periods of very intense rainfall. January 1974, with a monthly total of 872 mm, was the wettest January on record and the second wettest month ever, surpassed only by February 1893 in which 1,026 mm was recorded. Brisbane's mean January rainfall is 166 mm [5].

Although smaller in magnitude than the 1893 floods, the impact of the 1974 flood remains the most severe example of urban flooding in Australia. Damage in the 1974 flood was greater than in 1893 because Brisbane's population had grown from around 175,000 in 1893 to around 1 million and the greater exposure afforded by new buildings and infrastructure. Table 1 shows some comparative damage statistics from the 1893 and 1974 floods. Insured losses from the 1974 event ran to almost $2.29 billion (normalised to 2006 exposure, societal conditions, inflation and wealth) [6]. Given 2011 exposure, the normalized insured losses are of the order of $3.3 billion [7].

Table 1. Comparative damage statistics—February 1893 and January 1974 Brisbane River floods (Source: [8]).

Station	February 1893	January 1974
Deaths	35	16
Injuries	300	300
Persons left homeless	5,000	9,000
Persons evacuated		9,000
Buildings damaged	5,000	7,000
Cars damaged or destroyed		1,000

4. CAUSES (1): RAINFALL IN EASTERN AUSTRALIA: DECEMBER 2010 TO JANUARY 2011

4.1. Rainfall During December 2010 and January 2011

The second half of 2010 and early 2011 was characterized by one of the four strongest La Niña events since 1900 (Figure 4). Strong La Niña events are often associated with extreme rainfall and widespread flooding in eastern Australia [9].

An extremely wet spring (September to November) meant that catchments were already saturated before the December 2010 and January 2011 rains. Figure 5 shows total rainfall across Australia for November 2010 to January 2011. During this period, rainfall in the 600 to 1,200 mm range was widespread along most of the Queensland coast. Some stations north and west of Brisbane exceeded 1,200 mm during this same period.

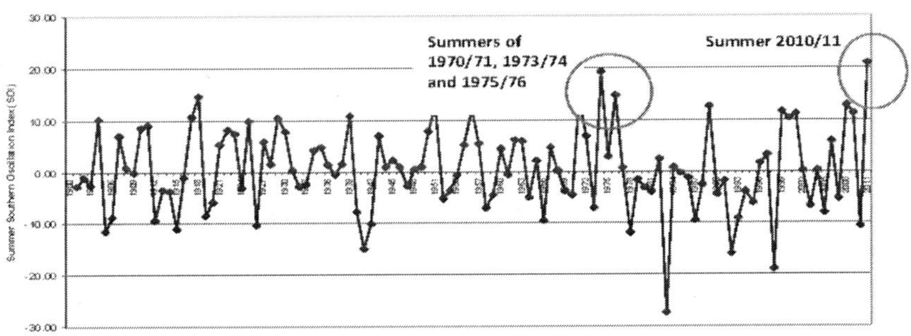

Figure 4. Average summer (October to March) Southern Oscillation Index (SOI), 1900/01 to 2010/11. (Source of data: [10]).

The Bureau of Meteorology warned the public and government about the likely strength of the La Niña event [12]. This did not, however, invoke any action on the part of the dam operators to draw down water levels from Full Water Supply to accommodate likely increased rainfall. We will address the question of dam operation during the flooding event in a later section.

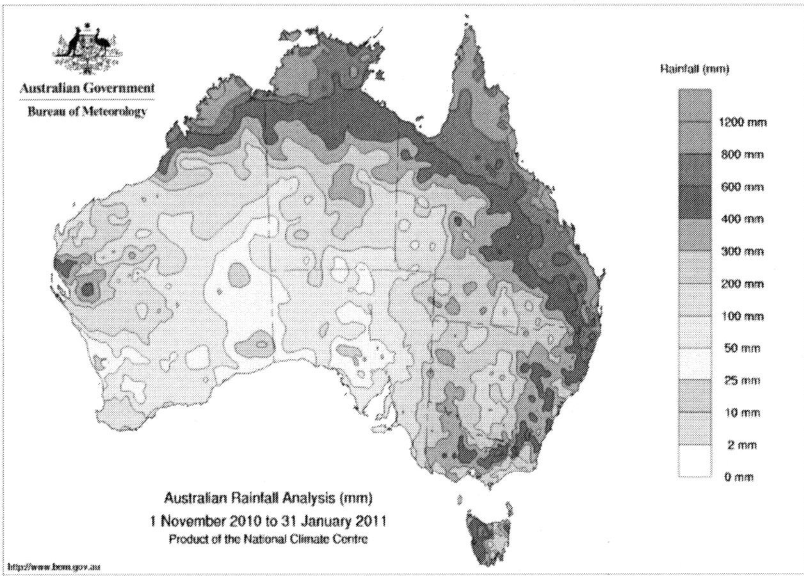

Figure 5. Total rainfall for the period November 2010–January 2011 (Source: [11]).

4.2. Flooding in Eastern Australia

The excessive rainfall over December and the first half of January resulted in almost every river in Queensland south of the Tropic of Capricorn and east of Charleville and Longreach recording major flood levels at some stage during this period [9]. Severe flooding also affected other parts of Queensland, northern and western Victoria, inland New South Wales and northern Tasmania (Figure 6).

4.3. Rainfall Meteorology in South East Queensland During 8th to 12th January 2011

The flooding in southeast Queensland during the second week of January was caused by the interaction of a low-pressure system situated off the mid and south Queensland coasts, and upper level and monsoonal troughs [13]. Between Friday 7th January and Sunday 9th January the low-pressure system moved in a southerly direction from east of Mackay (800 km north west of Brisbane) to an area north east of Fraser Island, before moving closer to the coast on Monday 10th January. Early on

Sunday 9th January the low-pressure system combined with an upper level trough to deliver heavy rains to southeast Queensland. This rainfall continued for much of the day.

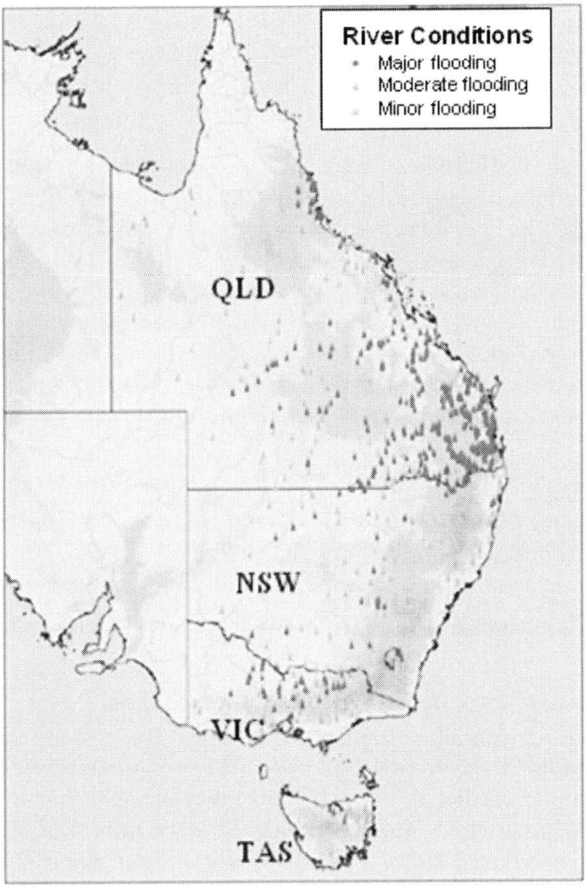

Figure 6. Flood peaks in eastern Australia over the period 26th November 2010–20th January 2011 (Source: [9]).

The upper level trough dissipated early on Monday 10th January, but the low-pressure system intensified and moved north to combine with a southward moving monsoonal trough. Warm moist air delivered into this trough system by a high pressure system located off New Zealand led to very heavy rainfalls across the southeast corner of Queensland from the evening of Monday 10th until the late afternoon of Tuesday 11th January, with rainfall tailing off on Wednesday 12th [13]. Direct rainfall for the period 9th–13th January into Wivenhoe and Somerset Dams was very heavy, with totals of 480 mm and 370 mm respectively.

The 10th–11th January rainfall led to very high water levels in the two dams. Levels in Somerset Dam peaked at 104.96 m AHD around 05:30 on Wednesday 12th January 2011. Water levels in Wivenhoe Dam peaked at 74.97 m AHD around 19:00 on Tuesday 11th January and remained above 74.9 m AHD for six hours from 18:00 on Tuesday 11th January [13].

Downstream of the dams, areas in the lower Brisbane River catchment also experienced significant rainfall, decreasing towards the south and east. In the Lockyer Creek catchment total rainfalls of around 450 mm were recorded close to its downstream end, with intensities having an Average Return Interval (ARI) greater than 100-years for durations greater than 3 hours. In the Bremer River catchment, 420 mm was recorded near the northwestern limit. The lowest rainfalls across the lower reaches of the catchment were recorded in the southern and eastern suburbs of Brisbane (110–160 mm) [13].

4.4. Comparing the Three-Day Rainfall Totals from the January 2011 Flood Event with Those of the January 1974

Figure 7 displays three-day rainfall totals from the period 10th–12th January 2011 and for the 25th–27th January 1974. Key points to emerge are:

(1) Peak rainfalls in 1974 were substantially heavier than those in 2011. A number of rainfall stations in southeast Queensland had three-day rainfall totals in excess of 1,000 mm in 1974 compared with a maximum of 648 mm recorded during the 2011 event. Over the Brisbane River catchment as a whole, the average three-day rainfall in 1974 was 349 mm, compared with 286 mm in 2011, and all four major sub-catchments were also wetter in 1974 than in 2011, although by small margins in the cases of the Bremer (442 mm *vs.* 417 mm) and Lockyer (331 mm *vs.* 292 mm) sub-catchments [9].

(2) The above observation also extends to metropolitan Brisbane where three-day and one-day totals of 600 mm and 314 mm in 1974 were significantly greater than the 166 mm and 111 mm recorded in 2011 [9].

(3) In 1974 the heaviest rains in south east Queensland occurred close to the coast, whereas in 2011 the heaviest rainfalls spread further inland, particularly on the western fringe of the Brisbane River catchment and on the Great Dividing Range.

Insufficient data exists for a full assessment of the 1893 event. However, available station data indicate that peak rainfalls in the region during the 1893 event were heavier than those during either the 1974 or 2011 events [9].

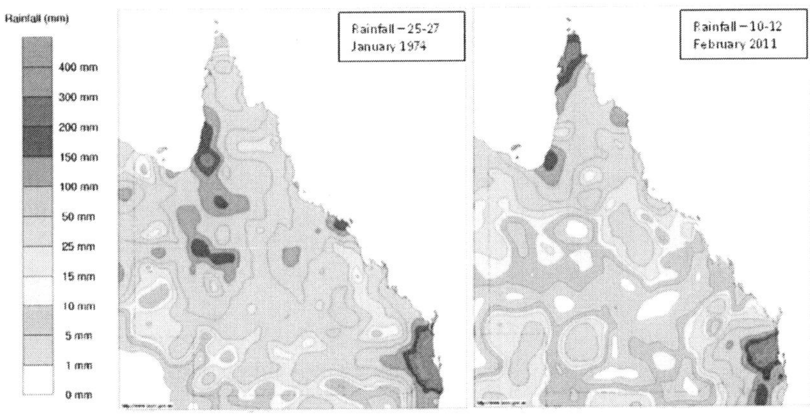

Figure 7. Three-day rainfalls for the periods from 25th–27th January 1974 (**left**) and 10th–12th January 2011 (**right**) (Source: [9], ©Commonwealth of Australia, 2011).

5. CAUSES (2): WATER RELEASES FROM WIVENHOE DAM

Hydrologists appointed by the Insurance Council of Australia to investigate events leading to flood damage claims in Brisbane, Ipswich, Toowoomba and the Lockyer Valley consider the Brisbane flood event to be a "dam release flood." In other words, in their view release of water from the Wivenhoe Dam was a key contributor to the flooding downstream over the period 11th–12th January 2011 [13]. This does not imply fault, but merely acknowledges that the water that caused much of the damage came from dam releases. In what follows we briefly discuss this issue.

Seqwater operates the Wivenhoe and Somerset Dams in accordance with procedures contained in its operations manual [14]. The manual provides objectives and strategies to guide operational decision-making during a flood event, and lists its operational objectives, in descending order of importance, as:

(1) ensuring the structural safety of the dams;

(2)providing optimum protection of urbanised areas from inundation;

(3)minimising disruption to rural life in the valleys of the Brisbane and Stanley Rivers;

(4)retaining the storage at Full Supply Level (for water supply purposes) at the conclusion of the Flood Event;

(5)minimising impacts to riparian flora and fauna during the drain down phase of the flood.

Apart from the obvious most important objective of ensuring the structural integrity of the dam itself, it can be seen that the dam is expected to serve two contradictory functions. On the one hand it serves as a buffer against drought, meaning that it is desirable to keep the dam as full as possible in case future rainfall is low (objective 4); on the other hand, the dam was specifically built to provide a buffer against floods, meaning it is desirable to keep it as empty as possible to maximize retention of flood waters (objectives 2, 3 and 5). Since keeping the dam full and empty are not simultaneously possible, Seqwater's engineers have to balance these conflicting objectives.

In January 2011, Wivenhoe Dam was called upon to accommodate massive inflows made up of surface runoff from the upstream catchment area, releases from Somerset Dam and direct rainfall into the lake. The greater the volumes and rate of inflow, the less effective are dams in mitigating flood flows, and the more constrained are management options for dam operators. According to a Seqwater report into the January 2011 flood event ([15,16]) operational decisions were made in accordance with the manual. Dam outflows were delayed until it was clear that no other option was available without risking the safety of the dams. Seqwater reported that:

Two distinct flood peaks entered Wivenhoe Dam during the flood event. The first flood into Wivenhoe Dam was similar in nature and magnitude to the comparable flood flows of the January 1974 event. The combined mitigation effect of Somerset and Wivenhoe Dams ensured that this first flood did not result in urban damage below Moggill (some 20 km upstream from Brisbane CBD), however achieving this result did cause significant filling of the dams' flood storage compartments. The second flood was also similar in nature and magnitude to the comparable flood flows of the January 1974 event. The flood compartments of the dams were filled to a high level by the first flood and there was not sufficient time to release this water prior to the second flood arriving. Accordingly, the second flood could not be completely contained without risking the safety of the dams. The resulting inflow of water into the Brisbane River, combined with floodwaters from Lockyer Creek, the Bremer River and the Lower Brisbane

River resulted in significant urban damage. However, the extent of this damage was greatly reduced by the operation of the dams.

The inflow of water into the Wivenhoe Dam, the outflows and the headwater elevation of the dam around the time of the 2011 flood are shown in Figure 8. The peak of the outflow from the dam (around 23:30 on 11th January) was approximately 40% lower than the peak of the inflows (at about 13:00 on the same day). Indeed, model estimates provided to Seqwater by Brisbane City Council indicate that without the mitigating effects of Wivenhoe the peak flood height measured at the City gauge near the Brisbane CBD would have been approximately 2.0 m higher than was actually experienced, and a further 14,000 properties impacted [15].

Figure 8 shows that water releases commenced only at 15:00 Friday 7th January when Wivenhoe Dam was at 109.6% and Somerset at 110.9% of FSL, with a total of 153,000 mL already in flood storage. In the week from 06:30 Tuesday 4th January, when Wivenhoe Dam was at 102.1% and Somerset at 102.9% of FSL and Seqwater were required under the manual to declare a Flood Event, there were only 14 hours during which the release rates from Wivenhoe Dam exceeded inflows.

The balancing act between retaining as much water as possible and releasing water to make space for further significant inflows has been automated to some extent with the operations manual directing that the water release strategy chosen at any point in time will depend on the *actual* water level in the dam, as well as modeling predictions about future water levels. The latter are to be made using the best available *forecast* rainfall and stream flow information [14]. In the aftermath of the January 2011 flood, a senior Wivenhoe engineer has stated before the Commission of Inquiry [17] that uncertainty in BoM rainfall forecasts and the paucity of rain gauges in the catchment immediately above the Wivenhoe Dam led them to conclude that the precipitation forecasts were not sufficiently reliable to form the basis for operational decision making [18]. Seqwater claim there were gaps in the information on which operational decisions needed to be made, despite them having available the best rain gauge of all-the dam itself!

With this in mind, and rather than employing data from forecasting models, Seqwater's modeling considered "no further rainfall" as the most appropriate scenario for predicting the future dam level. Thus less-than-perfect information contained in the BoM's forecast rainfall was given zero weight in decision-making. In other words, the dam operators chose a scenario guaranteed to be wrong over a forecast that was likely uncertain. From this outcome it is difficult not to conclude that the fact that Queensland had been in drought for most of the previous decade [19] anchored their decision making towards conserving FSL.

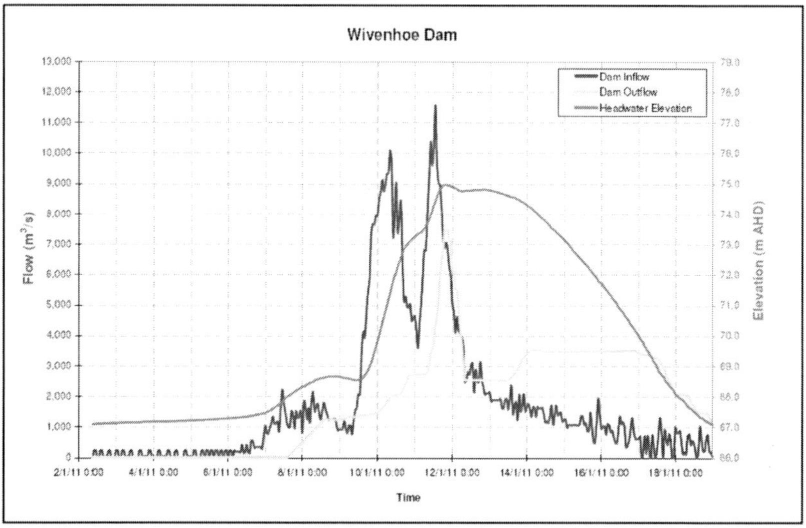

Figure 8. Inflow and release of water for the January 2011 flood event—Wivenhoe Dam (Source: [15]).

Many tools exist to facilitate decision making under uncertainty, e.g., [20]. None of these guarantee favourable outcomes, but we must learn to differentiate between bad outcomes and bad decisions. One example of a tool that may have proved useful to decision makers at Wivenhoe is the geostatistical technique of kriging [21,22]. This method is widely used to interpolate the value of a random variable, in this case rainfall, at an unobserved location from observations at nearby locations, a process that would seem very appropriate in the Wivenhoe rainfall problem.

Other tools such as decision trees [23] could have proved useful. Here uncertainties in random variables are expressed through probability density functions that describe the range of possible values and the relative likelihood of values in between. Rather than choosing a single scenario, the dam operators could have considered a spectrum of possible futures with various downstream flooding outcomes weighted by their likelihood of occurrence and contingent upon various rainfall scenarios. A decision tree would have allowed the operators to incorporate what was known or expected about future rainfall explicitly into the decision process. In the event, the dam operators chose not to do this and thus rather than making a decision under *uncertainty*, they chose to make a decision under *ignorance*.

How bad were the rainfall forecasts? Figure 9 compares the 24-hour Quantitative Precipitation Forecasts (QPFs) for the Wivenhoe Dam catchment, and the actual rainfall intensities observed for the first half of January 2011.

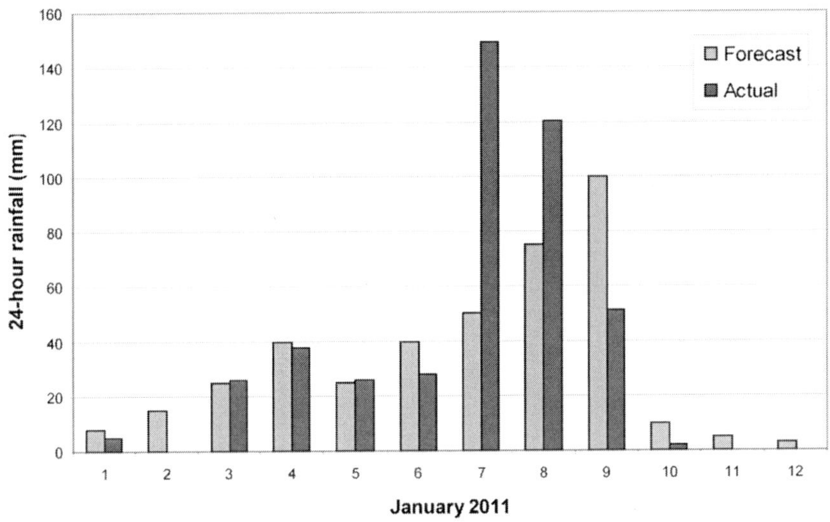

Figure 9. Forecast and actual rainfall intensities in the Wivenhoe catchment, January 2011 (Source of data: [15]).

Rainfall forecasts correspond reasonably well with actual average rainfall recorded up until 8th January 2011. Over the next two days the 24-hour forecasts significantly underestimated the average rainfall subsequently recorded. On 11th January the situation was reversed: the forecast was almost double that actually recorded. Thus extremely heavy rainfall was correctly predicted around the time of the flood, but the intensity of the main peak was underestimated, and its timing off by around 48 hours. Using this forecast as an input would allow the decision to be based on a probabilistic approach, which could be tested and implemented using Monte Carlo simulation or stochastic modeling. It would also be advisable to take account of the variability in spatial and temporal (as in Figure 9) patterns of rainfall forecasts. The accuracy of forecasts could be incrementally improved (that is, the variability of the forecast reduced) as more information became available. This is known as iterative "Bayesian updating".

The operators use a computer-based modeling tool, the so-called "real time flood model", to predict the reservoir water level. Two predictions are generated: a *with forecast* prediction that incorporates the BoM's rainfall forecasts and a *without forecast* that assumes zero rainfall. The former is tracked in blue in Figure 10 while the latter is shown in red. The *with forecast* prediction suggests that the level of the lake would exceed 74.0 m, the tipping point for the purposes of moving to a water release strategy where the primary consideration is the structural safety of the dam. This strategy, according to the manual, has no upper limit on the maximum release rate of water from the dam. The red line, however, the *without forecast*prediction, remains well below the 74.0 m threshold. In short, by failing to use all the information available to them (however incomplete), the dam operators made a sub-optimal decision, even if the dam was operated according to the operations manual.

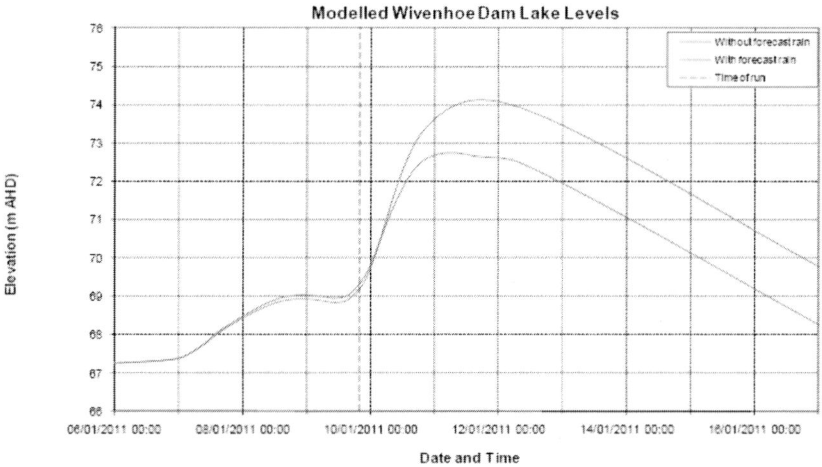

Figure 10. Modeled Wivenhoe dam lake levels—8:00 pm on 9 January 2011 (Source: [12]).

6. IMPACTS AND FLOODING

During the January 2011 event there were two major floods. The first of these was a severe flash flood that occurred on the afternoon of Monday 10th January in the inland city of Toowoomba and downstream in the Lockyer Valley (Figure 2). The Toowoomba/Lockyer Valley flash flooding was exacerbated by way a series of intense storms, with falls of 40–50 mm in 30 min, which moved over the already saturated catchment. Runoff was concentrated to maximum effect down gullies and streets in Toowoomba. Unable to cope with the volume of water, the creeks burst their banks, pushing a wall of water up to 11 m high through the city centre [24].

The speed of the flash flooding gave little time for warning. Flash flooding occurred at about 2:30 pm on 10th January; at 1:00 pm the Bureau of Meteorology issued a warning of localized flash flooding in the area and also alerted the State Disaster Coordination Centre. Informal warnings were blogged by the community at 1:40 pm, reporting that a wall of water was threatening Lockyer Creek.

The flood had reached up to 7 m by the time it struck with destructive velocity the town of Grantham, further downstream. Buildings adjacent to waterways were swept away in Murphy's Creek, Postmans Ridge and other locations. Conditions were especially hazardous at Grantham because of the number of dwellings constructed on the floodplain close to the river. Water depths of 2.0–2.5 m over the floodplain were realized in 10–15 min with velocities estimated to have been between 2 and 3 m/s. Nearly every house on the floodplain area of Grantham suffered major damage; a number were washed off their foundations and some totally destroyed. In total 23 people drowned in the Toowoomba and the Lockyer Valley flash floods, with nine others still missing, presumed dead.

The second major flood event occurred on 13th January when dam releases from Wivenhoe Dam flooded the city of Brisbane and surrounding areas. Figure 11 shows a hydrograph of the Brisbane River at the Brisbane City gauge between 8th January and 16th January.

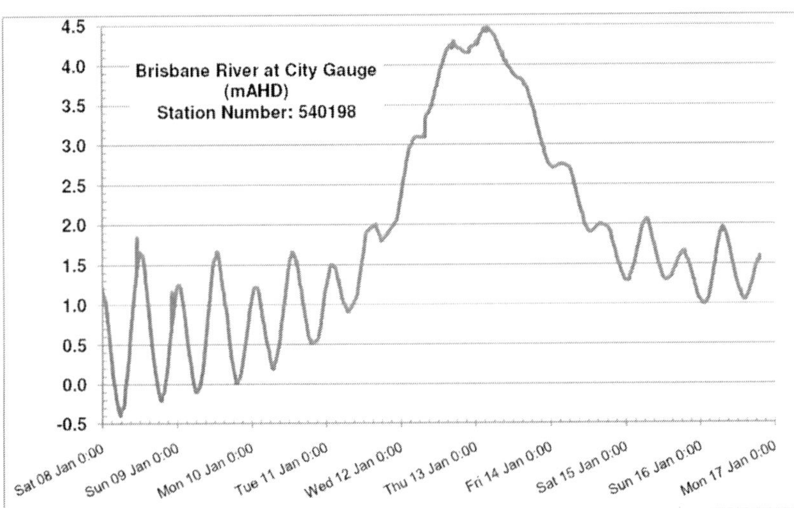

Figure 11. Height of the Brisbane River at the City Gauge between Saturday 8th January and Sunday 16th January (Source: [25], based on data from [26]).

Evident in Figure 11 are two peaks—a minor peak at 5:00 pm on Wednesday 12th January (4.30 m) and then later at 3:00 am on Thursday 13th January, when the river reached 4.46 m. Given a "major flood" at the Brisbane City gauge is defined as a gauge height of 3.50 m or greater, Brisbane City experienced a major flood from 10:00 am on Wednesday 12th January until 6:00 pm on Thursday 13th January, a period of 32 hours.

This 32-hour period at the City gauge spanned three high tides [25]:

- 14:15 Wednesday 12th January
- 02:34 Thursday 13th January
- 14:54 Thursday 13th January

The major peak of 4.46 mAHD at the City gauge corresponded almost exactly with the high tide at 2:34 am on 13th January. The minor peak of 4.30 mAHD at 17:03 Wednesday 12th January (Figure 11) probably represents the highest river flows. Brisbane felt the full force of these. More than 15,000 properties were inundated in metropolitan Brisbane [8] and some 3,600 homes evacuated.

Across Queensland approximately 12,000 people were accommodated in 34 evacuation centers managed by the Red Cross, with a number of other official and informal evacuation centers also operating. Altogether over 200,000 people were affected. Approximately 3,570 business premises were inundated, and commercial losses of approximately $4 billion were reported across the mining, agriculture and tourism sectors. Over 19,000 kilometres of roads were damaged, around 28 percent of the Queensland rail network damaged and three major ports significantly impacted. An estimated 28,000 homes would need to be rebuilt, while vast numbers of dwellings require extensive repairs [8]. According to the Insurance Council of Australia, almost 56,200 claims were received by insurers, with an insured cost of $2.55 billion [27].

The community response to assist with the aftermath of the floods was tremendous: over 55,000 volunteers registered to help clean up Brisbane's streets and homes, with thousands more unregistered volunteers. This overwhelming community involvement was a feature of the disaster response. Community Recovery Centres were set up in several locations, Red Cross and Lifeline provided a full range of services on the ground and eight specialised counselling teams from the Department of Communities were deployed across the Lockyer Valley and Toowoomba [28].

Inspection of the 2011 event shows that the extent of flooding in most areas of Brisbane City was very similar to that observed during the 1974 flood. The aerial photographs in Figure 12 and Figure 13 display the extent of flooding during the 2011 (blue line) event while the red line represents the flood extent in 1974. The four images depict different parts of Brisbane City. The point here is that while there may have been uncertainty in the Annual Return Interval of this flooding, where it flooded was largely predictable.

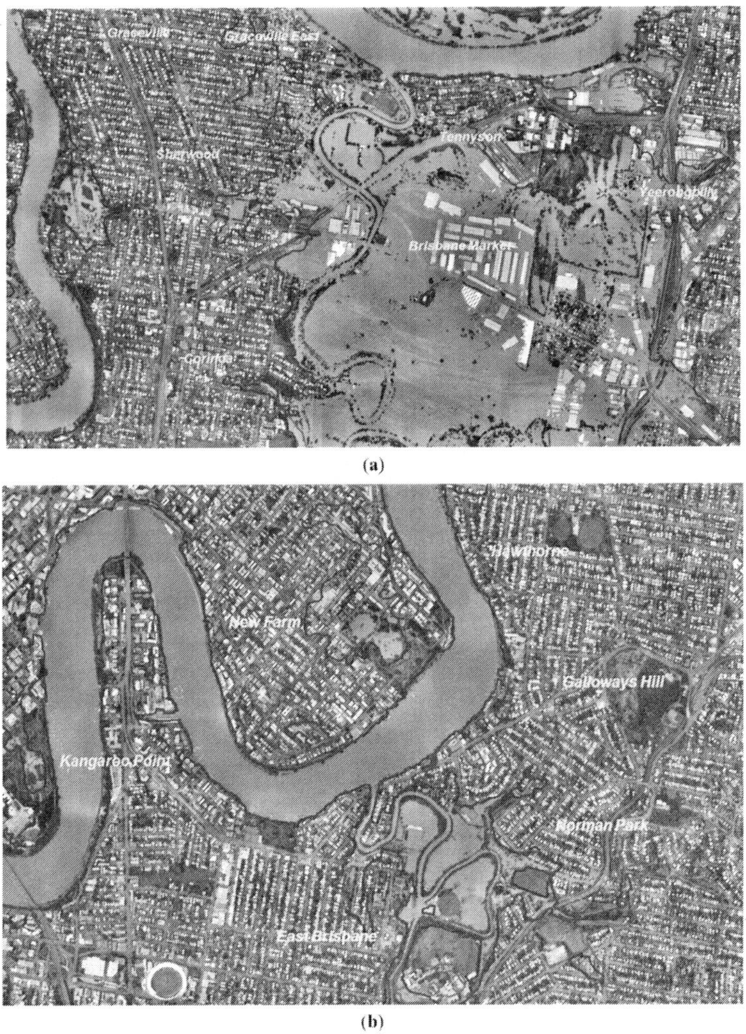

Figure 12. Flood extent comparison between (**a**) the 2011 flooding (blue polygon) and (**b**) the 1974 Brisbane flood inundation extent (red polygon). (Source: [29]).

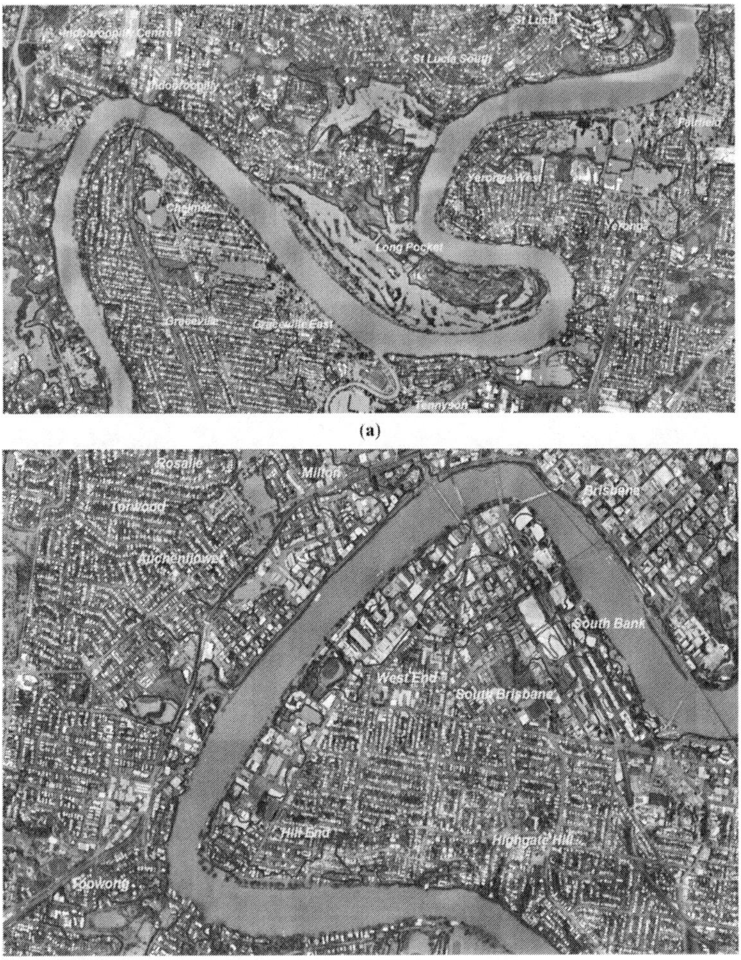

Figure 13. Flood extent comparison between (**a**) the 2011 flooding (blue polygon) and (**b**) the 1974 Brisbane flood inundation extent (red polygon). (Source: [29]).

7. IMPLICATIONS

7.1. Queensland Floods Commission of Inquiry

On Monday 17th January 2011 the Premier of Queensland, Anna Bligh, announced an independent Commission of Inquiry to examine the flood disaster. The terms of reference provide for an independent and thorough examination of the chain of events leading to the floods, and all aspects of the response and the subsequent aftermath. The Commission was tasked to deliver an Interim Report in August 2011, covering matters associated with flood preparedness to enable early recommendations to be implemented before the next summer's wet season. The Final Report is to be handed down by January 2012. The Interim Report [12] summarised submissions and testimony on the:

- preparation and planning for the flooding by governments at all levels, emergency services and the community;
- supply of essential services during the floods;
- adequacy of forecasts and early warning systems, with particular reference to Toowoomba and the Lockyer Valley;
- compliance with, and the suitability of, dam operational procedures for safety and flood mitigation.

A total of 175 recommendations were proposed, which were subsequently fully supported by the Queensland Government [30].

Most relevant to our study here is the Commission of Inquiry's view that Wivenhoe Dam had been operated according to the operations manual. Nonetheless, the Commission has recommended that the manual be reviewed including a re-evaluation of the design hydrology using a probabilistic approach, consideration of temporal and spatial variability in rainfall, and of the relative timings of inflows from the dams and downstream tributaries. It was also recommended that the reliability of rainfall forecasts and the use of weather radar be assessed and considered for decision-making. Furthermore the recommendations called for modeling of the consequences (in terms of risk to life and safety, and economic, social and environmental damage) of the water release strategies for a spectrum of historical, design and synthetic flood events. This is very much in line with the authors' view expressed earlier about the utility of decision trees and probabilistic methods in general to improve decision making in relation to dam releases of water during flooding situations.

The Commission also recommended that every local government with a flood risk should ensure that, before the onset of the next wet season (*i.e.*, before November 2011), it updates its local disaster management plan in accordance with Queensland's State-wide Natural Hazard Risk Assessment [31] and the *Disaster Management Act 2003*. Each local disaster management plan should address local circumstances and risks, be simple to use, and be published on the relevant local government's website. Training for disaster management personnel and volunteers should be provided, and community education programs regarding *inter alia* flood preparations, warnings, evacuation and emergency assistance instigated.

Matters relating to land use planning and insurance will be dealt with in the Commission of Inquiry's Final Report, and hearings will not be held until late in 2011. However information contained in public submissions provides the opportunity for some *a priori* discussion about issues and implications of these topics as they relate to flood risk management. We discuss these next.

7.2. Insurance Issues and Implications for Land-Use Planning

Following a natural disaster, the insurance industry plays a vital role in funding the rebuilding, repair or replacement of damaged homes, infrastructure and assets. In recent years, the Australian insurance industry has responded effectively to a range of natural disasters such as the Hunter Valley storms (2007), the 'Black Saturday' bushfires in Victoria (2009), the Perth and Melbourne hailstorms (2010), and Cyclone Yasi (2011) [32]. Some 88% of Australian homeowners have home and contents policies [32] that cover damage arising from most natural perils with the exception of riverine flood. Flash flooding, as a direct consequence of rainstorm runoff, following high intensity rainfall [33] is covered.

For the purposes of insurance in Australia, riverine flooding is defined as *the overflow from rivers and creeks following long duration rainfall over large catchment areas, or water rising up from flooding rivers, in contrast to water coming down from rain and storm* [33]. While some insurers offer cover for riverine flood, many currently do not, something that many policy holders discovered only after the January 2011 floods [34,35]. Thus whilst claims by insured victims of flash flooding in Toowoomba and the Lockyer Valley were generally settled by insurers, many residents along the Brisbane River had claims denied. Whilst the Insurance Council of Australia has put the proportion of denied claims at 15% [33,36], the perceived lack of performance of private insurers in assisting the recovery from the flood has led to a community backlash. Many people were unaware that they were at risk of flood (something that can and should not be blamed on insurers); others rejected flood insurance where offered because it was viewed as being prohibitively expensive [35].

To address the issue of availability and affordability of insurance for flood offered by the private insurance market to homeowners, the

Natural Disaster Insurance Review (NDIR) was announced by the Assistant Treasurer in the Australian Commonwealth Government on 4th March 2011. Key discussion points include the issue of a single unified and consistent definition of "flood" for insurance purposes, and, in the context of availability and affordability of flood insurance, non-insurance and under-insurance, consumer understanding and dispute resolution, flood risk measurement and mitigation, and some aspects of government funding of natural disaster relief and recovery.

Measuring flood risk by way of flood maps has been raised with the Review Panel by:

- insurers, as a necessary input to enable them to assess and price flood risk;
- homeowners seeking to assess their personal flood risk; and
- councils as part of their planning and risk management.

The Review Panel has produced an Issues Paper [33] for public discussion in which it identified two alternative models of flood insurance for the future, assuming that the status quo is no longer acceptable. Under the first model, cover for riverine flood would be automatically provided as part of home insurance policies. This would eliminate disputes about what constitutes a flood, but would significantly increase premiums for homes at risk of flooding. The second model allows a homeowner to 'opt out' of the flood portion of a home insurance policy. Both of these demand that high flood risk homes be identified, and the Review Panel has proposed that a discount be provided to owners of these high flood risk homes. How funding premiums for these homeowners would be achieved is an ongoing issue for discussion.

In the authors view, the fundamental question that needs addressing is not the insurability of flood risk, but how best to deal with the legacy of poor land-use planning decisions that has left some home owners in locations now designated as high risk. Development in the low-lying areas along the lower Brisbane River, for example, has occurred despite the city's history of severe flooding and has resulted in large concentrations of properties exposed (Figure 3). How to reduce this exposure to flooding should be the key policy objective. Flood risk management should aim to reduce a community's flood risk to acceptable levels, either by reducing exposure to flooding by prudent land-use planning, or by reducing the vulnerability of people and property to flooding [3]. The legacy issues are an issue that governments and the insurance sector are now grappling with. The authors believe that nothing will change until local councils are held accountable for bad land-use planning decisions [37]. Resolving the problem is not simple, but risk-adjusted insurance premiums must form part of the solution [38]. These provide a market signal and will act to incentivise risk-reducing behaviour by all parties. One very positive initiative of the local council in Grantham, involves a land swap for flooded residents. We return to this example in later discussion.

Insurance is not an alternative to risk management; it is a means of transferring the residual risk once risk management measures have been put in place. To actively contribute to flood risk management, insurance must act to reduce the number of homes at risk. The only mechanism available is to ensure that premiums reflect the real risk and thereby encourage homeowners, local councils and governments to undertake appropriate mitigation efforts and enact risk-informed land-use planning practices. Government may offer premium discounts to homeowners who have already been permitted to build in flood prone locations, but further development has to either be prevented or allowed only in a flood-resilient manner by choice of building materials and with raised floor heights. Any such construction must be undertaken with full disclosure of the flood risk. In discounting market premiums, it is critical that the government not interfere with actual pricing and so distort the price signal that can act to incentivise all parties to act to minimize risk.

Development in high-risk areas is certainly not unique to Brisbane [39] or just with respect to flood. Moreover, the practice of post-event handouts by government often encourages victims to rebuild in the same way and in the same location. This was the case immediately after the February 2009 "Black Saturday" bushfires in Victoria when the State Premier announced payments of up to $50,000 to homeowners whose principal residence was destroyed in the conflagration [40]. This was despite the knowledge that in the most severely affected communities of Marysville and Kinglake some 10% of destroyed structures were physically located within bushland, and some 60% within 10 m of the bushland boundary [41], and it having been well demonstrated that the likelihood of home survival in such extreme fires increases with increasing separation from the bush [42]. This is an example of government generosity not encouraging risk-reducing behaviour.

The situation is not all negative. A very positive example of sensible land-use planning to arise out of the 2011 flooding has occurred in the township of Grantham, devastated by flash flooding as discussed earlier. In March 2011, Grantham township was formally declared a "reconstruction area", which enabled the Lockyer Valley Regional Council to fast-track a master planning process, which was completed within four months. Part of the master plan includes relocation of residential homes from the floodplain to higher ground outside the flood-zone for flood-impacted residents from the townships of Grantham, Murphys Creek,

Postmans Ridge, Withcott and Helidon. It is expected that 70 to 80 percent of residents formerly living on the floodplain will move into new homes from December 2011. Nowhere else in Australia are we aware of a Local Government authority assisting a disaster-struck community in this way [43].

Another positive signal to come out of the 2011 flooding has been a decision by the Brisbane City Council to raise the Defined Flood Level (DFL) for rebuilding purposes using the January 2011 peak flood level in most areas. Prior to this the flood control levels were based on the notional (modeled) 1-in-100 year flood. The new minimum habitable floor level for residential development was changed to 500 mm above the DFL except where the existing level was higher. These specifications are now imposed upon all new residential, industrial and commercial developments [44], and will serve to reduce the city's vulnerability to future flooding. Needless to say the larger floods are still possible (Figure 3).

It seems axiomatic that if risk-reducing behaviours are to be encouraged, then homeowners must understand their flood risk. Similarly, it is difficult to see why insurers should offer flood cover if information that would allow them to assess and price risk is denied them. Unfortunately, such information is often not publically available even when flood studies by hydrologists and hydraulic engineers have been carried out and funded by public money. A recent study of publically available flood risk information on local government websites [45] identified that less than 50% of local councils provide flood risk information. Of those with a recognised flood risk, only 67% provided flood maps online.

Ideally insurers (and land-use planners) need modeled flood surfaces that can be interpreted to see how flood depths vary as a function of the Annual Return Interval (ARI) at given locations. When available, flood maps vary greatly in terms of information content with many only recording the extent of inundation in a modeled 1-in-100 year flood. This is not a measure of risk and says nothing about how flood damage may vary in larger floods. In the case of the Hawkesbury-Nepean River catchment, for example, many tens of thousands of homes have been built with planning approval just beyond the 1-in-100 year flood, and the difference in the height of inundation between the 1-in-100 year flood and the maximum possible flood is 9 m! [46]. This is a very different circumstance to some inland catchments where the difference may only be a meter or so. Again having a uniform ARI threshold for building approvals ignores the reality of risk, which is a function of the hazard, exposure and vulnerability.

The primary source of flood information currently available to the insurance industry in Australia is the National Flood Information

Database (NFID), which is developed from publically-available flood information and digital terrain maps, and which currently covers around 5.6 million individual addresses across the country [47]. The total number of addresses in Australia is around 12 million but many of these are not flood prone. Even so, NFID does not include every property at risk in Australia because not all catchments have been studied, and even where this has been done not all councils will make this information available, a situation that is most common in Queensland. In a submission to NDIR the Insurance Council of Australia (ICA) claim that this lack of completeness is one of the key factors limiting some insurers' ability to provide flood insurance in certain geographic areas [36].

The January 2011 floods have exposed the non-insurance or under-insurance of not just individual homeowners, they have also focused attention on the Queensland government's lack of insurance cover for its own infrastructure. State governments in Australia have a longstanding arrangement with the Commonwealth Government under the Natural Disaster Relief and Recovery Arrangements (NDRRA), whereby the Federal Government fund 75% of all infrastructure reconstruction costs in the event of a natural disaster. There is no compulsion at present for State Governments to take out insurance cover although Queensland is the only one to opt not to purchase reinsurance for its infrastructure assets, claiming that it was "not a value-for-money proposition" [48]. Thus 'self-insurance' by the State applies only to 25% of the reconstruction costs, with the Commonwealth covering the other 75%. To offset the NDRRA costs of the January 2011 flood event, the Federal Government will raise $1.8 billion by imposing a one-off flood levy on all Australian taxpayers with a taxable income greater than $50,000 and who were not in receipt of a government Disaster Recovery Payment for a flood event [49]. The impact of the flood will thus be felt by all Australians, not only those living in southeast Queensland.

8. CONCLUSIONS

This paper has examined the causes, impacts and some implications of the January 2011 Brisbane flood. The period December 2010 to January 2011 coincided with a strong La Niña event, often associated with extreme rainfall and widespread flooding in eastern Australia.

The flood has been termed a "dam release flood" by hydrologists appointed by the Insurance Council of Australia. This suggests that a release of water from the Wivenhoe Dam was a principal cause of flooding along the mainstream and tributaries of the Brisbane River downstream of the dam over the period 11th–12th January 2011. Whilst the dam operators were acting in accordance with the operations manual for the dam, their modeling did not take account of forecast rainfall in determining the predicted dam water level, and this resulted in a sub-optimal water release strategy. Employing tools for decision making under uncertainty would have resulted in a different water release strategy.

Many insurers do not provide riverine flood insurance due to a lack of information on which to determine and price the risk. An official Review of natural disaster insurance has proposed models to ensure more homeowners are covered by flood insurance. We argue that the real issue is how to deal with the legacy of past poor land use planning, which has left some home owners in locations now designated as high risk. Insurance premiums should reflect the real flood risk of a property, and insurers should have access to the best flood studies available on which to determine the cost of this risk.

Nevertheless, a positive example of sensible land-use planning to arise out of the 2011 flooding occurred in the township of Grantham. The plan drawn up after the floods includes relocation of residential homes from the floodplain to higher ground outside the flood-zone.

REFERENCES

1. 2011a Submissions; Floods Commission of Inquiry: Brisbane, QLD, Australia, 2011. Available online: http://www.floodcommission.qld.gov.au/submissions (accessed on 6 September 2011).
2. Flood Warning System for the Brisbane River below Wivenhoe Dam to Brisbane City; Bureau of Meteorology: Melbourne, Australia, 2011. Available online: http://www. bom. gov.au/hydro/flood/qld/brochures/brisbane_lower/brisbane_lower.shtml#PreviousFlooding (accessed on 31 March 2011).
3. Joint Flood Taskforce. Joint Flood Taskforce Report March 2011; Report prepared for the Brisbane City Council: Brisbane, QLD, Australia, 2011.
4. Known Floods in the Brisbane & Bremer River Basin, Including the Cities of Brisbane and Ipswich; Bureau of Meteorology: Melbourne, Australia, 2011. Available online: http://www.bom.gov.au/hydro/flood/qld/fld_history/brisbane_history.shtml (accessed on 1 April 2011).
5. Brisbane Floods—January 1974; Bureau of Meteorology: Melbourne, Australia, 1974. Available online: http://www.bom.gov.au/hydro/flood/qld/fld_reports/brisbane_jan1974.pdf (accessed on 31 March 2011).
6. Crompton, R.P.; McAneney, K.J. Normalised Australian insured losses from meteorological hazards: 1967–2006. Environ. Sci. Policy 2008, 11, 371–378.
7. Crompton, R.P.; Risk Frontiers, Macquarie University, NSW, Australia. Personal communication, 2011.
8. EMA Disasters Database; Australian Emergency Management Australia: Canberra, Australia, 2011. Available online: http://www.disasters.ema.gov.au/Browse/Search.aspx (accessed on 1 April 2011).

9. Frequent Heavy Rain Events in Late 2010/Early 2011 Lead to Widespread Flooding Across Eastern Australia; Special Climate Statement 24. National Climate Centre, Bureau of Meteorology: Melbourne, Australia, 2011.
10. SOI (Southern Oscillation Index) Archives—1876 to present; Bureau of Meteorology: Melbourne, Australia, 2011. Available online: http://www.bom.gov.au/climate/current/soihtm1.shtml (accessed on 31 March 2011).
11. Archive—Daily rainfall totals for Australia; Bureau of Meteorology: Melbourne, Australia, 2011. Available online: http://www.bom.gov.au/jsp/awap/rain/archive.jsp (accessed on 31 March 2011).
12. Interim Report, 1 August 2011; Queensland Floods Commission of Inquiry: Brisbane, Australia, 2011. Available online: http://www.floodcommission.qld.gov.au/ publications/interim-report (accessed on 23 August 2011).
13. Insurance Council of Australia. Flooding in the Brisbane River Catchment, January 2011; ICA Hydrology Panel: Sydney, Australia; 20; February; 2011.
14. Manual of Operational Procedures for Flood Mitigation at Wivenhoe Dam and Somerset Dam; Revision 7, November 2009. Seqwater: Brisbane, QLD, Australia, 2009. Available online: http://resources.news.com.au/files/2011/01/20/1225992/005512-110121-dam-manual.pdf (accessed on 3 May 2011).
15. Seqwater. January 2011 Flood Event: Report on the Operation of Somerset Dam and Wivenhoe Dam; Seqwater: Brisbane, QLD, Australia; 2; March; 2011.
16. The January 2011 Flood Event; Seqwater: Brisbane, QLD, Australia, 2011. Available online: http://www.seqwater.com.au/public/sites/default/files/userfiles/Factsheet-January_ 2011_FloodEvent.pdf (accessed on 3 May 2011).
17. Transcript of Proceedings; Queensland Floods Commission of Inquiry: Brisbane, QLD, Australia, 11 April 2011; pp. 77–96. Available online: http://www. floodcommission.qld.gov.au/__data./pdf_file/0018/3762/2011-04-11-QFCI-Day-2-Brisbane.pdf (accessed on 3 May 2011).
18. Ayre, R. Feasibility of Making Pre-releases from SEQWC Reservoirs; Seqwater: Brisbane, QLD, Australia, 2001.
19. Drought situation reports (Department of Employment, Economic Development and Innovation); Queensland Government: Brisbane, Australia, 2011. Available online: http://www.longpaddock.qld.gov.au/queenslanddroughtmonitor/queenslanddroughtre port/index.php (accessed on 6 September 2011).
20. Raiffa, H. Decision Analysis: Introductory Readings on Choices under Uncertainty; McGraw Hill: New York, NY, USA, 1997.
21. Krige, D.G. A Statistical Approach to Some Mine Valuations and Allied Problems at the Witwatersrand. Master Thesis, University of Witwatersrand, Johannesburg, South Africa, 1951.
22. Matheron, G. Principles of geostatistics. Econ. Geol. 1963, 58, 1246–1266.
23. Clemen, R.T. Making Hard Decisions: An Introduction to Decision Analysis; PWS-KENT: Boston, MA, USA, 1997.
24. Coates, L.; Haynes, K.; Gissing, A.; Radford, D. The Australian experience and the Queensland Floods of 2010–2011. In The Handbook of Drowning: Prevention, Rescue, Treatment, 2nd ed.; Bierens, J.J.L.M., Ed.; Springer-Verlag Publishing: Berlin, Germany, 2012; Chapter 10–17, in press.
25. O'Brien, M.J. Brisbane Flooding January 2011—An Avoidable Disaster; Queensland Floods Commission of Inquiry: Brisbane, Australia; 20; March; 2011.
26. Latest River Heights for Brisbane River at City Gauge; Station No 540198. Bureau of Meteorology: Melbourne, Australia, 2011. Available online: http://reg.bom.gov.au/fwo/IDQ65389/ IDQ65389.540198.tbl.shtml (accessed on 1 November 2011).
27. Historical and Current Disaster Statistics; Insurance Council of Australia: Sydney, Australia, 2011. Available online: http://www.insurancecouncil.com.au/ Default.aspx?tabid=1572 (accessed on 23 August 2011).
28. Flood Assistance; Lockyer Valley Regional Council: Gatton, Australia, 2011. Available online: http://www.lockyervalley.qld.gov.au/news-events/news/1270-lockyer-valley-road-closures-271210 (accessed on 9 September 2011).

29. Chen, K. Risk Frontiers; NearMap Pty Ltd.: West Perth, Australia, 2011. Available online: http://www.nearmap.com(accessed on 1 November 2011).
30. Queensland Government Response to the Floods Commission of Inquiry Interim Report; Queensland Government: Brisbane, Australia, 2011. Available online: http://www.premiers.qld.gov.au/publications/categories/reports./response-to-flood-inquiry.pdf (accessed on 23 August 2011).
31. Risk Frontiers. State-Wide Natural Hazard Risk Assessment; Report prepared for Queensland Department of Community Safety: Brisbane, QLD, Australia, 2011.
32. Insurance in Australia; Australian Trade Commission: Canberra, Australia, 2009.
33. Natural Disaster Insurance Review. quiry into Flood Insurance and Related Matters; Commonwealth of Australia: Canberra, Australia, 2011. Available online: http://www.ndir.gov.au/content/issuespapers/NDIRIssuesPaper.pdf(accessed on 23 August 2011).
34. Johnston, E.; Needham, K.; Tovey, J. Insurers Say no to Brisbane; Brisbane Times: Brisbane, Australia, 21 January 2011. Available online: http://www.brisbanetimes.com.au/environment/weather/insurers-say-no-to-brisbane-20110120-19y3m.html (accessed on 23 August 2011).
35. Williams, R. Victims Discover Insurance doesn't always Cover Flooding; Sydney Morning Herald: Sydney, Australia, 16 January 2011. Available online: http://www.smh.com.au/small-business/managing/victims-discover-insurance-doesnt-always-cover-flooding-20110117-19t31.html (accessed on 20 September 2011).
36. Insurance Council of Australia Response to the Natural Disaster Insurance Review; Insurance Council of Australia: Sydney, Australia, July 2011. Available online: http://www.ndir.gov.au/content/submissions/issues_paper_submissions/Insurance_Council_of_Australia.pdf (accessed on 12 October 2011).
37. Burby, R.J. Hurricane Katrina and the paradoxes of government disaster policy: Bringing about wise government decisions for hazardous areas. Ann. Am. Acad. Polit. Soc. Sci. 2006, 604, 171–191.
38. Roche, K.M.; McAneney, K.J.; van den Honert, R.C. Policy options for managing flood insurance. Environ. Hazards2010, 9, 369–378.
39. Crompton, R.P.; McAneney, K.J. Normalised Australian insured losses from meteorological hazards: 1967–2006. Environ. Sci. Policy 2008, 11, 371–378.
40. Victorian State Government. Bushfire Appeal Funds to Help Rebuild and Recover; Premier of Victoria: Melbourne, Australia, 2009. Available online: http://www.premier.vic. gov. au/ component/content/article/6508.html (accessed on 6 September 2011).
41. Crompton, R.P.; McAneney, K.J.; Chen, K.; Pielke, R.A., Jr.; Haynes, K.A. Influence of location, population and climate on building damage and fatalities due to Australian bushfire: 1925–2009. Weather Clim. Soc. 2010, 2, 300–310.
42. Chen, K.; McAneney, K.J. Quantifying bushfire penetration into urban areas in Australia. Geophys. Res. Lett 2004, 31, L12212.
43. Australian-First Deal to Relocate Grantham Residents; Lockyer Valley Regional Council: Gatton, Australia, 2011. Available online: http://www.lockyervalley.qld.gov.au/news-events/news/1318-australian-first-deal-to-relocate-grantham-residents (accessed on 9 September 2011).
44. 'Defined Flood Level' revised; Mullins Lawyers: Brisbane, Australia, 2011; Available online: http://www.mullinslaw.com.au/FeaturedItem.aspx?EntityId=106754 (accessed on 9 September 2011).
45. Box, P.; Thomalla, F.; van den Honert, R.C.; McAneney, K.J. How can residents know their flood risk? A review of online flood information availability in Australia. Aust Planner 2012. in submit.
46. Hawkesbury-Nepean Floodplain Management Committee. Achieving a Hawkesbury-Nepean Flood plain Management Strategy; Report prepared for the Hawkesbury-Nepean Flood Management Advisory Committee: Sydney, Australia, 1997.
47. Chen, K.; Risk Frontiers, Macquarie University, NSW, Australia. Personal communication, 2011.
48. Wilson, L. State Goes it Alone in Shunning Insurance; The Australian: Sydney, Australia, 3 February 2011. Available online: http://www.theaustralian.com.au/in-depth/cyclone-

yasi/state-goes-it-alone-in-shunning-insurance/story-fn7rj0ye-1225999067129 (accessed on 25 August 2011).
49. Flood Levy; Australian Government Treasury: Canberra, Australia, 2011. Available online: http://www.treasury.gov.au/contentitem.asp?NavId=022&ContentID=1949 (accessed on 25 August 2011).

CHAPTER 9

Dilemmas and Pathways to Dealing with Flood Problems in Twenty-First Century China

Guoyi Han1,* and Roger E. Kasperson2

1 Stockholm Environment Institute (SEI), 106 91 Stockholm, Sweden 2 Graduate School of Geography, Clark University, MA 01610, USA

ABSTRACT

With more than 60 percent of China's 1.34 billion population, about 50 percent of its most productive arable land, and the world's second largest and rapidly growing national GDP exposed to flooding of various kinds, China has an intractable flood problem. Envisioning potential impacts of climate change and continued intensification of floodplain development driven by rapid industrialization and urbanization, it is very likely that China will see a continued increase of exposure to floods this century. This over outlines and discusses fundamental dilemmas, plausible pathways, and key options for managing future flood risks in China in the context of rapid socioeconomic transition and climate change. Fundamental dilemmas are the embedded difficult trade-off choices, from balancing economic development with flood vulnerability reduction, to coordination and cooperation among increasingly diverse actors and across scales. Among plausible pathways, this article argues that a resilience strategy for managing flood risk is desirable. It would require human adjustment to flood, not by aiming for full protection and control but by adjusting our use of floodplains, integrating and experimenting with a wide range of flood risk management options, so that a dynamic balance is maintained between exposure and coping capacity and flood risk is contained at an acceptable level. Embracing variability and uncertainty lies at the heart of such a flood resilience centered paradigm. Reducing the flood toll cannot be had without trade-offs in economic development, food production, and agricultural productivity.

Keywords: China flood risk, dilemmas, flood management, pathways flood, risk management, flood resilience, strategy

1 INTRODUCTION

Globally, recent decades have seen rapid increases in losses due to natural disasters (UNISDR 2009). Despite progress in science and technology and the tremendous amount of attention and investment, vulnerability to flood disasters, in China and the world alike, has increased. There is a range of explanations for such a great paradox (White, Kates, and Burton 2001; IPCC 2011). Yet, in the case of flood losses, this also reflects the fundamental dilemma concerning flood risk management—the use and development of floodplains, while increasing the exposure to floods, are also vital for improving the general well-being of the populations and the economy of regions. The problem lies in the imbalance between exposure and coping and adaptive capacity. While coping capacity has also increased as countries have developed and both economi c conditions and governance have improved, the pace has not been sufficient to compensate for the increase in exposure and vulnerability, particularly in rapidly emerging economies (UNISDR 2009).

Into the twenty-first century, there has been a worldwide significant shift in the management of risks associated with river and coastal flooding (Harvey, Evans et al. 2009; Harvey, Thorne et al. 2009; Green 2010). The essence of this shift is moving away from control-oriented flood defense to a riskbased, more holistic approach, with an emphasis on system perspective and sustainability. There is now a strong recognition, at least within the flood management community, that sustainable flood mitigation solutions have to be achieved through integrated actions (White 1994; APFM 2003; Cheng 2005; UNDP 2004; UNISDR 2004). This shift can be traced back to the work of geographer Gilbert F. White and others who demonstrated convincingly the inadequacy of relying on engineering solutions alone to address flood problems. It was White who first argued that, instead of trying to control nature, society needs to learn to live more compatibly with the natural occurrence of floods (White 1945; Burton 1962; Burton, Kates, and White 1978).

Embracing variability and uncertainty lies at the heart of the new flood resilience centered paradigm. The fundamental and arguably most difficult change required is therefore for decision makers and society at large to accept that it is neither possible nor necessary to control all floods. A resilience strategy for managing flood risk would require human adjustment to flood, not by aiming for full protection and control but by adjusting our use of floodplains, integrating and experimenting with a wide range of flood risk management options, so that a dynamic balance is maintained between exposure and coping capacity

China has an intractable flood problem. Each year, China suffers billions in economic losses and hundreds to thousands of lives are lost. Looking into the future, the country's difficult flood problem is only going to become worse. By the middle of this century, China will be home to about 1.6 billion people and more than 70 percent of them will be living in cities on floodplains (Cheng 2008). If the current trends of socioeconomic development continue, China will become the largest economy in the world and close to 80 percent of its productivity will come from floodplains. Per capita GDP will at least quadruple its current value; thus the assets on the floodplains will keep on increasing exponentially. It is likely that climate change will also significantly increase both the intensity and the frequency of extreme weather events, making floods an even more severe threat (Shankman, Davis, and Leeuw 2009; IPCC 2011).

For millennia, China's flood management experience has predominantly centered on controlling floods through engineering structural measures, either by blocking or channeling the floodwater. For example, the legislative and institutional apparatus for flood management, such as the Flood Control Law of 1998, focuses primarily on controlling flood hazard and lacks almost completely any component of floodplain management, such as flood risk zoning and land-use regulation. To this day, there is little or no regional-scale floodplain planning (Shankman, Davis, and Leeuw 2009). Flood zoning and land-use planning measures in flood management remain at an embryonic stage (Cheng 2008).

As has been concluded by many analysts and observers, the increasing trend of flood losses, particularly during the 1990s, has stimulated a profound reflection and rethinking of flood management strategies in China (Cheng 2006; Harvey, Thorne et al. 2009). In 2003, the Ministry of Water Resources of China proposed to redirect flood prevention from flood control to flood management, which has led to a major institutional shift to enhancing understanding about interconnected systemic issues and risk awareness. Flood prevention work has thus shifted from attempts to eliminate floods to building capacity to live with and endure floods (GHD and IWHR 2006; Cheng 2008).

Three main aspects are central to this shift of overall strategy in flood management. First is the shift away from the notion of "full protection" to an acceptance that the flood defense system will have to be balanced with social and economic development on floodplains. Second is the shift from managing flood hazard to emphasizing the importance of floodplain management. Third is the approach to manage floods not only as a risk, but also as an opportunity and resource. This has involved a paradigmatic change from flood control to (ideally) controlled flooding. It is clear that this shift in China is consistent with and a reflection of an overall global rethinking and paradigm shift on humans' relationship with floods and how flood risk can be managed in a changing climate and an increasingly interconnected world. Embracing variability and uncertainty lies at the heart of such a new flood management paradigm.

While the shift in China from preventing flood events to managing flood risks is a fundamental step forward towards a more holistic and comprehensive approach to live with flood risk, it does not guarantee in any sense that the flood problem will be resolved. Fundamental dilemmas are entangled with flood problems in China. To what extent can vulnerability be significantly reduced without interfering with the progress in economic gains and improvements in overall well-being? To what extent are politicians as well as the general public prepared to accept the major investments that will be required for the transition to a more robust resilience strategy? What is an acceptable level of flood risk and how should it be determined? Understanding the fundamental dilemmas is essential for addressing the questions that are critical for devising a strategy that charts a path through complex science-policypractice interfaces and leads towards balanced, coordinated, and sustainable solutions.

The following two sections outline three fundamental dilemmas and three plausible scenarios of future pathways for managing flood risk in China. The article concludes with some remarks on the importance of embracing uncertainty in developing a resilience strategy.

2 THREE FUNDAMENTAL DILEMMAS FOR CHINA

A dilemma refers to a situation that requires a choice between options that are or seem equally unfavorable or mutually exclusive (American Heritage Dictionary). In essence, dilemmas are about making choices between alternative courses of action in which, in this case, trade-offs between risks and benefits are central and inescapable. Thus these dilemmas, by definition, defy solutions that appeal to all parties in that trade-offs exist in the course of achieving the resolution of desired societal wants or needs and these trade-offs must be balanced and negotiated

The contextual environment for flood risks in China is now undergoing rapid change due to global processes (such as globalization and global environmental change) and regional and local processes (such as rapid urbanization, industrialization, and social transformation). Deeply embedded in this changing context, China faces a set of fundamental dilemmas for managing its flood risk, where trade-offs are inevitably contentious yet must be made in balancing complex relationships between risk reduction and economic development.

2.1 "Space for Water" versus "Space for People"

By definition, floodplains are the land that could and would be flooded periodically. When floodplains are settled and developed, the competition for space between people and water (that is, floods) becomes the basic dilemma. With more than 60 percent of its 1.34 billion population, about 50 percent of its most productive arable land, and close to 80 percent of the world's second largest and rapidly growing national GDP depending on floodplains, China faces intense competition between "space for water" and "space for people"—a fundamental dilemma between flood risk reduction and economic development on China's floodplains.

Nothing illustrates this dilemma more vividly than the so-called flood detention / retention area dilemma in China. Since the 1950s, an important part of China's flood management strategy has been to establish designated flood detention and retention zones to accommodate excessive waters. Currently, China has 98 such designated zones, with a total area of 35,000 km2 and a total storage capacity of about 100 billion m3 . Yet, through decades of encroachment these zones now contain about 2 million hectares of cropland and more than 17 million people (Cheng 2005; Cheng 2008). Thus, in order for the designated zones to serve the intended flood adjustment function, China will have to find a way to resettle the 17 million people and change the ongoing settlement processes—not an easy dilemma to solve.

Furthermore, this difficult dilemma in all likelihood will only get worse. Within the floodplains, intensification of human activities and a shifting spatial concentration of both people and wealth will significantly increase China's exposure to floods in the coming decades. While China's population growth has slowed considerably due to the strict implementation of the one-child policy, its total population will likely grow to about 1.6 billion around 2030 before declining gradually. China will add the equivalent of the current U.S. population in the next twenty years, and the majority of this population increase will live in cities on floodplains. Even if this population intensification occurs at a slower pace than that experienced historically, the economic intensification on floodplains in the coming decades will still be tremendous and constitute a real challenge in Chinese history. It will shift the nature of the flood problem in China from a predominantly agricultural concern to an urban concern and food security threat.

This fundamental dilemma underlies the general trade-off between flood risk reduction and economic development—it is simply not possible to have both. Choices must be made, ranging from complete retreat from floodplains to bearing the full flood losses for economic gains. Anything in between is caught between irreconcilable goals. So far, the primary approach that China has employed to address this dilemma is to protect development from floods by extensive construction of flood defense infrastructures. Despite major investments, economic losses continue to increase (Cheng, Wu, and Wang 2004). The rising flood damage trend highlights the rapid increase of vulnerability due to intensification of floodplain development and assets accumulation. It also indicates the inherent limitations of the engineering approach.

Increasing flood vulnerability is essentially the result of the rapid development of China's floodplains. Vulnerability reduction thus requires a much broader approach to deal with the trade-offs between economic development and flood risk reduction in the context of multiple stresses and multiple vulnerabilities. Despite rapid development in recent decades, hundreds of millions of people in China are still living in poverty. Environmental deterioration and ecological destruction are widespread and devastating (Liu and Diamond 2005). China faces the dilemma of the need to achieve overall livelihood improvement while seeking to build societal resilience against floods. Flood risk reduction therefore will have to be integrated into or balanced against the broader context of poverty reduction, environmental and ecosystem management, improving overall well-being (longevity, income, education, health, and so forth), and sustainable development. It is a challenge that flood managers would like to avoid.

2.2 Cooperation versus Conflict

The second fundamental dilemma in flood risk management that China faces lies in the governance challenges where coordination and cooperation are central for resolving the inevitable conflicting interests among regions and social groups. Cooperation is a key potential resource needed for any viable future flood management approach. At the same time, economic decentralization, market reform and privatization, and growing regional and social disparities make cooperation increasingly difficult to achieve.

Reducing vulnerability is a complex task that involves an increasingly diverse range of actors, both public and private, concerning land-use decisions and the built environment (Hall 2008). All measures designed to reduce flood risks will have different spatial and social impacts. They typically involve spatial conflict (upstream and downstream, left and right riverbanks, urban and rural) and social tensions (among social groups, priorities, and so forth), thus demanding an enlarged capacity for balancing, coordination, and cooperation among regions and social groups. At the same time, with economic decentralization, the centralized command and control system in China has been weakened in its ability to "force" coordination and cooperation among regions and different stakeholders. The maneuvering space for coordination and cooperation is becoming increasingly small, while the actors and interests are becoming ever more diverse. A combination of balanced government and market instruments has yet to appear on the horizon. China still lacks the required and enabling institutional infrastructure. In the current Chinese flood management system, the roles of the various actors remain opaque and ill-defined and constitute major obstacles for effective coordination and cooperation.

The coordination and cooperation dilemma makes the flood risk management goal-setting process more challenging. What would be an acceptable level of risk reduction against which flood risk management should be measured, monitored, and evaluated? Setting such a goal is critical yet extremely difficult and contentious. An acceptable level of flood risk will always be problematic and uncertain since it rests on a wide range of processes, factors, and conflicting values. Often, as people get richer, their demand for safety correspondingly increases.

But the trade-offs between safety and wealth are highly sensitive. Making people safer and less rich may not be politically popular, because the risk reduction gains will be opaque and highly uncertain. Balancing tradeoffs between short-term and long-term goals is a difficult dilemma in determining an acceptable level of flood risk

Related to such goal-setting, a dilemma also exists with corresponding requirements for investment and capacity development. While flood response capacity will undoubtedly increase as overall development progresses, to keep pace with increasing exposure, an extra pace of capacity development is needed.

Yet who will pay for the extra pace of capacity development that will be required to stabilize the risk or even reduce it? What is the right balance between structural and nonstructural measures, and between government and market finances?

2.3 Coping with Changing Climate and Uncertainty

The third fundamental dilemma comes with the looming potential climate change impact, and with it, critical uncertainties about the patterns of future flood regimes (IPCC 2011). Planning and investment in flood protection infrastructures—such as dams and dykes—depend on predicting future climate variability. But climate change brings inherent unpredictability of future flood hazard patterns, adding a profound uncertainty for devising flood prevention and mitigation methods and investments. The climate change dilemma requires long-term planning and resolving a series of perplexing climate change uncertainties.

On a global scale, the Intergovernmental Panel on Climate Change (IPCC) has confirmed that climate change is now altering the predictability, intensity, and geographical distribution of many weather-related hazards through increased intensity of the water cycle and glacial melt and sea level rise (UNISDR 2009; IPCC 2011). At a national and regional scale, impacts of climate change on flood hazards are even more uncertain, while global trends seem to be moving towards an increasing frequency and intensity of extreme events (Milly et al. 2002; UNISDR 2009).

The average temperature in the Yangtze delta region, for example, is predicted to increase by between 2.3 and 3.3ºC by 2050 (NDRC 2007). This significant change in temperature is expected to lead to an array of changes in precipitation patterns, sea level rise, and coastal storm surges (Harvey, Evans et al. 2009).

From the perspective of human-environment interactions in coupled social-ecological systems, these dilemmas are all interlinked and characterized by great uncertainties, dynamic feedbacks, threshold effects, and nonlinear changes (Liu et al. 2007). They are fundamentally dilemmas between managing specific resilience (to floods) versus building a greater general resilience (Walker et al. 2004; Folke et al. 2010). These basic dilemmas place flood risk management decisions firmly in the domain of fundamental challenges to the governance of complex adaptive systems and sustainability, where coordination, cooperation, and social relations are essential areas that demand innovation.

With these dilemmas in mind, the next section discusses some key factors for future flood risk management, before outlining scenarios of plausible pathways.

3 KEY FACTORS FOR FUTURE FLOOD RISK MANAGEMENT

The looming climate change impacts and the rapidly changing socioeconomic conditions in China raise serious questions relating to future increases in flood risk. How will flood risk change in the coming decades, what number and level of flood events can we anticipate and prepare for? What policy interventions are available? To what extent will floods be a risk to China's long-term sustainable development? A complex array of factors shape and impact these issues.

Figure 1 is a schematic conceptual model of flood risk in the context of climate change and rapid socioeconomic transition. The key factors involve: (1) the major drivers, as well as the critical uncertainties associated with these drivers and their interactions; (2) the dynamics of vulnerability as determined by the covariance between exposure and response capacity that involves short-term coping and long-term adaptive capacity; and (3) the flood management system with both structural and nonstructural measures.

The interaction of all these factors and processes determines the dynamic pattern and change of flood risks and their impacts on society. Many have illustrated clearly the interactive nature of flood events and vulnerability in the context of a coupled human-environment system. Diverse drivers—natural, human, and technical—act as pressure or multiple stresses on the flood risk system, leading to dynamic changes of the control variables that affect and reshape both the pattern of hazard and the system's vulnerability to floods. The dynamics of vulnerability cannot be explained solely either by exposure (for example, population growth and intensification of economic activities) or response capacity (including both short-term coping and long-term adaptive capacities), but are the result of interactive change of both, or the covariance between them (Han 2011).

The key drivers for future flood risks in China are rapid socioeconomic transition, climate change, and the flood defense system itself. While the relative importance of these drivers will vary across river basins, they all are expected to experience major changes in the coming decades. The most challenging drivers for policy and decision makers are those associated with the highest levels of both risk importance and uncertainty, for example, potential climate change.

There is a general linkage between development stage and the pattern of losses from natural disasters. It is often the case that vulnerability tends to rise faster during the phase of transition (Sylves and Kershaw 2004). China today is in the midst of rapid urbanization and industrialization, which will almost certainly continue to increase human exposure to floods. Rapid urbanization, for example, acts as both pathway and receptor in terms of increasing flood risk. As a pathway, urban expansion increases the proportion of the impermeable surfaces and intensifies runoff. The general trend of current urban development decreases storage and drainage capacity for storm water, and heightens the vulnerability of cities to floods and waterlogging (Shi et al. 2005; Harvey, Evans et al. 2009).

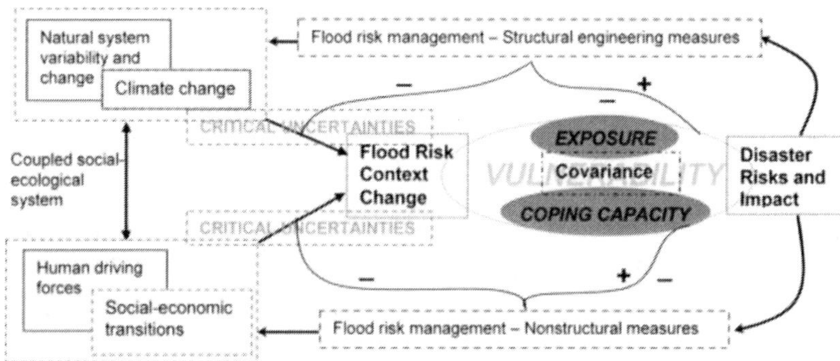

Figure 1. Conceptual model of flood risk in the context of climate change and rapid socioeconomic transition

In relation to the dynamics of vulnerability, the impact surrounding uncertainties arising from the changing socioeconomic transitions lies in whether and to what extent the growth of response capacity can match or exceed the increase in exposure (Kundzewicz and Menzel 2003). Examined globally, societal change and economic development are the chief culprits for increasing losses. If the current trends continue, global disaster losses will continue to outpace average economic growth (Bouwer et al. 2007).

In the coming decades, the transformation of the flood management system in China, spurred by ongoing institutional shifts, is a critical process that will determine both the direction and pace of flood response capacity change. The critical uncertainties in this respect reside not only with leadership and investment, but also in capacity development at all levels.

China's waterscape has been profoundly humantransformed and the transformation continues with massive water engineering projects such as the Three Gorges Dam and the ongoing South-North Water Transfer. From a system perspective, these transformations will continue to trigger new feedbacks that change the dynamics of the hydrological regimes, which in turn interact with the changing climate and social economic conditions to set the basic parameters for future flood risks in China. Integrating the above system dynamics from the interactions between climate and socioeconomic change with uncertainties to form scenarios is a key for robust, flexible, and sustainable risk reduction in the future (Schanze 2009).

4 PATHWAYS FOR THE FUTURE—SCENARIOS

With growing realization of the limited predictability of complex adaptive systems, scenarios are gaining recognition as a useful way for examining future changes. A series of recent scenario-based future risk studies have all confirmed that losses from weather-related disasters are projected to increase, due to a combination of the increasing exposure of people and assets and expected changes in the global climate. In the Netherlands, Bouwer, Bubeck, and Aerts (2010) found that due to socioeconomic change alone, annual expected losses may increase by between 35 and 172 percent by 2040, as compared with a baseline situation in 2000. If no additional measures are taken to reduce flood probabilities or consequences, by 2040 a combination of climate and socioeconomi c changes may increase expected losses by between 96 and 719 percent. In the United Kingdom, the Flood Foresight project concluded that expected annual flood damages may increase twentyfold by the 2080s if flood risk management policy were to remain as it is at present (Evans et al. 2004; Hall, Sayers, and Dawson 2005). In Japan, Ikeda, Sato, and Fukuzono (2008) developed various flood risk scenarios in their search for resilient social systems against flood disasters.

In China, similar research is currently under-way in the Taihu (Tai Lake) basini through a China-UK scientific cooperation project initiated in late 2006, which adopted the UK Foresight Project scenario approach (Harvey, Evans et al. 2009). While uncertainties exist in all the scenarios, given China's current development trend, there is little doubt that China's future flood losses will increase. This is not, however, to say that China's vulnerability to flood must also increase. The future state of flood risks in China will not only be influenced by critical uncertainties within the key drivers (for example, the speed and scale of climate and socioeconomic changes) (see Figure 1), but also by human choices that China makes today and along the way. Those choices will largely determine the pathway through which China's flood risk management system will evolve.

To highlight the different pathways, three plausible future flood risk management scenarios are outlined below. The underlying logic for constructing the scenarios is the prioritysetting on dealing with the trade-offs between flood risk reduction and economic development. In the first scenario, China maximizes economic development while continuing to deal with floods largely through engineering approaches. In the second scenario,

China prioritizes flood vulnerability reduction by developing and enforcing strict floodplain landuse regulations to limit and control floodplain development. The third scenario outlines a pathway of "living with floods" in which building resilience is achieved through embracing and internalizing, rather than controlling, variability and uncertainty in balancing flood risk reduction and economic development.

4.1 Controlling Floods—Continuation of the Engineering Approach

This is the more or less "business as usual" scenario in which China maximizes economic development while it continues its massive investment in structural measures to control floods. In the short term, economic gains and a technocratic bias will very likely make this pathway irresistible. But in the long term, the result of this pathway will lead China to be increasingly locked into a highly rigid system vulnerable to catastrophic floods, a likely "building up for disaster" scenario.

Over a span of more than half a century since 1949, China has constructed a massive flood control structure system, including more than 85,000 reservoirs of various sizes,ii a total of 286,900 km of dykes/levees of various standards,iii and close to 100 designated flood detention and retention (storage) zones with a total area of 35,000 km2 and a total volume of 103 billion m3 . Together, these measures form an extensive and effective flood defense system that has played a crucial role in flood control in China, reducing the annual inundation area by about 70 percent (as compared with that of 1949) and reducing loss of life dramatically. In the 1950s, the average annual casualties due to floods in China were 8976; this number was reduced to 1507 in the first decade of the twenty-first century (Cheng 2008).

While engineering protection has reduced the total area and frequency of inundation, the economic losses have grown enormously. Flood-related damage (yuan per hectare) has increased from 2190 in the 1960s to 12,120 in the 1980s and more than 20,000 in the 1990s on a national average. In rapidly developed areas such as the Tai Lake basin in the Yangtze delta, it has increased from 5565 in the 1960s to more than 30,000 in the 1980s (Shi and Pan 1997).

While structural approaches to flood control are essential to reduce flood losses, they have now reached a point where further expansion has become largely unsustainable. First, the "dyke syndrome" or the "dyke effect" (that is, dyke protection induces further development of floodplains that result in further demand for higher dykes) has been increasingly recognized as a driver for the erosion of long-term flood resilience of flood-prone human-environment systems. This has resulted in a continuously worsening discharge and flood level relationship.

Second, the cost-effectiveness of raising dykes/levees has been limited by economic, technological, and environmental constraints. Table 1 shows the investment feasibility calculated by the Institute of Water and Hydropower Research under the Ministry of Water Resources of China. It is extremely difficult to achieve the cost-effectiveness that is required (Xiang 2003).

Third, the existing flood control structures require tremendous investment in quality assurance and maintenance. More than one-third of the 85,000 reservoirs in China now require serious repairs.iv The levee systems along the major rivers are the result of centuries of construction, with extremely mixed quality and serious defects. Since 1998, more than 30 billion yuan have been invested in the levees in the middle and lower Yangtze River alone. Most of the investment has been in repair and infrastructure strengthening (Xiang 2003; Cheng 2006).

Lastly, China is running out of places to construct major water infrastructures. With the completion of the Three Gorges Dam on the Yangtze River, the Xiaolangdi Dam on the Yellow River, the Feilaixia Dam on the Pearl River, and the Nierji Dam on the Nen River in the Northeast, China has now basically run out of possible sites for major water control engineering constructions.

In addition, many other factors (such as resettlement and migration and ecological impacts) are now raising heightened concerns about new dam construction.

If China cannot break away from the current heavy reliance on its water engineering approach to flood control, one plausible future state of its flood risk system is an increasingly rigid system in which major investment is locked into the vicious circle of increasing and maintaining the protection standard of the dyke/levee system. At the same time, the natural flood carrying capacity of the rivers as well as the riverine ecosystems will remain in continuous decline, and waterlogging in cities will become intractable.

Table 1. Feasibility of flood loss reduction by raising levee protection standards

Current annual average river flood losses (billion yuan)	Loss reduction (%)	Loss reduction benefit (billion yuan)	Investment required for feasible engineering construction (billion yuan)
25	10	2.5	<30
	30	7.5	<90
	50	12.5	<150

Source: Xiang 2003.

While it sounds highly undesirable, this scenario is not only plausible but highly likely. This is the least-resisted pathway in the existing system because it simply requires doing business more or less as usual. In addition, compared to many developed countries, the standard of the flood control structures in China still has much catching up to do. For example, while Japan and China share a very similar percentage in population and assets on the floodplains, most cities in Japan are protected for 500- to 1000-year floods whereas most of the cities in China currently are only protected for 20- to 50-year floods. Furthermore, controlling floods is politically and socially much more straightforward than dealing with vulnerability reduction. Last but not least, many nested interests or "systemic institutional inertia" resist transformative changes (Ostrom 2004, 2007).

4.2 Reducing Vulnerability—Strict Regulation of the Floodplains

This is a scenario in which China prioritizes flood risk reduction and attempts to reverse the trend of increasing economic losses to floods by interfering and shaping developments on its floodplains. While a range of floodplain management options is available and need to be explored in China to strengthen the virtually nonexistent nonstructural measures for flood risk reduction, the biggest challenge for this pathway is how far it can go without interfering with economic development. Given China's extremely high dependence on its floodplains, a pathway that prioritizes flood risk reduction by limiting economic development will almost certainly encounter tremendous difficulties

Floodplain management for flood risk reduction is a particularly weak component in flood management in China. There are still many win-win solutions where integrative floodplain management can achieve both economic development and flood vulnerability reduction. Table 2 shows some initial results of the future flood risk scenario study in the Tai Lake basin, with a focus on the scope of risk reduction by careful land-use planning (Wang, Cheng, and Yu 2010). In all of the three scenarios they have analyzed—socioeconomic change only; climate change only; and combined impacts of socioeconomic and climate change—there is significant scope of potential flood loss reduction through careful risk mitigation-oriented land-use planning, which is nonexistent at present. Yet, in all three scenarios, even with careful land-use planning, expected annual flood losses will continue to increase, ranging from 3.64 to 8.01 billion yuan per year by 2050.

The positive results of this pathway would push China towards a more balanced approach between structural and nonstructural measures for flood risk management. But reversing the trend of increasing economic losses to floods in the future without major interference with the current trend of development intensification in the floodplains, both urbanization and industrialization, is highly unlikely.

4.3 Building Resilience—Living with Floods

The third potential pathway incorporates components from the two previous scenarios and rests on a series of changes that embrace variability and uncertainties, thus moving towards a strategy of building resilience and living with floods. Many regional flood studies have identified the need to break the feedback loops that have resulted in increasingly fragile and rigid systems with continuously declining capacity in coping with flood and waterlogging.

Resilience strategies, such as the "space for water" program to regain the system's natural flood adjustment capacity, are essential for achieving the sustainability of flood risk management in the future.

In this scenario, the current flood management institutiona l shift in China takes strong root in the coming years. China moves to adopt a resilience flood risk management strategy that seeks to restore the feedbacks required for a resilient and adaptable flood risk system in the future. A series of principal changes are critical for this scenario.

First, the objective in flood risk management is not solely to reduce flood losses but to ensure the good functioning of human-environment systems on the floodplains in terms of the overall benefits from human use and ecosystem services and functions, taking into account risks and uncertainties associated with flooding (Green 2010). Risk management focuses not only on flood losses, but on anticipating the "threshold" or "boundary" losses that could fundamentally undermine the sustainability of the system.

In this perspective, the absolute number of direct flood economic losses may not be as important as the ratio of such losses to GDP, since the latter can serve as a boundary or threshold variable that must be managed. Figure 2 shows the direct economic losses and their corresponding ratio to GDP in China from 1991 to 2004. It is clear that while the two are certainly correlated, their trends can be different depending on economic (GDP) development. For instance, while the direct economic losses in 1991 were roughly the same as in 2002, the loss to GDP ratio was significantly lower in 2002. The losses in 2002 were much less of a disturbance to the system than those in 1991.

Table 2. Expected future flood risks in the Tai Lake basin and benefit of land-use planning

Scenarios		2050– Socioeconomic change only	2050– Climate change only	2050– Socioeconomic and climate change
Expected annual damage (billion yuan)	Without land-use planning	6.69	7.14	13.23
	With land-use planning	3.95	3.64	8.01
Benefits of land-use planning (billion yuan/year)		2.75	3.50	5.23

Source: Wang, Cheng, and Yu 2010.

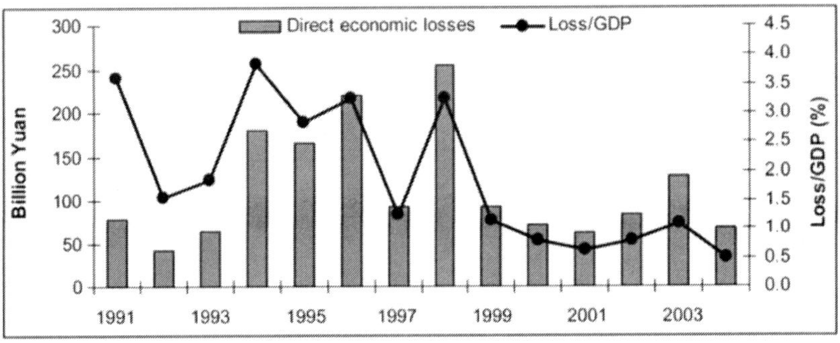

Figure 2. Direct economic losses due to floods and their ratio to GDP in China: 1991–2004 Source: Yang 2005.

In the near future, China could focus on two specific goals in this respect. One is to aim for a continued reduction in the loss of life due to floods. With continued socioeconomic development and improvement of general well-being, excessive loss of life due to floods will be increasingly unacceptable, socially and politically.vi The other goal should be to lower the ratio of flood economic loss to GDP, which on average stands around 1 percent in the early twenty-first century in China, considerably higher than that in most floodprone developed countries such as the United States (average ratio 0.03 percent) or Japan (0.2 percent) (Xiang 2003).

Second, effective flood management strategies are based on experimenting and learning to live with floods; land-use adaptation is essential (Klijn, van Buuren, and van Rooij 2004). A full range of floodplain management tools can and should be used to address the flooding problems in China, but the combination and integrated application of those tools will vary among places and communities. Their effectiveness will have to be assessed locally (White 1945, 1974, 1994). Since 1998, the implementation of the "space for water" program in China has encountered tremendous difficulties, partly due to the strict and inflexible top-down scheme that overlooks local dynamics and regional variations.

Third, "controlled flooding" can be applied innovatively as a central strategy to deal with and embrace natural variability and uncertainty. It seeks to manage floods not only as hazards but also as a resource. Controlled flooding seeks to allow places to be flooded when there is excessive water but is flooding by design in terms of locations and sequences. It is a way to accommodate and retain the natural river dynamics and, at the same time, to minimize flood losses and maximize the benefit of natural floods. In the coming decade, major investment can be made to modify and readjust the existing massive flood defense structures so they are adapted from flood control to controlled flooding. The designated flood detention and retention areas (with a total area of about 5 percent of the flood-prone areas) would be the priority areas to initiate such experiments. One critical enabling condition to make the controlled flooding approach plausible is the rapid technological advancement in recent decades in flood forecasting, real-time monitoring, and transformative changes in communication technology (White 1998).

While a resilience flood risk management approach is possible, implementation of such a strategy often is beset with major socioeconomic dilemmas, such as balancing and accepting major investment in the short term while seeing benefits only in the long term. Shifting the short-term, smallscale rationality to the long-term and large-scale rationality that a resilience strategy requires is not only a challenge for Chinese society, but globally a major obstacle for the pursuit of sustainability (Klijn, van Buuren, and van Rooij 2004; Turner 2010). As illustrated by the Dongting Lake case (Han 2011), or the Poyang Lake case (Shankman, Davis, and Leeuw 2009) and Tai Lake (Taihu) case (Harvey, Thorne et al. 2009), a resilience strategy requires many trade-offs, each of which embodies a range of difficult social, economic, and political issues including investment, compensation, social justice and social stability, poverty reduction, ecosystem services, and biodiversity.

Looking into future flood risks in China, a broader approach is needed, which integrates comprehensively the vulnerability components and aims for building resilience. Until that is done, the "dyke syndrome" will remain and the vicious circle of "higher dyke—more development—even higher dyke" will continue. Ultimately, it is a hard choice to make between less return and more resilient and more short-term gains and less resilient in the long run. So far, few modern societies have chosen the resilience path. To do so requires nothing less than a paradigm shift to embrace rather than control and eliminate variability and uncertainty. That requires fundamental change in the way we perceive the world, what we believe, and the ways we act (Holling 1996, 2001).

5 CONCLUSIONS

By outlining three fundamental dilemmas and three plausible pathways for managing future flood risk in China, this article concludes:

As the development on floodplains in China continues and intensifies in the coming decades, climate change will very likely increase both the intensity and the frequency of extreme weather events. Accordingly, China will need to learn to live with an increasing vulnerability to flood risk in the foreseeable future. To reduce human vulnerability to floods, multiple pathways exist but all involve difficult trade-offs. Hard choices are inevitable and inescapable.

China needs a new flood management strategy—one that incorporates vulnerability reduction and resilience building. Compared with the traditional engineering approach, a resilience approach requires not only changing the focus from modifying flood hazard events to reducing vulnerability. But it is also essential to embrace and internalize variability and uncertainty in decision making. The latter is essential for changing the management bias from community preparedness achieving full protection or zero flood risk. While vulnerability reduction must be integrated into overall development planning, society will always be vulnerable as long as people live in and develop floodplains.

A resilience strategy for managing flood risk thus requires human adjustment to flood, but not an approach aiming for full protection and control but for adapting human uses of floodplains, integrating, and experimenting with a wide range of flood risk management options. In this way, a reasonably dynamic balance, while not perfect, can be achieved between exposure and response capacity, and flood risk can be contained at a level judged to be socially and economically acceptable

While a resilience strategy embodies a wide range of opportunities for long-term sustainable flood risk prevention and mitigation, it also raises major challenges. China faces major dilemmas in managing its future flood risks. Trade-offs between economic development and flood vulnerability reduction and between cooperation and tensions across diverse actors and scales cannot be escaped.

Floods are part of Chinese culture, and for many in the floodplains a way of life. Moving from flood control to controlled flooding is a strategic choice towards living with floods in its true meaning. It requires a broad flood risk management approach that is centered on vulnerability reduction and resilience building of the social-ecological system within which the overall goal of sustainability is considered paramount and overall well-being is pursued.

An effective flood risk strategy will only emerge as a result of social learning and innovation and through human interactions with the flood risk system. Embracing variability and uncertainty will be the departure point. Given the past thirty years of transformative changes in the country, China may be a place where such a policy experiment and innovation are not only timely but also embraced.

ACKNOWLEDGMENT

This work was partly supported by the Risk and Hazard Science Base (111 Project) at the State Key Laboratory of Earth Surface Processes and Resource Ecology, Beijing Normal University, China.

NOTES

i Taihu (Tai Lake) is the third largest fresh water body in China, with an area of over 2000 km2 . The drainage basin of the Tai Lake covers 36,985 km2 , including some of the most economically developed areas in the delta region of the Yangtze River.

ii In 1949, China had only six large reservoirs. The 42,000 km levees were in poor condition.

iii In comparison, this number in the United States is about 40,000 km. In Japan it is less than 10,000 km, in the Netherlands it is about 4800 km, and in India 20,000 km.

iv According to the Ministry of Water Resources, there were 30,413 (36 percent of the total) "sick and dangerous" reservoirs in China in 2009: of these 143 were large reservoirs, and 1118 were mediumsized reservoirs. These reservoirs now present a severe source of flooding risks threatening hundreds of millions of people and huge areas of cropland.

v Current annual average flood economic losses amount to less than 2 billion yuan in the Taihu basin; in the 1990s the annual average flood economic losses were around 70 billion yuan nationwide (Xiang 2003).

vi The annual average loss of life due to floods in China is still above 2000 people.

REFERENCES

1. APFM (Associated Programme on Flood Management). 2003. Integrated Flood Management: Concept Paper. Technical Document No.1. Geneva: The Associated Programme on Flood Management.
2. Bouwer, L., P. Bubeck, and J. Aerts. 2010. Changes in Future Flood Risk due to Climate and Development on a Dutch Polder Area. Global Environmental Change 20(3): 463–471.
3. Bouwer, L. M., R. P. Crompton, E. Faust, and P. Höppe. 2007. Confronting Disaster Losses. Science 318(5851): 753.
4. Burton, I. 1962. Types of Agricultural Occupance of Flood Plains in the United States. Research Paper No. 75. Chicago: Department of Geography, University of Chicago.

5. Burton, I., R. W. Kates, and G. F. White. 1978. The Environment as Hazard. New York: Oxford University Press.
6. Cheng, X. T. 2005. Changes of Flood Control Situations and Adjustments of Flood Management Strategies in China. Water International 30(1): 108–113.
7. —. 2006. Recent Progress in Flood Management in China. Irrigation and Drainage 55(s1): 75–82.
8. —. 2008. Recent Progress in Flood and Drought Management Research. Journal of the Institute of Water Resources and Hydropowe r Research of China 6(3): 191–198 (in Chinese).
9. Cheng, X. T., Y. C. Wu, and Y. Y. Wang. 2004. On the New Concepts of Flood Management and Flood Security Systems. Beijing: China WaterPower Press (in Chinese).
10. Evans, E. P., R. Ashley, J. W. Hall, E. Penning-Rowsell, P. B. Sayers, C. R Thorne, and A. Watkinson. 2004. Foresight. Future Flooding. Scientific Summary: Volume I—Future Risks and their Drivers. London: Office of Science and Technology.
11. Folke, C., S. R. Carpenter, B. Walker, M. Scheffer, T. Chapin, and J. Rockström. 2010. Resilience Thinking: Integrating Resilience, Adaptability and Transformability. Ecology and Society 15(4). Article 20. http://www.ecologyandsociety.org/vol15/iss4/art20/.
12. GHD and IWHR (Institute of Water Resources and Hydropower Research). 2006. National Flood Management Strategy Study. Zhengzhou: Yellow River Water Press (in Chinese).
13. Green, C. 2010. Towards Sustainable Flood Risk Management. International Journal of Disaster Risk Science 1(1): 33–43.
14. Hall, J. 2008. Editorial. Journal of Flood Risk Management 1(3): 131–132.
15. Hall, J. W., P. B. Sayers, and R. J. Dawson. 2005. National Scale Flood Risk Assessment of Current and Future Flood Risk in England and Wales. Natural Hazards 36(1–2): 147–164.
16. Han, G. Y. 2011. Understanding Regional Dynamics of Vulnerability: A Historical Approach to the Flood Problem in China. Ph.D. Dissertation. Graduate School of Geography, Clark University, Worcester, MA, USA.
17. Harvey, G. L., E. P. Evans, C. R. Thorne, and X. T. Cheng. 2009. Scenario Analysis Technology for River Basin Flood Risk Management in the Taihu Basin: A China-UK Scientific Cooperation Project. Discussion Paper 44. Nottingham: China Policy Institute, University of Nottingham.
18. Harvey, G. L., C. R. Thorne, X. T. Cheng, E. P. Evans, S. Han, J. D. Simm, and Y. Wang. 2009. Qualitative Analysis of Future Flood Risk in the Taihu Basin, China. Journal of Flood Risk Management 2(2): 85–100.
19. Holling, C. S. 1996. Surprise for Science, Resilience for Ecosystems, and Incentives for People. Ecological Applications 6(3): 733–735.
20. —. 2001. Understanding the Complexity of Economic, Social and Ecological Systems. Ecosystems 4(5): 390–405.
21. Ikeda, S., T. Sato, and T. Fukuzono. 2008. Towards an Integrated Management Framework for Emerging Disaster Risks in Japan. Natural Hazards 44(2): 267–280.
22. IPCC (Intergovernmental Panel on Climate Change). 2011. Summary for Policymakers. In Intergovernmental Panel on Climate Change Special Report on Managing the Risks of Extreme Events and Disasters to Advance Climate Change Adaptation, edited by C. B. Field, V. Barros, T. F. Stocker, D. Qin, D. Dokken, K. L. Ebi, M. D. Mastrandrea, K. J. Mach, G.-K. Plattner, S. Allen, M. Tignor, and P. M. Midgley. Cambridge and New York: Cambridge University Press.
23. Klijn, F., M. van Buuren, and S. van Rooij. 2004. Flood-Risk Management Strategies for an Uncertain Future: Living with Rhine River Floods in the Netherlands? Ambio 33(3): 141–147.
24. Kundzewicz, Z., and L. Menzel. 2003. Flood Risk and Vulnerability in the Changing World. Paper presented at the International Conference: Towards Natural Flood Reduction Strategies, 6–13 September, Warsaw, Poland.
25. Liu, J. G., and J. Diamond. 2005. China's Environment in a Globalizing World. Nature 435(7046): 1179–1186.

26. Liu, J. G., T. Dietz, S. Carpenter, M. Alberti, C. Folke, E. Moran, A. N. Pell, P. Deadman, et al. 2007. Complexity of Coupled Human and Natural Systems. Science 317(5844): 1513–1516.
27. Milly, P. C. D., R. T. Wetherald, K. A. Dunne, and T. L. Delworth. 2002. Increasing Risk of Great Floods in Changing Climate. Nature 415(6871): 514–517.
28. NDRC (National Development and Reform Commission). 2007. China's National Climate Change Programme. Beijing: National Development and Reform Commission, People's Republic of China.
29. Ostrom, E. 2004. Book Review of Panarchy: Understanding Transformations in Human and Natural Systems: Lance H. Gunderson and C. S. Holling (Eds.), Island Press, Washington, DC, 2002, ISBN: 1559638575, 450 pp. Ecological Economics 49 (4): 488–91.
30. —. 2007. A Diagnostic Approach for Going beyond Panaceas. PNAS 104(39): 15181–151887.
31. Schanze, J. 2009. Editorial. Journal of Flood Risk Management 2(3):149–150.
32. Shankman, D., L. Davis, and J. de Leeuw. 2009. River Management, Landuse Change, and Future Flood Risk in China's Poyang Lake Region. International Journal of River Basin Management 7(4): 423–431.
33. Shi, P. J., Y. Ge, Y. Yuan, and W. P. Guo. 2005. Integrated Risk Management of Flood Disasters in Metropolitan Areas of China. Internationa l Journal of Water Resources Development 21(4): 613–627.
34. Shi, P. J., and Y. Z. Pan. 1997. Flood, Typhoon and Storm Surge Disaster in China. Paper presented at the Seminar on Natural Perils and Insurance/Reinsurance Techniques. September 1997, Beijing, China.
35. Sylves, R., and P. J. Kershaw. 2004. Reducing Future Flood Losses: The Role of Human Actions (Summary of the Disasters Roundtable, March 2, 2004). Washington, DC: National Research Council and National Academies Press.
36. Turner, B. L., II. 2010. Vulnerability and Resilience: Coalescing or Paralleling Approaches for Sustainability Science? Global Environmental Change 20(4): 570–576.
37. UNDP (United Nations Development Programme). 2004. A Global Report: Reducing Disaster Risk—A Challenge for Development. New York: United Nations Development Programme.
38. UNISDR (United Nations International Strategy for Disaster Reduction). 2004. Guidelines for Reducing Flood Losses. Geneva: United Nations International Strategy for Disaster Reduction.
39. —. 2009. Global Assessment Report on Disaster Risk Reduction: Risk and Poverty in a Changing Climate. Geneva: United Nations International Strategy for Disaster Reduction.
40. Walker, B., C. S. Holling, S. R. Carpenter, and A. Kinzig. 2004. Resilience, Adaptability and Transformability in Social-Ecological Systems. Ecology and Society 9(2). Article 5. http://www.ecologyandsociety.org/vol9/iss2/art5/.
41. Wang, Y. Y., X. T. Cheng, and C. Q. Yu. 2010. The Benefit of Land Use Planning of Taihu Basin. Workshop Presentation at the International Policy-Oriented Discussions—International Flood Risk Management Approaches: From Theory to Practice, November 30–December 1, 2010, Washington, DC. http://www.nfrmp.us/ifrma/docs/pre/summary/YanyanPaper.pdf.
42. White, G. F. 1945. Human Adjustment to Floods: A Geographical Approach to the Flood Problem in the United States. Research Paper No. 29. Chicago: Department of Geography, University of Chicago.
43. —, ed. 1974. Natural Hazards: Local, National, Global. New York: Oxford University Press.
44. —. 1994. Testimony and Prepared Statement for the "Midwest Floods of 1993: Flood Control and Floodplain Policy and Proposal: Hearing before the Subcommittee on Water Resources and Environment," October 27, 1993, 103d Congress, 1st Session, House Committee on Public Works and Transportation. Washington, DC: US Government Printing Office.
45. —. 1998. Looking toward the Horizon: Prospects for Floodplain Managers. Paper Read at Floodplain Management in a Multifaced World—Proceedings of the 21st Annual Conference, April 28–May 2, 1998, Little Rock, Arkansas.

46. White, G. F., R. W. Kates, and I. Burton. 2001. Knowing Better and Losing Even More: The Use of Knowledge in Hazards Management. Environmental Hazards 3(3–4): 81–92.
47. Xiang, L. Y. 2003. Comparative Study of Flood Management Strategies between China and Selected Countries. Water Resources Development Research 3(5): 12–18 (in Chinese).
48. Yang, K. 2005. Flood Emergency Management in China. Paper presente d at the ICID 21st European Regional Conference 2005, May 15–19, Frankfurt (Oder) and Slubice, Germany and Poland.

CHAPTER 10

Momentum Attraction by Flood Plains in Compound Channel

Mehdi Behdarvandi Askar[1], and M Fathi Moghadam[2*]

[1]Ph. D. Graduate, Water Sciences Engineering Faculty, Shahid Chamran University, Ahvaz, Iran.

[2]Professor, Water Sciences Engineering Faculty, Shahid Chamran University, Ahvaz, Iran.

ABSTRACT

Most natural rivers have flood plains that extend laterally away from the main river channel at a gentle gradient or in a series of terraces. In certain cases multistage channels are deliberately formed in order to increase conveyance capacity in times of flood and to have recreational land available at other times of the year [1]. Two-stage channels thus consist typically of a main river channel in which there is some discharge all of the time and flood plains, which are dry for most of the time yet perform a vital function in times of flood. Since flood alleviation schemes are the focus of much engineering work, the prediction of the conveyance capacity, velocity distribution and boundary shear stress distribution in such channels is clearly important. The boundary shear stress distribution is a prerequisite for studies on bank protection and sediment transport. The prediction of these parameters in two stage or compound channels is complicated by the lateral exchange of momentum that takes place in the shear layer that forms between the generally faster moving water in the main river channel and the slower moving water on the flood plain. The superposition of high lateral shear on bed-generated turbulence and longitudinal secondary flow structures is an intriguing problem in fluid mechanics. In the context of river channels with flood plains, the problem is usually further complicated even for moderately straight channels by the complex geometry of the cross-section and the heterogeneous nature of the boundary roughness [2].

Keywords: main channel, flood plain, dredging, bank protection.

1. INTRODUCTION

Compound channels are hydraulic sections that are formed of main channel and floodplain channels. The main channel is part of the lower floor is level and generally have a rectangular or trapezoidal cross-section. This section, pass the typical runoff and base flow in rivers. The floodplain channel, in contrast, is composed of sections of the floor level is higher than the main channel and is located is located one side or both sides of the main channel [5]. This section has normally no role in flow transfer and action is in flood conditions. Generally, its hydraulic roughness is much greater than the main channel. Given the above definitions it is clear that the compound section, hydraulic section is typical of many rivers in alluvial plains. That is where he has long been the cradle of human life and civilization. Certain geometric conditions cause the compound levels are specific hydraulic conditions which require special justification is in these sections. Sections of special geometry combined with a significant difference in roughness of main channel and the flood plain Cause differences between the following sections of this channel is immediately noticeable.

This difference in velocity and depth of interaction in turn led to the formation of a boundary area between the sections the energy perspective, the exchange of internal stresses in this area will cause significant energy loss. Also, from another perspective, the striking velocity gradient between the levels of tension between the inner shear flow is in the following section In addition to the bed and wall shear stress is the channel. It is commonly called the apparent shear stress. Regular or irregular channel as to what is considered to be low in energy or longitudinal friction mostly due to the action involved with the channel bed and the solid wall. However, the friction factor and low power portion of the longitudinal the compound channels are included and the corresponding effects on the inner sections cannot be ignored [3].

The main differentiation stages compound with regular and typical sections of the same phenomenon is also hence the use of conventional hydraulic relationships associated with the regular channels, the compound channels, in combination with a significant incidence of errors.

So resulting in a hydraulic study of the sections started decades ago and continues today. review the boundary shear stress not only for discussion is very important in sediment transport and coastal protection it can be keep in check the compound sections in comparison main channels and flood plains to infer the nature of the transverse momentum transfer phenomena and their effects on the distribution of boundary shear stress and the hydraulic properties used [7].

2. RESEARCH BACKGROUND

Topics discussed with momentum transfer near the shear stress and shear force issues range of research on this issue directly and indirectly shaped the more it will be referred to Sellin [14]. He developed a technique for obtaining information on the mechanism and nature of the phenomenon can be used to exchange momentum. He has a high reflecting power of aluminum powder on the surface with a compound channel unraveled and with a camera that was installed just above the water level of the surface pattern was created by the filming and photograph. For more accuracy and velocity under the camera with the same average velocity of the secondary flow at the junction of Main channel and the floodplain channel were moved. Tamai and Kawahara [15] used similar techniques for calculating the momentum transfer in the channel width and length were used. They also calculate the flow rate of hydrogen bubble technique was used in coastal channels.

Zheleznyakov [16] in his research for a compound symmetrical cross section which consists of a Main channels to the part geometry floodplain surrounded by both parties that the same was done.he understood that average velocity and position in Main channel in the shallow floodplain will suffer loss. However, the local velocity at the junction of main channels and floodplain increases at the same time. He also managed to introduce a critical factor in his subsequent articles and compound sections was effective that it Were the Main channels depth to the depth of the channel at the flood plain $n = \frac{y_c}{y_f}$; later called the effective depth [13]. Rapid changes in compound channel of Quadrant First experiments in channels with Main section and two on a broad floodplain In the latter case, their experiments in a channel with Main section and a broad flat floodplain made and results in a logarithmic vertical velocity distribution reached. Myer and Elsawy [10] examined shear stress distribution in symmetric and asymmetric channels. They stated that Transverse momentum transfer greatly reduces shear stress in Main channel and increases shear stress in the flood plains. The momentum transfer between the floodplain and Main section came to the parameter of shear stress. Dimetrio and Knight [7] applying a shear force of shear stress was obtained. Effects of momentum transfer between floodplain and main channel of the transverse momentum Quadrant and achieved remarkable results. They also face two types of vertical and horizontal shear stresses were examined. Posey [11] presented a different approach in calculating the compound sections discharge. The sections were divided into smaller sections after calculating each stated discharge they found the total discharge is obtained from different sections discharge. Finally they found that any error in their results refer to irrespective of momentum transfer between main sections and flood plains. They also found in high transverse momentum transfer mode reduces the bank full discharge and flow rate are compared with the regular channels. They studied apparent shear stress for several hypothetical interaction between main channels and floodplain. They were also introduced an index, so it was considered important indicators of this effect was less than 2.

3. THE PREVAILING THEORY

In turbulence due to the turbulent flow due to velocity fluctuations in the three directions of movement also Stresses that occur in otherwise known as the apparent shear stress.

Continuity and momentum equations with respect to x along the flow we have:

$$\frac{\partial v}{\partial y}+\frac{\partial w}{\partial z}=0 \tag{1}$$

$$\rho\left(v\frac{\partial u}{\partial y}+w\frac{\partial u}{\partial y}\right)=\rho g S+\frac{\partial \tau_{yx}}{\partial y}+\frac{\partial \tau_{zx}}{\partial z} \tag{2}$$

That ρ is Density of water, g is the acceleration of gravity, s is Channel slope, τ_{yx} τ_{zx} Shear stresses along the x axis, and respectively zx and yx are in the pages. Momentum equation can be written as follows:

$$\frac{\partial(\rho uv)}{\partial y}+\frac{\partial(\rho uw)}{\partial z}=\rho g S+\frac{\partial \tau_{yx}}{\partial y}+\frac{\partial \tau_{zx}}{\partial z} \tag{3}$$

Methods of measuring the shear stress can be divided direct and indirect [12]. Of the various methods of measuring the shear stress, the Preston tube can be named as the most common method of measuring the shear stress.

The shear stress is calculated using the following equations to calculate it and get help from the apparent shear stress or the momentum transfer,

$$\frac{\rho g A_{mc} S_f}{2} = \int \tau_0 dp + ASF_v \tag{4}$$

Preston tube is in fact a Pitot tube modified by Preston [12] to measure the shear stress indirectly been developed and placed on the wall or bed.

The main drawback of long computation in one hand and the nature of its high computational errors due to the large errors and is not accountable for the rough beds. Because in rough Beds due to collide in the boundary layer and not in the Preston tube into proper position Extractive numbers of Preston was not reliable. In this study, the stress measured using the drag force and momentum absorption technique has been used. In this method, the fluid drag forces on submerged objects using a variety of power meters, including dynamic load cell is recorded. Fathi Moghaddam and Kowen [6] and Kouwen and Fathi Moghadam [8] used this technique to measure the momentum and the drag of the fluid used by the vegetation elements. Since this method the energy method, the measurement is done in one spot, much less measurement error and precision of vector data is higher than the other methods. In this study, this method for measuring the momentum and friction force taken by the roughness of the channel will be used for floor and wall. In order to excite shear stress in this study used four different types of roughness size, which will be discussed in detail.

DIMENSIONAL ANALYSIS: For the direct channel and the water smooth, steady and uniform flow and no surface wave motion and bed load, the parameters affecting the shear stress can be said:

Sensor placement for dynamic load inertia at the start of KEF flume flow, before the experiment is to channel some slight slope, the slope parameter is ignored. Therefore the effective factors in determining the average velocity shear walls and floor can be expressed as follows.

$$f(\overline{u}_{*b}, \overline{u}_{*w}, \overline{u}_{*}, \rho, \upsilon, g, V, h, H, b, K_{s},) = 0 \tag{5}$$

Using Buckingham theorem above equation can be transformed into relations with several dimensionless parameters. The combination of these three variables, ρ ,V ,Y and repeated with the other parameters and variables Following relationship to estimate the total average shear velocity, average velocity and average bed shear wall shear velocity are obtained:

$$f\left(\frac{\overline{u}_*}{V},\frac{\overline{u}_{*b}}{V},\frac{\overline{u}_{*w}}{V}\frac{\upsilon}{VH},,\frac{gH}{V^2},\frac{b}{h},\frac{k_s}{H},\frac{H-h}{H},\right)=0 \tag{6}$$

$$\beta=\frac{H-h}{H} \tag{7}$$

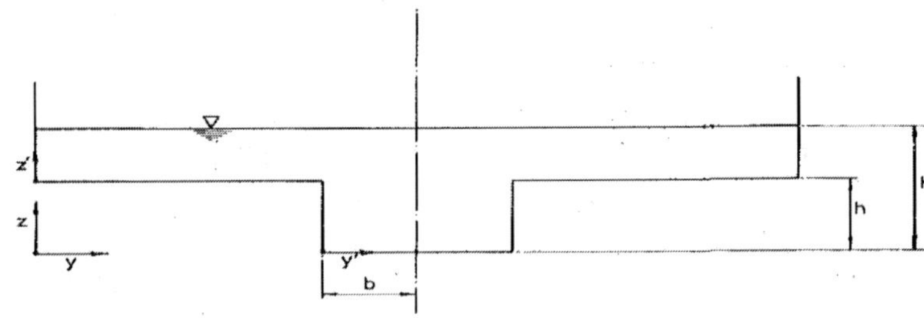

Figure 1: Schematic cross section of the compound channel

Due to the fact that the tests will be performed completely turbulence. The Reynolds number of judgments on the situation will be postponed to later experiments. However, this parameter is omitted due to rough conditions with low bed roughness is permitted in front of viscose.(2)

4. THE TESTED PARAMETERS:

In order to achieve adequate and reliable data and after consideration of various sources, was appointed to review the following parameters.

As **TABLE 1** show β is the parameter called the relative depth which comes of dividing the depth of shallow floodplain to the depth of main channel. Because of more than 0.5 has been neglected is that higher than it actually in compound channels there is not any, hydraulic mean .and main channel, will operate such as a rectangular hole in the middle of the channel. As noted earlier the floor within the main channel and the floodplain is fixed in all experiments.

While the height of the main channel and the floodplain in four different sizes of 6, 9, 12 and 15 cm will change. another test is to measure the roughness Shear stress excite absorption feature in order to show momentum rough ground than other methods that are used almost powerless.

Table 1: Tested Parameters

The tested parameters	The tested parameters value
$\beta = \frac{H-h}{H}$	0/1 . 0/2 . 0/25 . 0/3 . 0/35 .0/4 .0/5
$\delta = {}^{b}/_{h}$	0/66 . 0/83 . 1/11 . 1/66
$D50$	1/31 . 4/6 . 9/36 .12/5

5. MATERIALS AND METHODS

As was mentioned earlier, this study aims to measure the shear stress and momentum transfer between main channels and floodplain composed of rectangular cross section with momentum absorption technique is used. Momentum measure imposed on the channel bed and walls, changes in the method proposed by Fathi-Moghadam[6] were applied as a table knife edge and a new method was developed under the knife edge of the flume. Experiments was done in a flume with a net width of 80 cm, depth 550 cm and a length of 8.3 meters, including 4.1 meters upstream channel, a knife edge of the flume meter (measuring range) and 3.2 meters downstream channel. Controls of the downstream level and width to depth ratios of the different conditions in the main channel and flood plains of the discharge, a valve at the end of the flume have been pre nose. The total stress of the sensor will use to measure the dynamic load. To measure the dynamic loads of the power converter is used. This tool is used to connect moving part of the channel to the fixed part and the friction force changes recorded on the walls and floor of the mobile channel. Note that for the calibration of these devices are used in different weights.

In this study, the channel was transformed into a rectangular and all the necessary force that causes the longitudinal momentum in the direction of flow in a rectangular channel can be the body of a rough and a rough floor in a state with more than 100 readings was tested. The different modes consist of channels (about 112 cases) were tested. Then, the longitudinal momentum measurements in the compound channel by the readings of a rectangular channel for 7 per share were calculated on the wet and The size of the direct measurement of momentum in the compound channel were compare Subtracting them from the transverse momentum transfer and the nature of the apparent shear stress, respectively.

It is worth noting that the experiments in rectangular channels for four different roughness sizes 1.31, 4.6, 9.36 and 12.5 mm for the roughness of a rough bed mode and a rough wall mode for different depths of the compound channel. Then, compound channel for four different tests that were the height of the main channel sizes 6, 9, 12 and 15 cm for the height of the roughness at different depths was tested in four states were continued. Then, as was mentioned earlier, the comparison of rectangular and compound mode and the momentum transfer between main channel and the floodplain and how the changes were studied.

6. RESULTS AND DISCUSSION

After various tests and the following assumptions are:

Flume bed and the wall shear stress distribution in the central axis of symmetry of the flume The use and application of research findings in practice, is large enough scale simulation is, Assumed to be fully developed channel flow. The change in longitudinal position measured on the surface does not lead to a change in friction coefficient. As can be seen in **FIGURE 2** to **5** the horizontal axis represents the percentage of apparent shear force % ASF and the vertical axis represents β is the relative depth. In **FIGURE 2** for all four types of wall roughness and the height of 6 cm is shown in the main channel.

$$\frac{\rho g A_{mc} S_f}{2} = \int \tau_0 dp + ASF \tag{8}$$

$$\%\, ASF = \left(\frac{ASF}{\rho g A_{mc} S_f}\right) 100 \tag{9}$$

Figure 3 for the main channel wall height of 9 cm and **FIGURE 4** and **5** respectively for main channel 12 and 15 cm high will be used.

According to **FIGURE 2** Can be concluded that in all different roughness, resulting in high relative depth of the shallow floodplain channel is very much on minimum transverse momentum transfer occurs. But in lower relative depth of the transverse momentum transfer more until in β=0.25 Transverse momentum transfer in the highest percentage in the roughness of all falls occur. The reason for this is that, in the depth of the above, the compound channels are actually losing their property and The main channel acts like a hole in the floor of a rectangular channel, It is important to note in **FIGURE 2** that with increasing Roughness size, the amount of transverse momentum transfer of longitudinal momentum towards increased so most transmission of the apparent shear force (momentum transfer) of about 10% and the particle size is 12.5 mm As shown be At shallow depth (h = 6)is not much difference between momentum transfer to the particle size 1.31 and 4.6 mm.

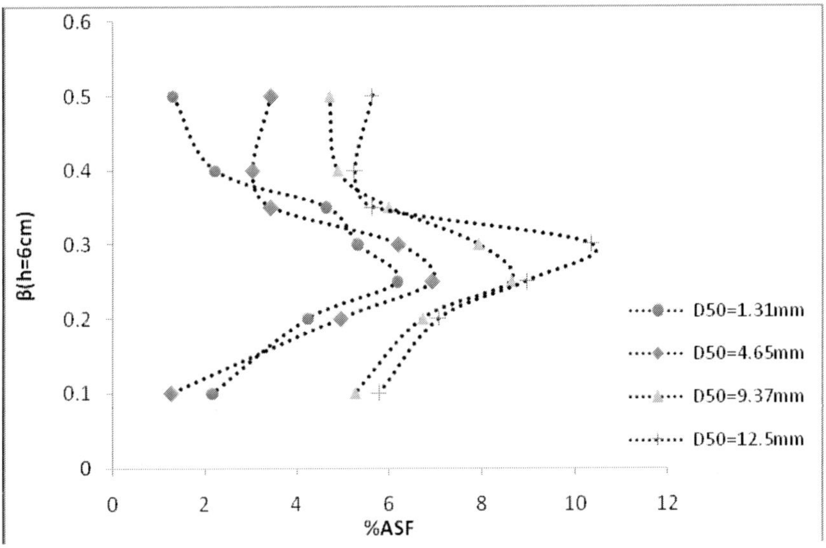

Figure 2: Distribution of shear force (momentum transfer) vs the relative depth h = 6cm

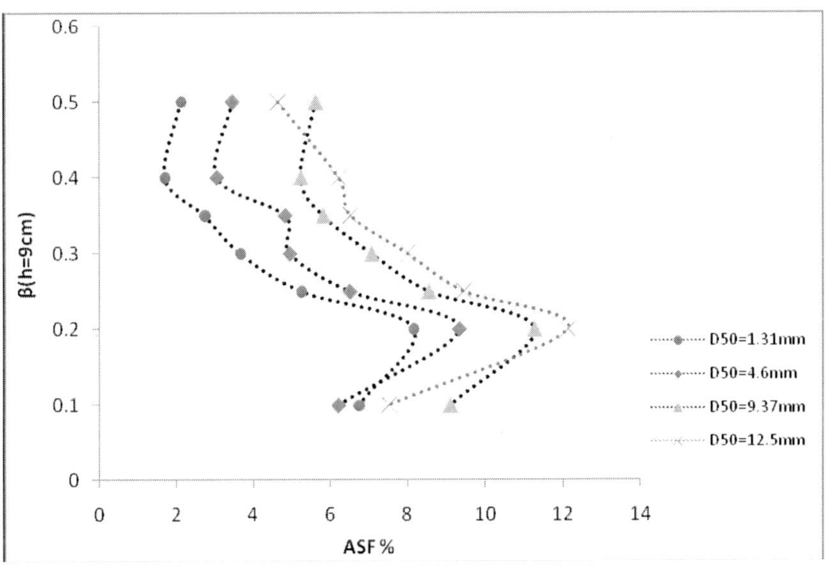

Figure 3: Distribution of shear force (momentum transfer) vs the relative depth h = 9cm

In this case given the very broad momentum transfer range of different roughness (from 1% to 10%) are. according to **FIGURE 3**, which corresponds to the case (h = 9cm)that is Roughness of all, the high relative depth, which caused high depth on floodplain channel. minimum transverse momentum transfer occurs but in Lower relative depth, the transverse momentum transfer more and more increases until that in β=0.2 Transverse momentum transfer in the highest percentage in the roughness of all falls occur.

Most transmission of the apparent shear force (momentum transfer) of about 12.5% and the particle size is 12.5 mm. As can be seen increasing the wall height of the main channel, compared to the previous mode and clear separation between the surface roughness lower shear force values (1.31 and 4.6 mm sizes) will emerge. In this case The momentum transfer range of different roughness (from 3% to 10%) . In **FIGURE 4** can be found in the state, (h = 12cm) high relative depth all the different roughness, which caused high depth floodplain channel is on. Minimum transverse momentum transfer occurs but in Lower relative depth, the transverse momentum transfer more until in β=0.25 Transverse momentum transfer in the highest percentage in the roughness of all falls occur.

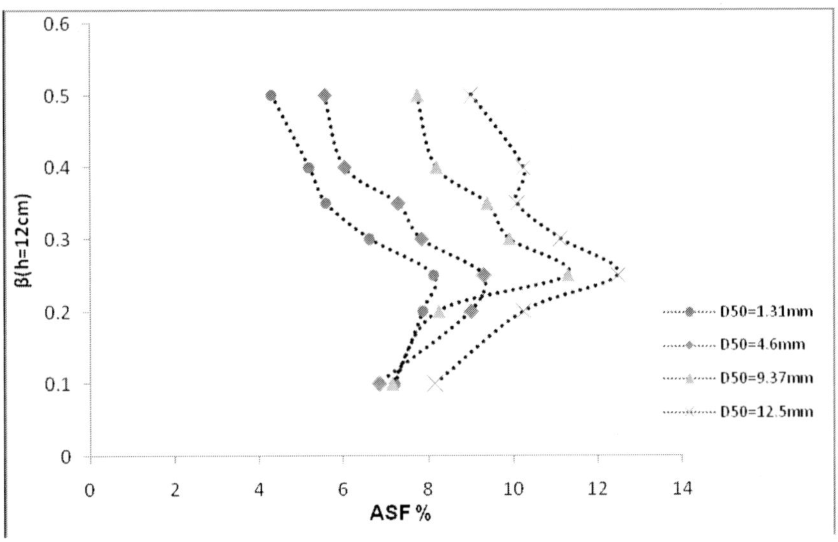

Figure 4: Distribution of shear force (momentum transfer) vs. the relative depth h = 12cm

In this case most transmission of the apparent shear force (momentum transfer) of about 14% and the particle size is 12.5 mm. In **FIGURE 5**, which corresponds to the maximum wall height is the main channel Most transverse momentum transfer, the roughness of 9.36 mm in the third place which is about 15% in this case Momentum transfer range can be from 2% to 15% .

It should be noted that Maximum momentum transfer for the roughness of 12.5 mm occur in β=0.25 and the other in roughness will occur in β=0.2. Total shear force in all states and all the relative roughness increases with increasing depth that is evident in **FIGURES 6** and **7**, **8**and **9**. It should be noted that maximum shear stress in the range (h = 15) occurs between 1 to 9 of 9 Newton and the maximum roughness of about 12.5 mm. Shear force corresponding to the lowest range (h = 6) is its value between 0.5 to 3.5 is the Newton.

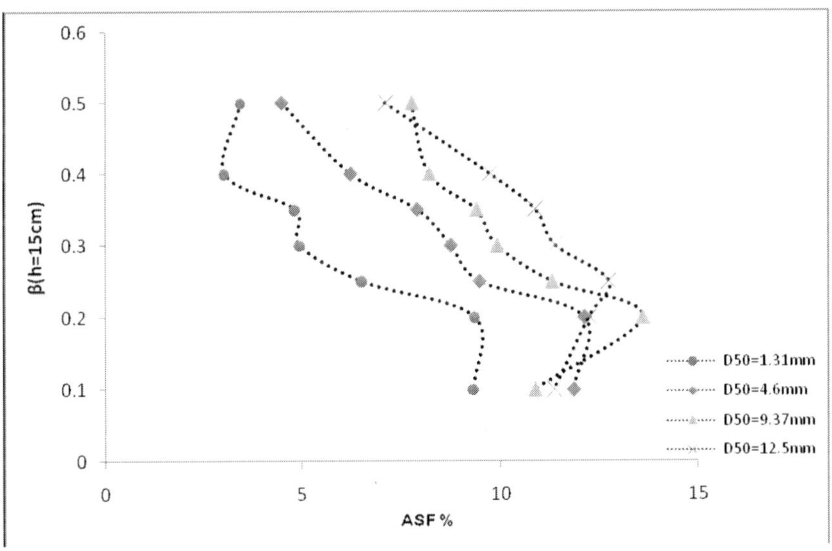

Figure 5: Distribution of shear force (momentum transfer) vs the relative depth h = 15cm

As the **FIGURES 6** to **9** are show Increase in shear force after β=0.5 with lower slope increases., at the high depth of floodplain the characteristics of compound channel were reduced. Despite the apparent shear force that decreases with increasing relative depth can be found that the shear force will increase with increasing relative depth.

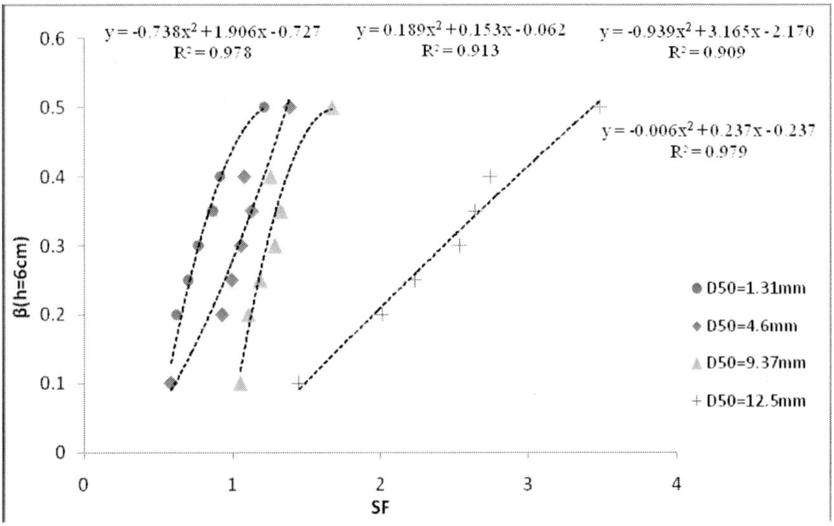

Figure 6: longitudinal shear force at different roughness in (h = 6cm)

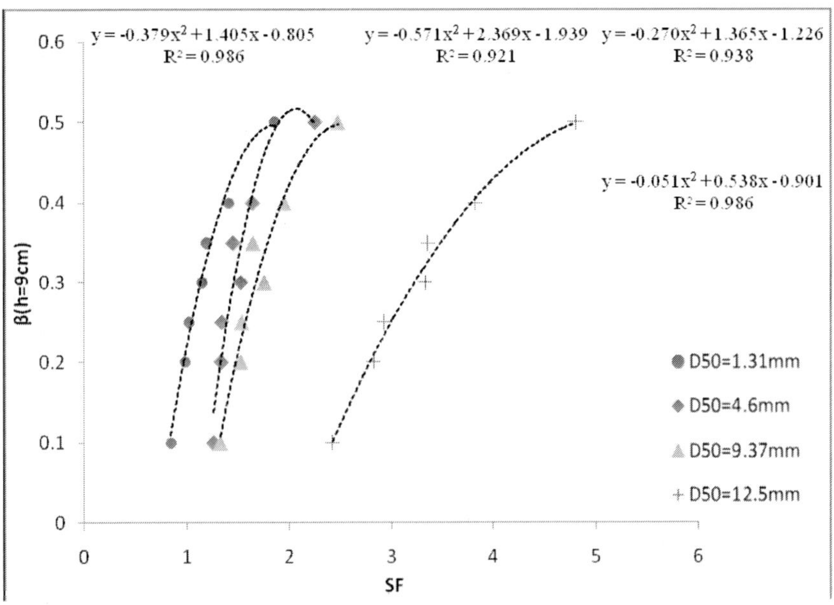

Figure 7: longitudinal shear force at different roughness in (h = 9cm)

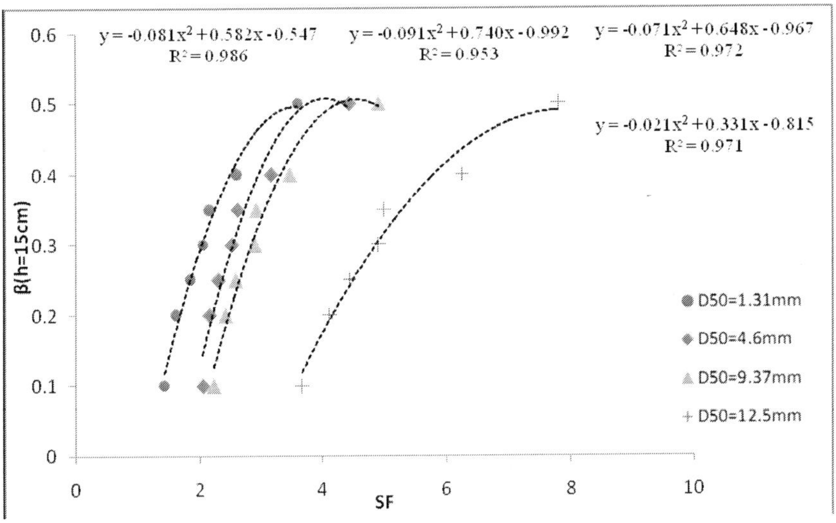

Figure 8: longitudinal shear force at different roughness in (h = 12cm)

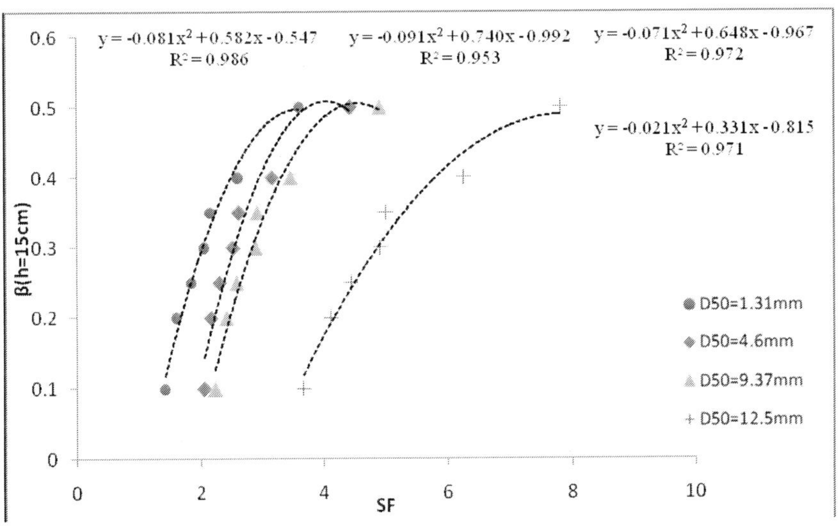

Figure 9: longitudinal shear force at different roughness in (h = 15cm)

Generally these are the results of the study can be summarized as follows:

Changes of the apparent shear force (transverse momentum transfer) is to increase the depth relative to the high ratio of β=0.25 Reduced relative to depth ratio of β=0.2 in the following sections, the compound of the apparent shear force (transverse momentum transfer) is reduced.

In β=0.25 and β=0.2 usually most of the apparent shear force (momentum transfer cross) occurs. This means that flow in floodplain, slowing the movement of flow in the main channel or in other words flow in main channel takes the flow in floodplain that is very important In river and coastal engineering and organization. Will be considered in β=0.5 When exactly is the water depth in the floodplain, equal to the main channel wall height so in compound sections in rivers and natural watercourse about conditions that may be due to dehydration of water depth in the channel is very little more attention be paid to floodplain because This may increase the apparent shear force and consequently the secondary vortex and consequently will lead to coastal erosion and degradation.

The compound sections, the relative depth greater than β= 0.5 are losing their property be the compound and the rules governing the flow of hydraulic fluid channel does not comply with the compound sections. More shear force in the apparent roughness belongs to larger average particle size. Also, the height of the wall of the main channel is more, and the amount of apparent shear force will be more. Maximum frame rate of change of shear force occurs at the lowest height of the wall of the main channel. That in River and coastal projects in the economic arguments, it is important to discuss the importance of dredging the river makes it clear.

The steep increase in the total shear force at h = 12cm were much worse than other states that Is illustrated that at the height of the main channel wall shear rate increases force overall trend of the relative depth to be faster.

ACKNOWLEDGEMENTS

The authors would like to acknowledge Chamran University of Ahwaz, the Centre of Excellence on Operation Management of Irrigation and Drainage Networks for financial support and facilitation of the experiments.

REFERENCES

1. Abril JB, Knight DW. Stage-discharge prediction for rives in flood applying a depth-averaged model. J Hydraulic Res. 2004; 42(6): 616–629.
2. Ackers P. 1991, Hydraulic design of straight compound channels. SR Report 281, HR Wallingford, UK, Vol. 1 & 2, pp 1-130 & 1-140.
3. Ackers P. Hydraulic design of two-stage channels. Proceedings of ICE Journal for Water, Maritime and Energy. 1992;96:247-257.
4. Ackers P. Flow formulae for straight two-stage channels. J Hydraulic Res. 1993; 31(4):509-531.
5. Alavian V, Chu VH. 1985, Turbulent exchange flow in shallow compound channel. Proceedings of 21st Congress of IAHR, Melbourne. Vol. 3, pp. 446
6. Fathi Moghadam M, Kouwen N. Nonrigid, nonsubmerged vegetative roughness on floodplains. J Hydraul Eng. 1997;123(1), 51–57.
7. Knight DW, Demetriou JD, Homed ME. Boundary shear in smooth rectangular channels. Agric Water Manage.110(4),405-422at the concave bank. Hydraulic Proc. 1984;12:1323-1338.
8. Kouwen N, Fathi-Moghadam M. Friction Factors for Coniferous Trees along Rivers. J Hydraulic Eng ASCE. 2000;126(10): 732-740.
9. Lashkar-Ara B, Fathi-Moghadam M, Shafai-Bajestan M, Jael A. Boundary shear stresses in smooth channels. J Food Agri Environ. 2010;8(1):132-136.
10. Myers WRC, Elsawy EM. Boundary shear in channel with floodplain , Journal of the hydraulics Division. ASCE. 1975;101:933-946
11. Posey CJ. Computation of discharge including overbank flow ,.civil engineering. ASCE. 1963:62-63
12. Preston JH. The determination of turbulent skin friction by means of pitot tubes. J Royal Aeronautical Soc. 1954;l54:109-121
13. Rajaratnam N, Ahmadi RM. Interaction between main channel and flood plain flows journal, of the hydraulics division. ASCE. 1979;105:573-588
14. Sellin RH. A Laboratory investigation into the interaction between the flow in the channel of a river and that over its flood plain, La Houille Blanche, Grenoble,France,7,1964,pp793-802.
15. Tamai N, Kawahara Y. Resistance Law of a flow in composite cross sections, 3rd congress Asian and pacific regional division of IAHR Boundary ,1982,pp.42-51
16. Zheleznyakov GV. Interaction of channel and floodplain streams. Proceeding of the international association for hydraulic research congress. 1971;5:144-148.

CHAPTER 11

Effective and Sustainable Flood and Landslide Risk Reduction Measures: An Investigation of Two Assessment Frameworks

Yvonne Andersson-Sko¨ld[1,2], Lars Nyberg[2]

1. Swedish National Road and Transport, Research Institute (VTI), Gothenburg, Sweden

2. Centre for Climate and Safety, Karlstad University, Karlstad, Sweden

ABSTRACT

Natural events such as floods and landslides can have severe consequences. The risks are expected to increase, both as a consequence of climate change and due to increased vulnerabilities, especially in urban areas. Although preventive measures are often cost-effective, some measures are beneficial to certain values, while some may have negative impacts on other values. The aim of the study presented here was to investigate two frameworks used for assessing the effectiveness and sustainability of physical and nonphysical flood and landslide risk reduction measures. The study is based on literature, available information from authorities and municipalities, expert knowledge and experience, and stakeholder views and values. The results indicate that the risks for suboptimization or maladaptation are reduced if many aspects are included and a broad spectrum of stakeholders are involved. The sustainability assessment tools applied here can contribute to a more transparent and sustainable risk management process by assessing strategies and interventions with respect to both short- and long-term perspectives, including local impacts and wider environmental impacts caused by climate change, for example. The tools can also cover social and economic aspects. The assessment tools provide checklists that can support decision processes, thus allowing for more transparent decisions.

Keywords: Flood risk, Landslide risk, Risk reduction, measures, Sustainability, assessment tools, Sweden

1. INTRODUCTION

Floods and landslides can have severe and even disastrous consequences with fatalities, diseases, construction and infrastructure failures, and can damage or completely destroy land (Dai et al. 2002; Srivastava and Laurian 2006; Holcombe and Anderson 2010; Singh 2010). The risks related to a flood or landslide can be described as the potential for loss, damage, or destruction of an asset as the result of a hazard exposing a vulnerability related to the event. The risk is a function of the probability, magnitude, and other characteristics of an event and the consequences of the event. The consequences depend on the exposure and the vulnerability characteristics of the elements at risk (humans, landscape and ecosystems, buildings and constructions, the social structure, and other values in the area at risk). The vulnerability is a function of the susceptibility (the likelihood of being exposed) and the adaptive and coping capacity. Both the probability and the consequences of floods and landslides are expected to increase in the coming decades, as a result of climate change and increased vulnerabilities, especially in urban areas (Poussin et al. 2012; IPCC 2013). The consequences may be damages caused directly or indirectly by a flood or landslide. An example of an indirect consequence is delays due to road or railroad damages (Holcombe and Anderson 2010; Suh et al. 2011). Strategies can be developed to reduce either the probability of an event or the consequences, or both (Dai et al. 2002; Brooks 2003; Sarewitz et al. 2003). Both physical and nonphysical measures can be implemented to reduce risks. The measures can be decided and implemented at different levels, from an individual level to national and international political and management levels.

At the national and international levels, measures to reduce risks can be enforced by policies, directives, legislation, frameworks, and other guidelines, and by economic incentives such as taxes and subsidies. International directives and agreements such as the EU Flood Directive (Directive 2007/60/EC), the EU Inspire Directive (European Directive 2007/2/EC), the EU Seveso Directive (Directive 2012/18/EU)), the Hyogo Framework for Action 2005–2015 (UNISDR 2005), and the Sendai Framework for Disaster Risk Reduction 2015–2030 (UNISDR 2015)—influence the development of national strategies and legislation. National measures also include information campaigns and education (Bormann et al. 2015) and programs such as the Room for the River program in the Netherlands.[1] National hazard and risk mapping, as well as governmental investigations and inquiries, are also important nonphysical, national-level measures to reduce risks (Schuster and Highland 2007). Physical and nonphysical risk reducing measures can also be applied at local and catchment levels, for example.

Examples of physical measures are classic flood risk reducing methods such as barriers, dikes, embankments, walls, and dams (Goltermann and Marengwa 2012). Landslides can be prevented by soil reinforcements and erosion-prevention measures in the most landslide-prone areas. Physical measures can be implemented from a catchment perspective, or from a political regional perspective, but can also be taken to a local and estate or household level. Regardless of the level at which the risk reduction strategy is implemented, it may result in consequences that involve two or more provinces and even countries.

Some of the more important nonphysical measures are education, information campaigns, communication with landowners, subsidies, increased preparedness, and the establishment of risk management networks (Glaas et al. 2010; Glavovic et al. 2010; Holcombe and Anderson 2010; Andersson-Sköld et al. 2013; Bormann et al. 2015). Other measures can be recommendations and restrictions in master plans and in detailed spatial planning (Srivastava and Laurian 2006; Holcombe and Anderson 2010).

An appropriate and sustainable risk reduction strategy needs to be defined, not only for the current risk but also for the protection requirements and the effectiveness of the risk reduction measures (Dai et al. 2002; Holcombe and Anderson 2010; Singh 2010). Risk reduction requirements can be defined on the basis of risk acceptance, for example, protection of a city in relation to certain flood levels. They can also be based on an analysis of the costs of risk reduction activities in relation to the consequences in the case of an event, preferably taking into account environmental and social aspects (Roberts et al. 2009; Anderson et al. 2010; Zeng et al. 2012). Developing the most sustainable risk reduction strategy requires assessing different alternative measures (or combinations of measures) in the context of short- and long-term perspectives. Previous studies have shown that the more open and transparent the decision-making process is, the more effective the risk reduction can be expected to be (Sharma et al. 2012).

1.1 Monetary Cost-Benefit Assessments in Risk Management

Monetary assessments can be powerful tools within risk management. A study of socioeconomic impacts and risk reduction strategies in urban landslide-prone areas in the United States showed that improved planning codes and professional practice reduced the monetary losses from landslides by more than 90% (Schuster and Highland 2007). The actions taken ranged from nonphysical measures, such as restrictions against new developments, implementation of construction codes, installation of warning systems, and the possibility to insure private property, to physical measures, such as excavation, grading, and other physical measures for existing buildings (Schuster and Highland 2007).

Poussin et al. (2012) examined the cost-effectiveness of different adaptation strategies at a local/regional level to reduce future flood damage in the Meuse river basin, including changes in land use and climate between 2000 and 2030. The assessment included calculated risk reduction in monetary terms of houses and inhabitants impacted by floods (water depth, flow velocity, and duration). The impact assessment was done by applying exposure, sensitivity, and adaptive capacity functions as described by Ernst et al. (2010). The strategies examined were dry-proofing (for example, sandbags, cofferdams, panels on doors and windows), wet-proofing (all semistructural and nonstructural measures that can be implemented to adapt the exterior and interior of a house), and a combination of both (Poussin et al. 2012). The results of the study showed that land use planning played a more important role than the impacts of climate change for the increasing flood risks in the Meuse river system.

However, many ecological and social factors such as disturbed ecosystem functions, social unrest, or psychological stress are difficult to evaluate in quantitative or monetary terms (Ismail-Zadeh and Takeuchi 2007). Another limitation of a cost-benefit approach is related to impacts and potential maladaptation due to differing spatial and temporal scales. Patterson and Doyle (2009), for example, found that county-based planning and monitoring may result in unwanted and costly impacts elsewhere in a catchment area. Taking all those aspects into account in the cost-benefit analysis requires broader evaluation methods.

1.2 Integrated Assessments in Risk Management

An alternative to cost-benefit analysis is analysis that integrates monetary and nonmonetary assessments and valuations. Multi-criteria models, or multi-criteria analyses (MCA), are being developed as tools for illustrating potential conflicts of interest and unwanted outcomes of planning strategies. To achieve wanted and positive, and to avoid unwanted and negative, impacts of interventions, short- and long-term as well as wider consequences need to be assessed and valued (Barnett and O'Neill 2010; Nyberg et al. 2014).

Multi-criteria analyses (MCA) are frequently used for integrated assessments. Their foci and purposes vary, but costs are often valued in relation to technical performance and/or impacts on the environment (Renn 2005; Volchko et al. 2014). In general, the methodology also encourages an integrative role of stakeholders. The outcome of applying such models could be that stakeholders with differing experiences, needs, views, and interests find that they interpret the risk differently (Renn 2005; Bormann et al. 2015). A study by Gamper and Turcanu (2009) showed that a broad participation in the MCA process can improve the risk reduction and decision-making process by increasing common understanding, serving as a means of conflict resolution, creating a win–win solution, and acting as a complementing instrument in land use planning and risk management. Two MCA methods have recently been developed for assessing flood and landslide risk management strategies—the benefit value tree (BVT) by Bana e Costa et al. (2004), and the matrix-based decision support tool (MDST) by Andersson-Sköld et al. (2014a).

From the perspective of sustainable development, the functionality of measures (level of protection, time and cost for construction, and maintenance), their impacts on the environment (local-scale effects as well as use of resources and contribution to climate change), and social and socioeconomic aspects need to be taken into account, both from a short- and a long-term perspective (Barnett and O'Neill 2010; Nyberg et al. 2014; Andersson-Sköld et al. 2015). The short-term perspective applies mainly to the implementation phase of a measure, while the long-term perspective applies when a measure is in use, as well as to a post-use phase.

1.3 Aim of the Study

The aim of this study was to investigate the potential for assessing the effectiveness of risk-reducing measures based on a sustainability perspective. The following two methods were used:

1) the benefit value tree (BVT) by Bana e Costa et al. (2004), where the costs (in monetary terms) are weighed against a selection of environmental and social benefits or negative impacts; and
2) the matrix-based decision support tool (MDST) by Andersson-Sköld et al. (2014a), where monetary cost is one aspect among others.

The BVT and the MDST allow the inclusion of both quantitative and qualitative information in the assessment. The two methods were applied in two Swedish case studies to evaluate the risk reduction and sustainability of flood and landslide risk reduction measures by using available case information, literature, and stakeholder views, experiences, and perceptions.

2. BENEFIT VALUE TREE (BVT) AND MATRIX-BASED DECISION SUPPORT TOOL (MDST): TWO METHODS TO ASSESS EFFECTIVENESS AND SUSTAINABILITY OF RISK REDUCTION MEASURES

In this study the BVT and the MDST are applied on two case studies. Both models are stepwise processes to assess and value the impacts of risk reduction measures. The assessments can be done using qualitative, quantitative, or semiquantitative ratings, and the valuation can be done in different ways. The main difference between the two approaches focused on here is the sustainability aspects considered in the assessment of impacts caused by the risk reduction measures.

2.1 The Benefit Value Tree (BVT)

The BVT for the evaluation of flood control measures was developed by Bana e Costa et al. (2004). It is a stepwise method initiated by an analysis of the problem context, followed by identification of options, assessment, and analysis of the options' impacts, and finally sensitivity and value analysis (Bana e Costa et al. 2004).

In the assessment, the costs of one or more alternative flood-control measures are compared to the risk reduction effectiveness as well as other impacts of the risk reduction measures under investigation. The BVT is based on expert views and consists of three main non-economic components (key benefit dimensions)—environmental impacts, social impacts, and technical effectiveness—that are analyzed in relation to the costs of the flood-control measure being assessed. In total, 16 non-economic ("benefit") components (Bana e Costa et al. 2004) are compared to the costs of the measure:

Environmental impacts

- Water
  - Time of inundation of the riverine zone.
  - Risk of discharge obstruction due to sedimentation.
  - Quality of surface water after a flood event.
  - Piezometric level of aquifer.
  - Quality of groundwater.
- Soil
  - Area of agricultural soil.
  - Soil contamination.
- Fauna and flora
  - Nature conservation interest.
- Landscape
  - Urban integration.
  - Enhancement of landscape.
- Social impacts
  - Perception (concern/anxiety) of flood (landslide) risk.
  - Effects on the social fabric.
  - Effects on public health.
- Technical effectiveness
  - Technical complexity of the intervention.
  - Complexity of maintenance.
  - Level of protection.

The impact assessment and analysis of the potential risk reduction measures can be done through qualitative descriptions combined with simple ranking, such as pros and cons. This can also be followed by, or done by, a more advanced and detailed information and assessment procedure, for example through the MACBETH impact assessment and weighting procedure as described in Bana e Costa et al. (2003, 2012). For a subsequent sensitivity analysis the VISA model can be used, for example, to calculate the aggregated benefit scores and perform the sensitivity analysis of the potential different flood risk reduction measures (Bana e Costa et al. 2004). Finally, the total cost (investment and other initial costs) versus overall benefit analysis is done taking into account the estimated variations in weights and uncertainties (Bana e Costa et al. (2004).

In this study, we use a semiquantitative scale. The scale ranges from −2 (very negative impact on the "benefit" component, that is, significantly more negative than −1) to +2 (very positive impact on the component, that is, significantly more positive than +1), where 0 implies no, or no significant, impact. The main reason for choosing this scale is that the study focuses on the 16 benefit dimensions in relation to the aspects regarded in the MDST described below. The study does not aim to make a detailed assessment but rather an aggregated assessment, since the uncertainties in the different assessments are so numerous that fine assessments are not relevant.

2.2 The Matrix-Based Decision Support Tool (MDST)

The MDST is a generically applicable tool (Andersson-Sköld et al. 2014a). It consists of a stepwise approach for analyzing the possible effects of different measures, for example, measures to reduce the consequences of natural hazards such as floods, climate change adaptation measures, as well as the consequences of strategies for contaminated land and general land use planning. A matrix-based, multi-criteria approach first defines the problem, then suggests possible risk reduction measures, and finally identifies and assesses their current and future social, economic, and environmental local and global impacts. The tool involves impact identification and assessment as part of the procedure. The impact categories considered cover health and environment, use of resources, social and economic aspects such as investment costs, and well-being and perceived welfare. Flexibility is also added as an individual impact category to be assessed. In total, 11 aspects are assessed regarding suggested measures or land use strategies:

- Environment
  - Global warming—emissions of greenhouse gases, carbon sequestration.

  - Air—emissions of toxic gases, emissions of particles, airborne bio-accumulative substances, emissions that contribute to eutrophication, acidification, oxidants, and formation of ground layer ozone.
  - Water—ecosystem status and drinking water quality, including biodiversity, fisheries, marine and limnological properties of high conservation value, eutrophication through leaching.
  - Soil—terrestrial impacts, such as soil quality and soil pollution load, impacts on terrestrial biodiversity, ecosystems, and properties of high conservation value.
- Resources
  - Energy—energy consumption.
  - Raw material—raw material acquisition.
  - Landscape and land resources.
- Social and economic aspects
  - Well-being/perceived welfare—perception such as concern or anxiety about flood risk/landslide risk, perception of other aspects of doing nothing or the intervention such as aesthetics, attachment, perceived disturbances by construction, intervention, maintenance, and so on.
  - Socioeconomic costs and revenues—infrastructure, cultural environment/heritage, accessibility, business activity, jobs, recreation and health (other than covered under environment due to global warming, air, water, and soil impacts), and so on.
  - Direct costs/revenues—investment, maintenance, potential revenues (in the short- and long-terms), and so on.
- Flexibility
  - High flexibility implies no-regret solutions, and reversibility of the system.

The assessment should also comprise both short- and long-term impacts. The tool should be applied early in a decision-making process and the process benefits from iterative use. The tool has been applied in practical cases, for example, brown-field site developments in Gothenburg, Sweden (Andersson-Sköld et al. 2015). The tool encourages discussion, and the systematic view of sustainability increases awareness of the holistic perspective (Jonsson et al. 2012; Andersson-Sköld et al. 2015). The discussion when applying the tool draws attention to institutional adaptation measures as a complement to pure physical measures.

The analysis can be done at different levels of specification (qualitative description, semiquantitative, or based on quantitative estimates). It can, as in this study, be used for assessing the impacts, and the process can further be continued by including weighting, which can be done through different standard methods. The results should also undergo a sensitivity analysis, for which different methods can be applied. In this study the result of the assessment is presented through applying a scale ranging from −2 to +2. No weighting system was applied.

3. METHODOLOGY

The study was performed by applying two different MCA methods—the BVT (Bana e Costa et al. 2004) and the MDST (Andersson-Sköld et al. 2014a)—for the assessment of two different risk-reducing measures. The material is based on literature, available information from authorities and municipalities, expert knowledge and experience, and stakeholder workshops. In order to have a common basis for the stakeholders' views and values, a case study area approach was applied. The study aims to test the two assessment methods, as well as to describe the effectiveness of different risk reduction measures based on the available information.

3.1 Case Study Areas

A case study approach offers a common basis for understanding and interpreting the aspects studied. It provides opportunities to build an in-depth understanding of complex social, environmental, and economic interactions (Jonsson et al. 2012) and to realize stakeholder-oriented research that may have practical implications in the case (Johansson 2013). Two case study areas in Sweden were included in the investigation: the flood-prone Ljusnan–Voxnan river system and the landslide-prone municipality Lerum (Figure 1).

Municipalities have large responsibilities in the Swedish risk management system (Jonsson et al. 2005). In Sweden, as in many other modern societies, different health, safety, and security matters are normally handled separately and in parallel planning and management processes by different sectors in the municipal administration (Johansson 2008, 2013; Norén 2016). Under the Planning and Building Act, the municipalities are responsible for the master plans and detailed spatial plans in the municipality (Johansson et al. 2006; Johansson 2013; Norén

2016). The municipality is responsible for planning and coordinating crisis management involving national agencies and authorities, nongovernmental organizations (NGOs), industry, and other organizations within the geographical area of the municipality.[2]

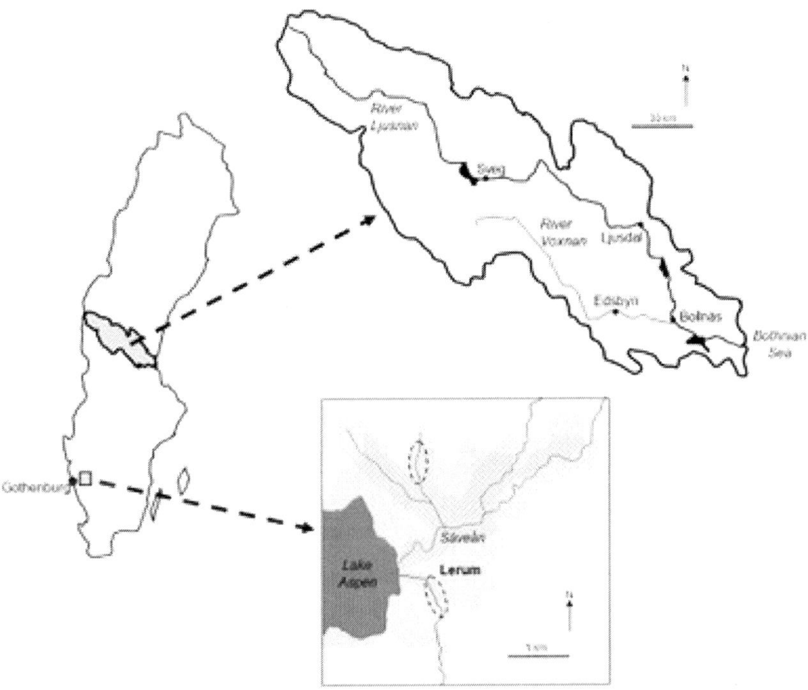

Figure 1. Location of the two case study areas in Sweden: the Ljusnan–Voxnan River (*top*) and the municipality of Lerum (*bottom*). Edsbyn is the seat of Ovanåker Municipality. In the map of Lerum the *light grey areas* represent densely populated areas, and the *hatched areas* represent landslide-prone clay areas. The two areas where preventive measures were undertaken are marked by *dashed lines*

Currently there are hardly any systematic, and no common, approaches applied among Swedish municipalities for risk management, following the full chain from risk identification to assessing measures for flood risk reduction (Norén 2016). Municipalities often have some type of flood risk maps, but there is no systematic analysis of risk reduction measures or integrated assessments of pros and cons of potential measures (Norén 2016).

3.1.1 The Ljusnan–Voxnan River System

The Ljusnan–Voxnan river system (Figure 1) is a flood-prone area that includes five municipalities of more than 10,000 inhabitans (Härjedalen 10,200; Bollnäs 26,900; Ljusdal 19,000; Ovanåker 11,600; and Söderhamn 25,800), five larger towns/municipality centers of 2500–13,000 inhabitants (Sveg 2500; Ljusdal 6200; Bollnäs 12,800; Edsbyn 4000; and Söderhamn 11,800), and around ten smaller towns/villages of 200–2200 inhabitants (Färila 1300; Järvsö 1400; Vallsta 300; Arbrå 2200; Segersta 300; Kilafors 1100; Marmaverken 400; Ljusne 1900; Ovanåker 200; Alfta 2200). Weather-related flooding events with costly consequences have occurred repeatedly.

Examples of preventive physical measures undertaken are the establishment of dikes and new pumping stations, dredging, and the removal of a river neck (T. Wannqvist, Head of the Rescue Services Department, personal communication, 2013). In this study the effectiveness of previous preventive measures were investigated through focus group discussions. The clearing of a river grove was further assessed by applying the BVT and MDST.

3.1.2 Lerum Municipality

Lerum Municipality (about 40,000 inhabitants, whereof 17,000 in the city), is located in western Sweden, just east of Gothenburg (Figure 1). The river Säveån runs through Lerum Municipality and the town (with the same name). The name Lerum means clay (ler) village (um) and describes the geological conditions in this landslide-prone area dominated by sensitive quick clayey soils. Recently measures to prevent landslides have been implemented and at the time of this study new measures were planned.

Examples of previous preventive measures are culverting of streams and ditches, excavation and filling, tree removals, erosion control, road reinforcements, and slope reinforcements (U. Lundgren, co-ordinator of landslides, flooding, and geotechnical issues, Lerum municipality, personal communication, 2013). The effectiveness of previous measures was discussed in focus groups. In addition, the effectiveness of soil reinforcements and erosion reduction in one part of the stream was assessed by applying the BVT and MDST.

3.2 Focus Groups

Three focus groups were recruited, representing different networks and organizations involved in the preventive and/or operational risk management in the two case study areas. Two stakeholder groups represented the local level, one for Lerum Municipality (6 participants), one for Bollnäs and Ovanåker Municipalities (9 participants), and one group represented the stakeholders of the Ljusnan–Voxnan river system (14 participants).

The groups were composed of civil servants with expertise in different sectors including technical experts on the built environment, geological information, physical planning and infrastructure, technical services, as well as experts on environment, health, social services and administration, education, child care, and rescue services. In two of the focus groups (Lerum and the Ljusnan–Voxnan river system) representatives of the county administrative boards participated. In the Ljusnan–Voxnan focus group energy producers and water regulators also participated. All three groups had taken initiatives regarding risk analyses, planning, and implementation of risk-reducing measures, and internal and external networking and communication. Their expectations were dominated by a view that their groups had made important interventions to reduce risks, and had played important roles in the risk management system.

During the focus group meetings the impacts of the different risk-reducing interventions were discussed based on the impacts included in the two methods. The discussions were based on available information on taken or planned measures in the case study areas. The information varied from well described as a basis for national co-funding (Lerum) to rather little information provided due to lack of documentation. During the focus group discussions questions on the measures' effectiveness, investment and maintenance costs, pros and cons regarding environmental impacts (local to global scale), and social and socioeconomic aspects were discussed aiming to achieve information that could be applicable in the assessments.

The assessments are not based on averages of individual gradings or scaling but on the experiences and expectations of the expert respondents. The grading presented in the result tables (Tables 1, 2) was determined by the authors based on the qualitative impacts and their relative magnitudes described in the focus group discussions.

The reason for the semiquantitative and the discussion-based approach is the lack of relevant information, and especially in-depth and quantitative information, due to lack of structured processes and documentation already in the decision and follow-up processes, which has also been found as a general obstacle in the Swedish risk management (Andersson-Sköld et al. 2013).

Table 1. Results from the BVT assessment based on expert judgments in the focus group discussions—clearing of a 10–15 km river grove (Ljusnan–Voxnan river system), and landslide prevention by soil reinforcements and erosion reduction along a stream stretch (Lerum)

Aspects considered in addition to costs in the BVT	Clearing of 10–15 km river grove. Reduces water level, and thereby the inundation zone		Landslide prevention. Soil reinforcements and erosion reduction along a stream stretch. Aims to completely reduce the landslide risk at the site (negligible hazard probability)	
Environmental	Effect of intervention	Value	Effect of intervention	Value
Water				
Time of inundation of the riverine zone	Aims to completely reduce the inundations, but not designed for specific protection level. Therefore, estimated high impacts (+1)	1	Not directly relevant	0
Risk of discharge obstruction due to sedimentation	The intervention may result in sedimentation elsewhere in the river system (−1)	−1	The intervention per se may result in sedimentation elsewhere in the river system during a very short time	0
Quality of surface water after a flood event	Low agricultural activities (no significant leaching of nutrients), that is the intervention has no/low impact on water quality. The intervention itself causes short time re-suspension. Erosion in river not significantly impacted by intervention	0	Low agricultural activities (no significant leaching of nutrients), that is the intervention has no/low impact on water quality. The intervention itself causes short time re-suspension. Erosion in river not significantly impacted by intervention	0
Piezometric level of aquifer	Very low/no significant impact on aquifer	0	Not directly relevant	0
Quality of groundwater	Large inundation areas in case of no intervention. The river water status: high levels of nutrients, some ongoing activities and some previously contaminated sites (reduced accumulation due to fewer flood events and less time)	1	No impact on groundwater quality	0
Soil				
Area of agricultural soil	Large areas of agricultural land that will be less impacted by flood events	1	No impact on agriculture	0
Soil contamination	See quality of groundwater	1	No impact on soil contamination	0
Fauna and flora				
Nature conservation interest	The intervention per se has an impact on the natural environment (its geology and habitat) due to the reduced riverine area and the dredging	−2	The intervention per se has an impact on the natural environment, but only for a limited stretch	−1
Landscape				
Urban integration	No/very low impact	0	No/very low impact	0
Enhancement of landscape	No known impacts and no focus group responses	0	The focus group respondents were in general slightly positive, but did not provide any strong positive or negative response. Two of the participants did, however, provide information that one of the landowners finds that the intervention has a high negative impact on aesthetics	−1
Social				
Perception (concern/anxiety) of flood risk	The focus group found that the intervention has contributed to reducing the flood risk (+1)	1	The focus group highly values the risk reduction in a long-term perspective (+2), but not all stakeholders agree (−2)	0
Effects on the social fabric	No known impacts	0	No known impacts (0) apart from that the interventions will result in costs that one landowner finds not acceptable (it is perceived to have impact on the owners' well-being and may cause conflicts)	−1
Effects on public health	No known health effects before or after intervention	0	No known health effects before or after intervention	0
Technical				
Technical complexity of the intervention	Excavation/dredging, relatively low technical complexity (+1)	1	Excavation and slope reinforcement, relatively low technical complexity	1
Complexity of maintenance	Dredging, low complexity, relatively low frequency (+1)	1	No or low complexity and frequency	2
Level of protection	Some level of protection (+2)	1	High-hazard probability reduction	2
Total		+5		+2

Valuing/ranking scale: −2 very negative/not wanted, −1 negative/not wanted, 0 no (significant) impact, +1 positive/wanted, +2 very positive/wanted

Table 2. Results from the MDST assessment based on expert judgments in the focus group discussions—clearing of a 10–15 km river grove (Ljusnan–Voxnan river system), and landslide prevention by soil reinforcements and erosion reduction along a stream stretch (Lerum)

MDST		Clearing of 10–15 km river grove. Reduces water level, and thereby the inundation zone		Landslide prevention. Soil reinforcements and erosion reduction along a stream stretch. Aims to completely reduce the landslide risk at the site (negligible hazard probability)	
Health and environment		Effect of intervention	Impact value	Effect of intervention	Impact value
Global warming	Emissions of greenhouse gases, carbon sequestration	The intervention demands machineries and transportation of the dredged sediment and excavated masses, which will contribute to emissions of greenhouse gases. The activity is, however, only temporary and the contribution to GHG emissions is therefore limited over time (−1). No impacts on sequestration expected due to the activity	Short-term −1 Long-term 0	The intervention demands machineries and transportation of the excavated masses, which will contribute to emissions of greenhouse gases. The activity is, however, only temporary and the contribution to GHG emissions is therefore limited over time (−1). No impacts on sequestration expected due to the activity	Short-term −1 Long-term 0
Air quality	Emissions of toxic gases, emissions of particles, airborne bio-accumulative substances, emissions that contribute to eutrophication, acidification, oxidants, and ground layer ozone formation	The intervention demands machineries and transportation of the masses, which will contribute to emissions. The impact is, however, only temporary, and will not last over a long time	Short-term −1 Long-term 0	The intervention demands machineries and transportation, which will contribute to emissions. The impact is, however, only temporary, and will not last over a long time	Short-term −1 Long-term 0
Water	Ecosystem status and drinking water quality, including biodiversity, fisheries, marine and limnological properties of high conservation value, eutrophication through leaching	The intervention may result in sedimentation elsewhere in the river system, which may have impacts on the biodiversity and ecosystems (−1). The intervention itself causes short-term re-suspension, but erosion not significantly affected (0). Very low/no significant impact on aquifer (0). Large inundation areas in case of no intervention. The river water status: high levels of nutrients, some ongoing activities and some previous (contaminated sites) result in reduced accumulation of contaminants in groundwater due to fewer flood events and shorter duration (+1)	Short-term −1 Long-term 1	The intervention itself causes short-term re-suspension and possibly minor erosion (−1) in the stream, but long-term erosion will be reduced at the site (+1). The intervention may impact fish in the stream, but if done according to regulations, the impacts will be low or negligible (0). Very low/no significant impact on aquifer (0)	Short-term −1 Long-term 1
Soil	Terrestrial impacts, such as soil quality and soil pollution load, impacts on terrestrial biodiversity, ecosystems, and objects of high conservation value	Large areas of agricultural land that will be less impacted by flood events in a long-term perspective (+1) and over a long period, reduced accumulation of contaminants in soil (+1). The intervention per se has an impact on the natural environment (its geology and habitat) due to the reduced riverine area and the dredging (focus group response) long- and short-term response (−2)	Short-term −2 Long-term 0	The intervention per se has an impact on the natural terrestrial environment (its geology and habitat) due to excavation and reinforcements but the area affected is very limited (−1)	Short-term −1 Long-term −1
Resources					
Landscape and land resources	Use of land, housing, enhancement of landscape	No known impacts and no focus group responses	Short-term 0 Long-term 0	In addition to the impacts mentioned under environment (that is, minor changes of landscape). No known impacts and no focus group responses	Short-term 0 Long-term 0
Energy	Energy consumption	The intervention demands machineries and transportation of the masses, which will use energy. The activity is, however, only temporary and the use of energy is therefore limited over time (−1). The intervention has no impact on the hydropower dam system (0)	Short-term −1 Long-term 0	The intervention demands machineries and transportation of the masses, which will use energy. The impact is, however, only temporary, and will not last over a long time (−1)	Short-term −1 Long-term 0
Raw material	Raw material acquisition	The raw material acquisition depends on the energy source in use	Short-term −1 Long-term 0	The raw material acquisition is mainly the energy source in use	Short-term −1 Long-term 0
Social					
Well-being/ perceived welfare	Perception (concern/anxiety) of flood risk, aesthetics	The focus group found that the intervention has contributed to reducing the flood risk and will do so even in future (+1). No negative impacts mentioned by the focus group participants (0)	Short-term 1 Long-term 1	The impact value depends on the respondent. The focus group highly values the risk reduction in a long-term perspective (+2), while one of the landowners finds that the intervention has a high negative impact on the aesthetics and will result in high costs impacting the landowners' well-being (−2)	Short-term −2 Long-term 2
Direct costs	Investment, maintenance, revenues, and so on	The intervention cost was €17,000, fairly high for small municipalities despite low complexity (−1). No or very low maintenance costs (0)	Short-term −1 Long-term 0	The intervention cost is estimated at €700,000, fairly high for small municipalities and landowners despite low complexity (−2). No or very low maintenance costs (0)	Short-term −2 Long-term 0
Socioeconomic costs	Infrastructure, cultural, accessibility, business activity, jobs, recreation, and health	The aim is to completely reduce the inundations, but the intervention is not designed for a specific protection level, therefore estimated to have high impact on reducing socioeconomic impacts over the long term (+1), providing minor jobs in the short-term (+1). No known effects on the social fabric (0)	Short-term 1 Long-term 1	Aims to completely reduce the landslide probability for several houses and inhabitants, therefore estimated to have very high impact on reducing socioeconomic impacts over the long term (+2). No known effects on the social fabric (0)	Short-term 0 Long-term 2
Flexibility	High flexibility implies no-regret solutions, reversibility of the system, and high resilience	The focus group mentioned that the intervention would not have been implemented today, since the ecosystem was affected too much by long-term impacts (−1)	Short-term −1 Long-term −1	The intervention will alter a very small stretch of a stream, but otherwise no significant positive or negative impacts on flexibility. No other nearby interventions expected (0)	Short-term 0 Long-term 0
Total—short term			−7		−9
Total—long term			+2		+4
Total			−5		−5

Valuing/ranking scale: −2 very negative/not wanted, −1 negative/not wanted, 0 no (significant) impact, +1 positive/wanted, +2 very positive/wanted

4. RESULTS

The results of the assessments of costs and benefits from a monetary perspective as well as the wider cost and benefit analyses (BVT and MDST) based on the experiences and expectations of the expert respondents are presented in Sect. 4.1, and an analysis and synthesis of the results is presented in Sect. 4.2.

4.1 Costs and Benefits of Physical Measures and Land Use Planning

The perception among all the focus group participants was that flood risk preventive measures can be cost-effective. In the focus group discussions related to the Ljusnan–Voxnan river system, the measures undertaken in the river system up-to-date were regarded as both relevant and cost-effective by the participants. There was one exception: the clearing of the river grove in the Ljusnan–Voxnan system, which, most likely, would not have been implemented today due to conflicting ecological aspects, mainly related to regulations according to the EU Water Framework Directive.

The cost-effectiveness of flood risk reduction measures, however, was found difficult to estimate in quantitative terms, and to evaluate, because of the rarity of events and the fact that measures may have both wanted and unwanted, and often unexpected, impacts on the river system. Due to the lack of recent events, the current vulnerability was also hard to predict.

The focus group discussions made clear that preventive measures were perceived as cost-effective also in the landslide-prone municipality Lerum. This is in agreement with monetary estimates for recent and planned preventive measures. One example is a recently conducted preventive landslide risk reduction measure, including culverting, excavation and filling, tree removal, and erosion control, in a residential area called Torpadal in Lerum (5 houses accommodating 13 people). The preventive measures added up to a cost of 700,000 Euros. The expected costs in the case of an event were estimated at 5 million Euros, based on road and house damages, fatalities and injuries, and a maximum of one casualty based on an expected average exposure and sensitivity (Andersson-Sköld et al. 2014b). The expected loss of life, which also was valued in monetary terms, was based on 24 h averages and would be expected to be much higher in the case of a nighttime event without warning.

Another investigated, but not yet implemented, measure in Lerum refers to a residential area with private properties where preventive measures included slope reinforcement (drainage, embankment, and excavation). The expected costs in the case of a small landslide event ($p = 0.5\ a^{-1}$) due to reconstructions and renovations are up to 100,000 Euros. A larger landslide ($p = 0.1\ a^{-1}$) would result in costs of up to 0.5–1 million Euros. The costs include road damages and related reconstruction, damages on houses and infrastructure (water, sewage, IT, and so on), personal injuries, and loss of life (Andersson-Sköld et al. 2014b). In the case of either a small or large landslide, salmon living in the Säveån River (smolt)[3] may be affected depending on the time of year of the event. The preventive measures planned were slope reinforcement (drainage, embankment, and excavation), and the budget was less than 450,000 Euros. The expected loss of life again was based on daytime averages and would be expected to be much higher in the case of a nighttime event without warning.

A cost-benefit analysis that only includes aspects such as degree of protection and the cost of investment and maintenance will, however, not reveal if some aspects are in conflict with the measure. Conflicts of interest, such as aesthetic aspects, were also mentioned with regard to landslide preventive measures. In general, budget issues were found to restrict preventive measures. In addition, unequal benefits and the fact that those financing the measures may not be the ones benefit the most were seen as conflicts of interest. An example is interventions upstream, at least partly financed by and having an impact on landowners not under flood risk themselves, that significantly may reduce the impacts downstream. Otherwise the costs for the ones at risk may be significantly higher despite that the cause is to be found upstream in the catchment area.

4.1.1 Application of the Benefit Value Tree (BVT)

The BVT method (Bana e Costa et al. 2004) was developed specifically to enable a more transparent and systematic assessment of the costs and benefits of flood risk reduction measures. Here, the method was applied to assess the clearing of a river grove in the Ljusnan–Voxnan river system and landslide prevention through soil reinforcement and erosion control. The selection of the river grove was made because this intervention was the most highlighted and discussed in general in the catchment area, and therefore the current knowledge and awareness of its pros and cons were most reliable. It caused also most discussions among the experts and focus group participants.

The application of the BVT to assess the clearing of a river grove in the Ljusnan–Voxnan system provided a summary of how the flood risk reduction measure will impact the inundation time, water and soil quality aspects, the fauna and flora, and landscape, as well as social and technical aspects (Table 1). The measure will shorten the time of inundation, contribute to improve the water quality and result in less negatively impacted agriculture and reduced flood risk-related concerns among citizens. The measure is also regarded as not being complex in either the construction phase or with regard to maintenance, and it is perceived to provide protection (Table 1). The intervention is, however, also expected to result in unwanted, possibly severe, impacts on the geology, and the intervention will have negative impacts on the ecosystem and habitat due to the reduced river area and the dredging (Table 1). In total, the result is slightly positive (+2) if all environmental and social values are summarized and even higher (+5) taking into account the level of protection and the low technical complexity of the intervention. There is, however, a potential serious impact on the ecosystem that needs further analysis with regard to the value of impacts and/or regarding potential alternatives.

The BVT was also tested with the landslide preventive measure, that is, soil reinforcements and erosion reduction along a short stream stretch in Lerum. Despite being developed for flood risk reduction measures, many of the aspects considered were also relevant with regard to landslides. The aspects “time of inundation of the riverine zone” and “piezometric level of aquifer” were neglected as not directly related to landslides (Table 1). These two aspects are indirectly relevant, since a landslide may cause flooding and piezometric change, if occurring along water courses. In particular, landslides in quick clay areas may cause severe flooding, large waves, and even new meandering, as well as direct consequences not included in the BVT. The landslide impacts are more site-specific and less generically quantifiable compared to the time of inundation and the piezometric level of aquifer. The suggestion is to differentiate between “direct impacts” and “indirect impacts” of a landslide, and the values of the impacts need to be assessed and estimated based on the site-specific context and its potential consequences.

The other aspects included in the BVT are all relevant for landslides. In a short-term perspective, landslide risk reduction measures may result in increased suspension and sedimentation in the system and changes in water quality (surface and groundwater) as well as soil quality. For example, the risk of contaminant spreading may be reduced due to reduced risks of erosion and landslides. But under some conditions, for example, due to excavation of contaminated sites, the risks during the construction phase may increase. The aspects are therefore relevant to and applicable to landslides. In the same way, aspects related to fauna, flora, and landscape are also applicable to landslides. The impact depends on the method, design, and scale of the intervention and where the intervention is done. The social aspect "perception" was interpreted as concern or anxiety of flood risk. All technical and social aspects are also relevant and applicable to landslides.

Applying the BVT in this way on the landslide risk reduction measure resulted in a slightly positive outcome (+2) (Table 1). All the technical aspects, including level of protection, were positive (+2 or +1), and if there had not been a conflict of interest regarding aesthetic landscape values among civil servants and one of the landowners (−2) the result would have been even more positive. The interventions will impact the fauna and flora, but only to a limited extent (−1) (Table 1).

Applying the BVT method compiles and illustrates the pros and cons in a transparent way. It helps to illustrate the impacts of a measure, and whether there is a need for alternative strategies, a need for a more in-depth analysis, or a need for assessing the acceptance of the risk and the impacts of the intervention.

4.1.2 Application of the Matrix-Based Decision Support Tool (MDST)

The MDST also provides a summary of the impacts of the interventions (Table 2). The results differ as MDST was developed to assess the sustainability aspects of land use planning strategies as well as risk reduction measures. The rating is provided in relation to the situation if no changes are made. The method differs from the BVT in that the economic aspects are included among the other aspects. In addition, the economic aspects are divided into both direct costs and revenues, and socioeconomic impacts. The method further includes both short- and long-term impacts, impacts of relevance to the global, regional, and local scales, as well as other environmental aspects that are local/site-related. The flexibility related to the measure is considered (Table 2).

When applying the MDST, both interventions (clearing of the river grove and landslide prevention) have a high negative sustainability impact (−5). The impact varies, however, depending on which time perspective is considered. In a short-term perspective, there is a large negative impact (−7 and −9) for the clearing of the river grove and the landslide prevention respectively. The reason is that both interventions contribute to the emissions of greenhouse gases and other air pollutants, to affecting the water quality negatively initially, to the use of energy and resources, and to disturbing the ecosystems.

In a long-term perspective, however, both measures will contribute to improved water quality and will have positive impacts on well-being and socioeconomic aspects (Table 2). The results show a positive impact of both measures (+2 and +4) for the clearing of the river grove and the landslide prevention respectively.

The results show that neither of the interventions can be regarded as acceptable in comparison to the current situation if both short- and long-term perspectives are taken into account and valued equally. It is important that the impacted stakeholders or representatives evaluate long-term versus short-term impacts separately. Also, the importance of the different aspects considered by the MDST has to be valued by a weighting procedure and alternative solutions should be considered.

4.2 Analysis and Synthesis

Although both methods include aspects of the three sustainability dimensions (social, economic, and environment), the two methods consider different aspects and different temporal and spatial scale perspectives. Therefore, the two methods yield different and even opposite results. Both interventions yield positive (+5) or slightly positive (+2) results when applying the BVT, while both interventions are negative (−5) when applying the MDST.

Previous studies show that broader environmental aspects are rarely taken into account in daily decision making (Johansson 2008; Andersson-Sköld et al. 2014a, 2015). The MDST offers a structured and transparent method for such analyses. For analyzing the pros and cons of flood and landslide preventive measures, the MDST lacks some of the local/site-specific detailed analyses provided by the BVT. Therefore, regarding decisions to contribute to sustainable development, we suggest that the two methods be merged.

As the measures' effectiveness is essential we suggest this to be the starting point of the analysis. The functionality, how well-functioning the flood risk reduction measure is, in a short- and long-term perspective can be related to the technical benefit aspects in BVT. In order to cover both the local and global scales of environmental, landscape, and social impacts we suggest to merge those as a combination of impacts from the two models. The direct costs can be divided into short-term investments and longer-term maintenance costs. As there are large uncertainties related to flooding and landslides, both due to the socioeconomic developments and climate change impacts, flexibility in a long-term perspective should also be considered. The resulting combination based on the two models includes the following aspects (in a short- and long-term perspective) for assessing measures to reduce risks related to flooding and landslides:

Functionality

- Level of protection with regard to

- direct impacts of both landslides and flooding: consequences for life and personal injuries, buildings, and infrastructure,
- indirect consequences of landslides, for example, potential flooding/energetic waves, meandering, and the related consequences for life, buildings, and infrastructure.
- Technical complexity of the intervention (short-term), and complexity of maintenance (long-term)
- Environmental aspects
- Global warming—impacts on emissions of greenhouse gases and carbon sequestration.
- Air—emissions of toxic gases, emissions of particles, airborne bio-accumulative substances, emissions that contribute to eutrophication, acidification, oxidants, and formation of ground layer ozone.
- Water quality, including quality of surface and groundwater (during construction in the short-term, and due to flood or landslide events in a long-term perspective)Soil quality and terrestrial impacts, such as potential discharge of nutrients or contaminants from agricultural soil.
- Ecosystem functions—fauna and flora (biodiversity, impacts on ecosystem functions and services such as fisheries, terrestrial, marine and limnological properties of high conservation value).
- Flooding—time of inundation, piezometric level of aquifer.

Resources and landscape

- Energy consumption and raw material acquisition.
- Urban integration.
- Enhancement of landscape.

Social

- Perception, such as concern or anxiety of flood risk/landslide risk, perception of other aspects of doing nothing or the intervention, such as aesthetics, attachment, perceived disturbances of construction, intervention, maintenance, and so on.
- Socioeconomic aspects (not considered under direct and indirect impacts) such as effects on the social fabric, jobs, business activity.
- Health and recreation.

Costs

- Investments short-term.
- Maintenance long-term.

Flexibility

- High flexibility implies no-regret solutions, and reversibility of the system.

The list of aspects now includes both local and large-scale impacts, as well as other aspects of relevance for sustainable development (social, environment, economy). An example of how to apply those merged aspects based on the results from Tables 1 and 2 is provided in Table 3. The corresponding result for the full list of aspects is a slightly positive (+3) impact for the flood risk reduction intervention and a neutral (0) impact for the landslide intervention. For both interventions, there was a negative result in a short-term perspective (−3 for the flood risk reduction and -8 for the landslide risk reduction intervention) due to disturbances during the construction and use of resources, and so on, and a positive result in a long-term perspective (+6 for the flood risk reduction and +8 for the landslide intervention) due to the increased protection level and related social aspects. Both interventions result in more positive total impacts compared to the MDST method. The neutral result for the landslide risk reduction measure when not taking into account any weighting indicates that the intervention may be relevant. If weights had been applied the result may have differed depending on the perceived importance of long- versus short-term impacts. The results indicate further that the short-term impacts on environment and resources are high and that a cautious approach needs to be taken—for example, less energy and material demanding methods need to be considered during the intervention work.

The results of the analysis of expert experience and knowledge are useful for illustrating the pros and cons. The results are especially useful for the comparison of potential alternative strategies and to identify which strategies provide the most benefits compared to negative impacts.

The results of any of the methods (BVT, MDST, merged method) do not take into account the stakeholders' and decision makers' valuation of the importance of considering different aspects. The perceived importance or value of the individual sustainability aspects may vary among different decision makers, experts, and stakeholders. Therefore, if the perceived value is expected to vary among the individual sustainability aspects, we suggest an additional step in which the value of the impact is included through a weighting procedure, for example, a ranking or more advanced methods such as monetary valuations. The value of each of the individual sustainability aspects will then equal the product of the assessment estimate and the weight of the impact (Figure 2).

Table 3. Examples of applying the merged BVT and MDST methods based on the results presented in Tables 1 and 2—clearing of a 10–15 km river grove (Ljusnan–Voxnan river system), and landslide prevention by soil reinforcements and erosion reduction along a stream stretch (Lerum)

Aspects considered	Clearing of a 10–15 km river grove. Reduces water surface level, and thereby the inundation zone		Landslide prevention. Soil reinforcements and erosion reduction along a stream stretch. Aims to completely reduce the landslide risk at the site (negligible hazard probability)	
Functionality	Effect of intervention	Impact value	Effect of intervention	Impact value
Technical complexity of the intervention and complexity of maintenance	Excavation/dredging relatively low technical complexity (short-term impacts). Dredging, low complexity, relatively low frequency (long-term impacts)	Short-term 1 Long-term 1	Excavation and slope reinforcement relatively low technical complexity (short-term impacts). No or low complexity and frequency (long-term impacts)	Short-term 1 Long-term 2
Environmental aspects				
Water quality, including quality of surface and groundwater (during construction in the short-term and due to flood or landslide events in a long-term perspective)	The intervention may result in unwanted (−1) sedimentation elsewhere in the river system. No significant leaching of nutrients (0). Reduced contaminant leaching and accumulation due to fewer flood events and less time (+1)	Short-term −1 Long-term 1	The intervention itself causes short-term re-suspension and possibly minor erosion (−1) in the stream, long-term erosion will be reduced at the site (+1)	Short-term −1 Long-term 1
Ecosystem functions—fauna and flora (biodiversity, impacts on ecosystem functions and services such as fisheries, terrestrial, marine, and limnological properties of high conservation value)	The intervention per se has an impact on the natural environment (its geology and habitat) due to the reduced riverine area and the dredging (focus group response) long- and short-term response (−2)	Short-term −2 Long-term −2	The intervention per se has an impact on the natural terrestrial environment (its geology and habitat) due to excavation and reinforcements but the area affected is very limited (−1)	Short-term −1 Long-term −1
Social				
Perception, such as concern or anxiety of flood risk/landslide risk, perception of other aspects of doing nothing or the intervention, such as aesthetics, attachment, perceived disturbances of construction, intervention, maintenance, and so on	The focus group found that the intervention has contributed to reducing the flood risk and will do so even in future (+1). No negative impacts mentioned by the focus group participants (0)	Short-term 1 Long-term 1	The impact value depends on the respondent. The focus group highly values the risk reduction in a long-term perspective (+2), while one of the landowners finds that the intervention has a high negative impact on the aesthetics and will result in high costs impacting the landowners' well-being (−2)	Short-term −2 Long-term 2

The result of the weighting is context-dependent and the perceived values may vary over time. Therefore, the weighting has to be done in a transparent and well-documented way. For successful risk management, the process needs to be continued by following up activities, decisions, and detailed planning (Figure 3).

4.2.1 Expert and Stakeholder Involvement in the Assessment Process

To open the decision process up for additional potential solutions early on, the weighting process must not only include experts but also the decision makers and preferably also the general public.

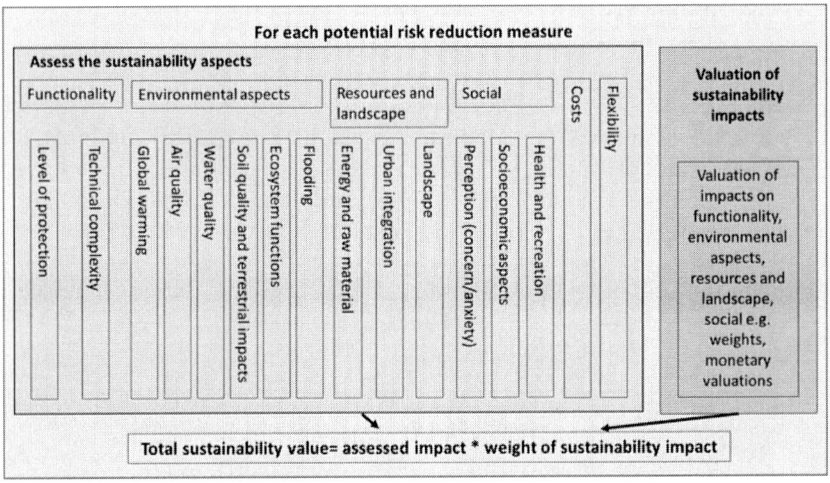

Figure 2. Valuation of sustainability aspects, a function of assessed impact and -weight of the sustainability impact

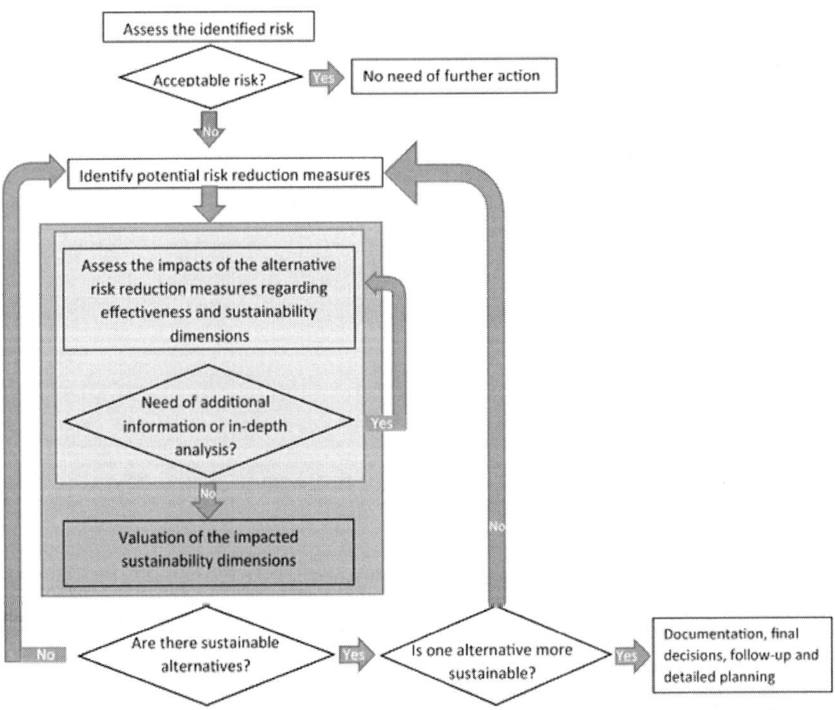

Figure 3. Iterative risk analysis and management process

The risk analysis and management process should preferably also be a dialogue-driven iterative process where the valuation procedure includes relevant stakeholders. This increases the acceptance of the decision and contributes to the identification of new, even more sustainable measures, as well as increased knowledge and understanding among the involved participants.

The process of using a group of stakeholders and complementary experts for assessment and weighting provides not only increased awareness among the participants but also information on where there are disagreements and knowledge gaps that may need further in-depth analysis before a decision on the most favorable strategy is made. For example, the landslide risk reduction measure needs further analysis before a decision can be taken. One important aspect is that one of the landowners is neither willing to accept, nor contribute to the intervention. A next step in the process could be to include the landowners in discussions of the proposals as well as alternative solutions, their costs, negative impacts, and benefits to achieve a common understanding of the alternatives among the landowners and the civil servants.

The application of either method illustrates that there are some aspects that benefit from the interventions and others that do not. For example, both interventions will have negative and unwanted impacts on ecosystem functions and the environment, while they will reduce the risk, which is the main aim. Therefore, further analyses and communication are needed, in which the pros and cons are valued in relation to each other. The valuing can be done through workshops, questionnaires, or web-based activities, as well as through already existing networks.

4.2.2 Networks Offer Effective Stakeholder Involvement

One important aspect of the assessment methods applied here is how to reconcile the use of expert opinions with the need for participation and involvement of stakeholders and practitioners within and outside the responsible authorities. During the study of Ljusnan–Voxnan river system and the city of Lerum several of these aspects were raised in the focus group discussions.

The results of the focus group discussions showed that well-functioning networks, the Ljusnan–Voxnan network,[4] for example, provide large benefits and a basis for effective communication. According to the respondents, well-functioning networks need a clear goal and aim, as well as continuity of concrete and relevant activities. Such activities can, for example, be to communicate with the public, politicians, and administrations.

The benefits of the Ljusnan–Voxnan network are that it has increased the knowledge and smoothness of contact routes in case of an event, increased the understanding among stakeholders, increased dialogue, exchange of experience, and knowledge transfer and thereby also competence among the stakeholders. The network has moved towards greater consensus despite the increase in number of perspectives, views, and new ideas, and a better understanding of values and valuing among the network members.

The network also resulted in increased ability to keep an overview and control (for example, on the runoff in the system), and increased preparedness and safety/security, including keeping flood risk on the agenda within the river system. Another benefit was increased cooperation and more preventive activities, including risk identification and prognoses and increased ability to achieve funding for preventive measures. The cost of maintaining the network was regarded as very small (a modest number of working days per year for each member, including 2–3 days travelling) compared to the identified benefits.

There were existing networks also in the case study municipalities. They were regarded as well-functioning but were informal and depended on the individuals involved. The internal organization and communication were deemed well-functioning within the networks, but the cooperation and communication between a larger group of stakeholders, the general public, as well as between the municipality departments, could be increased. To this end, the suggestion was to introduce and maintain routines. This would also increase the effectiveness of introducing new staff in the organizations.

5. DISCUSSION

In agreement with previous research, this study found that physical landslide preventive measures are cost-effective in built-up areas with high risk (Plate 2007; Schuster and Highland 2007; Hinkel et al. 2010; Zeng et al. 2012; Andersson-Sköld et al. 2014b). Physical measures to reduce flood risk can be cost-effective as well.

Physical preventive measures may, however, involve large-scale changes, and they can entail complex and costly investments that require a long-term perspective. They also demand a solid and broad knowledge base since such investments may increase the risk of maladaptation (Patterson and Doyle 2009). This can be a challenge, especially in areas where flood events are rare and the probability, exposure, and consequences with and without the measure may be difficult to assess.

In addition, physical measures use resources, impact the local environment, add emissions to air, soil, and water, and may alter the risks of nutrient and soil contaminant leaching. In this study, we developed the BVT and MDST methods into a merged method that provides a way to include these aspects. The BVT contributes to assessing the local-scale impacts of flooding and risk reduction, and the functionality and effectiveness (level) of the protection. The MDST contributes to a wider environmental assessment and also includes air quality in addition to the terrestrial and water focuses in BVT. In the merged version, as in both BVT and MDST, individual well-being, socioeconomic aspects, and the impacts of investment and maintenance are included. Such broad perspectives in the local planning process are of great importance in addressing urgent and serious challenges such as climate change. Otherwise we have to face a situation in which measures to reduce risks such as flooding will generate new problems as an unwanted effect and waste scarce resources.

The results of applying the MDST and the merged method indicate the importance of taking both short- and long-term perspectives into account. The time perspective illustrates not only the investment costs in relation to the monetary costs and benefits in a long-term perspective, but also the environmental and social impacts during the investment and construction phase as well as over the measure's in use time. The information is useful, not least in the weighting process as the short-term impacts may be valued as more important than the long-term perspectives. This is, for example, often the case when applying discount rates in cost-benefit analyses. Investment budgets may be restricted, and the short-term socioeconomic impacts may be severe (for example the impact on an area's attractiveness) and important at the time. In this study, we have developed a method designed to take short- and long-term perspectives into account.

Conflicts among different stakeholders could be decreased through increased transparent documentation and increased communication among different interests and stakeholders (Johansson et al. 2006; Andersson-Sköld et al. 2013). An important issue in the risk management process is to decide for whom the intervention should be effective and who should be responsible for funding. Some measures are beneficial to many and others only to an individual landowner. The municipality or landowner who may need to finance a risk reduction measure may benefit less than others in the river system or in a landslide-prone area, and for others the measure may even cause negative impacts. It is important to evaluate such impacts in the planning and risk reduction strategies.

The merged sustainability tool, based on the MDST and the BVT, should preferably be applied initially by experts as a basis for the communication and involvement of key stakeholders in the decision process. Further communication and discussions are needed to obtain stakeholders' values, needs, and views. The application of the merged method is accordingly recommended to be iterative, taking into account new options, views, and values.

The assessments in this study have been based on results from focus group discussions with experts. For more in-depth results more advanced methods such as catchment and hydraulic modelling, stakeholder and expert questionnaires, and/or in-depth interviews, as well as refined sensitivity- and weighting analyses can be applied. The merged method, as applied here, is however applicable as a structured first preliminary assessment, to reveal pros and cons.

Not least the method can be applied as a checklist and a tool to increase the understanding, learning process, and the communication among different stakeholders. However, a deeper understanding and quantification of the consequences (in a short- and long-term perspective and regarding impacts on local to larger scales) would further improve the assessment. It would also contribute to better and more robust valuations of risk reduction alternatives. This would both need more thorough investigations and follow-up of previous events and ex ante studies and research.

The current level of assessment, applying the merged method developed here, is useful in the decision and planning process as decision makers can consider the different measures in an understandable and transparent way. It also offers a method to eliminate the usual problem of taking the broader environmental aspects into account in the local management processes. Based on the structured method (checklists) a wide range of stakeholders can be involved in the process of decision making.

Communication and stakeholder involvement are, however, challenging and can be time-consuming. Networks including relevant stakeholder representatives are effective in overcoming some of those barriers. The results of the focus group discussions showed that well-functioning networks (with clear goals and aims, continuity, and relevant activities) also provide a well-functioning base for communication, not only among its members but also with external stakeholders and the general public. The results of this study indicate that such networks or organizations can be relevant in leading or facilitating the communication process.

In some cases, there may be certain additional stakeholders that need to be involved in the process such as impacted landowners. These need to be involved early in the iterative process (Figure 4).

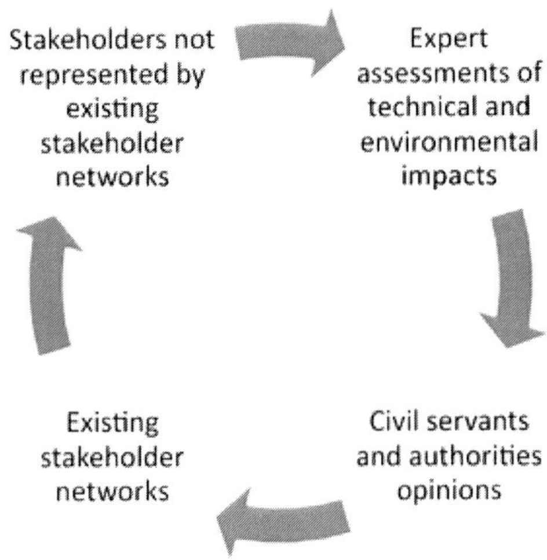

Figure 4. Iterative processes including stakeholders, civil servants/ authorities, and experts

6. CONCLUSION

The effectiveness of a flood or landslide risk reduction measure should be viewed in a wide perspective. An important question is for whom the measure should be effective and who should be responsible for funding. The merged BVT and MDST method provides a comprehensive and integrated assessment of flood and landslide risk reduction strategies, including economic, social, and local aspects as well as global environmental impacts and use of resources in a short- and long-term (future generation) perspective. The methods are to be used as checklists for discussion and as frameworks for decisions to improve the possibility of more sustainable decisions. The results of this study show that the risks of sub-optimization and maladaptation can be reduced by including many aspects in short- and long-term perspectives and involving a broad spectrum of stakeholders.

FOOTNOTES

1 Room for the River program in cooperation with UNESCO-IHE 2013. https://www.unesco-ihe.org/sites/default/files/13270-rvdr-brochure-governance-engels_def-pdf-a.pdf.

2 Swedish law act 2006:544. Law (2006:544) of municipalities and County Council actions before and during extraordinary events in peacetime and preparedness (in Swedish), Swedish Code of statute 2006:544.

3 Salmon in the stage of its first migration to the sea.

4 Further information about the existing river system networks in Sweden in general (in Swedish) can be found at https://www.msb.se/sv/Forebyggande/Naturolyckor/Oversvamning/Alvgrupper/ and http://www.uio.no/forskning/tverrfak/demokrati/aktuelt/arrangementer/konferanser/2012/papers/paper---norkom-olausson.pdf.

ACKNOWLEDGEMENTS

This work was supported by the Swedish Civil Contingencies Agency (MSB), the Swedish Government through the Centre for Natural Disaster Science (CNDS) and the Faculty of Health, Science and Technology at Karlstad University. Thanks are due to the reference group of the project, to the participants in the focus group discussions, and the two anonymous reviewers of the manuscript.

REFERENCES

1. Anderson, M.G., E. Holcombe, M. Esquivel, J. Toro, and F. Ghesquiere. 2010. The efficacy of a programme of landslide risk reduction in areas of unplanned housing in the Eastern Caribbean. *Environmental Management* 45(4): 807–821.
2. Andersson-Sköld, Y., R. Bergman, M. Johansson, E. Persson, and L. Nyberg. 2013. Landslide risk management—A brief overview and example from Sweden of current situation and climate change. *International Journal of Disaster Risk Reduction* 3: 44–61.
3. Andersson-Sköld, Y., P. Suer, R. Bergman, and H. Helgesson. 2014a. Sustainable decisions on the agenda—A decision support tool and its application on climate-change adaptation. *Local Environment* 21(1): 85–104.
4. Andersson-Sköld, Y., S. Falemo, M. Tremblay. 2014b. Development of methodology for quantitative landslide risk assessment—Example Göta river valley. *Natural Science* 6(3): 130–143.
5. Andersson-Sköld, Y., S. Thorsson, D. Rayner, F. Lindberg, S. Janhäll, A. Jonsson, U. Moback, R. Bergman, and M. Granberg. 2015. An integrated method for assessing climate-related risks and adaptation alternatives in urban areas. *Climate Risk Management* 7: 31–50.
6. Bana e Costa, C.A., J.-M. De Corte, and J.-C. Vansnick. 2003. *MACBETH. LSEOR* 03.56. London: Department of Operational Research, London School of Economics and Political Science.
7. Bana e Costa, C.A., J.-M. De Corte, and J.-C. Vansnick. 2012. MACBETH. *International Journal of Information Technology & Decision Making* 11(2): 359–387.
8. Bana e Costa, C.A., P.A. Da Silva, and F.N. Cirreia. 2004. Multicriteria evaluation of flood control measures: The case of Riberia do Livramento. *Water Resources Management* 18(3): 263–283.

9. Barnett, J., and S. O'Neill. 2010. Maladaptation. *Global Environmental Change* 20(2): 211–213.
10. Bormann, H., R. van der Krogt, L. Adriaanse, F. Ahlhorn, R. Akkermans, Y. Andersson-Sköld, C. Gerrard, N. Houtekamer, G. de Lange, A. Norrby, N. van Oostrom, and R. Sutter. 2015. Guiding regional climate adaptation in coastal areas. In *Handbook of climate change adaptation*, ed. W.L. Filho, 337–357. London: Springer.
11. Brooks, N. 2003. Vulnerability, risk and adaptation: A conceptual framework. Working paper 38. Norwich, UK: Tyndall Centre for Climate Change Research and Centre for Social and Economic Research on the Global Environment (CSERGE).
12. Dai, F.C., C.F. Lee, and Y.Y. Ngai. 2002. Landslide risk assessment and management: An overview. *Engineering Geology* 64(1): 65–87.
13. Ernst, J., B.J. Dewals, S. Detrembleur, P. Archambeau, S. Erpicum, and M. Pirotton. 2010. Micro-scale flood risk analysis based on detailed 2D hydraulic modelling and high resolution geographic data. *Natural Hazards* 55(2): 181–209.
14. Gamper, C.D., and C. Turcanu. 2009. Can public participation help managing risks from natural hazards? *Safety Science* 47(4): 522–528.
15. Glaas, E., A. Jonsson, M. Hjerpe, and Y. Andersson-Sköld. 2010. Managing climate change vulnerabilities: Formal institutions and knowledge use as determinants of adaptive capacity at the local level in Sweden. *Local Environment* 15(6): 525–539.
16. Glavovic, B.C., W.S.A. Saunders, and J.S. Becker. 2010. Land-use planning for natural hazards in New Zealand: The setting, barriers, 'burning issues' and priority actions. *Natural Hazards*54(3): 679–706.
17. Goltermann, D., and J. Marengwa. 2012. SAWA Final Report summary, Hamburg. www.sawa-project.eu/uploads/documents/SAWA_Final_Report_Summary_% 20draft 25052012.pdf. Accessed 1 Oct 2016.
18. Hinkel, J., R.J. Nicholls, A.T. Vafeidis, R.S.J. Tol, and T. Avagianou. 2010. Assessing risk of and adaptation to sea-level rise in the European Union: An application of DIVA. *Mitigation and Adaptation Strategies for Global Change* 15(7): 703–719.
19. Holcombe, E., and M. Anderson. 2010. Tackling landslide risk: Helping land use policy to reflect unplanned housing realities in the Eastern Caribbean. *Land Use Policy* 27(3): 798–800.
20. IPCC (Intergovernmental Panel on Climate Change). 2013. *The physical science basis*. Cambridge: Cambridge University Press.
21. Ismail-Zadeh, A., and K. Takeuchi. 2007. Preventive disaster management of extreme natural events. *Natural Hazards* 42(3): 459–467.
22. Johansson, A. 2013. Societal risk and safety management—Policy diffusion, management structures and perspectives at the municipal level in Sweden. PhD thesis, Karlstad University, Sweden.
23. Johansson, A., I. Svedung, and R. Andersson. 2006. Management of risks in societal planning—An analysis of scope and variety of health, safety and security issues in municipality plan documents. *Safety Science* 44(8): 675–688.
24. Johansson, M. 2008. Barriers and bridges—Communicative conditions in the Swedish environmental objective performance. Ph.D. thesis No. 469, Linköping Studies in Arts and Science, Linköping University, Sweden.
25. Jonsson, A., I. Danielsson, and A. Jöborn. 2005. Designing a multipurpose methodology for strategic environmental research—The Rönneå catchment dialogues. *AMBIO* 34(7): 489–494.
26. Jonsson, A., M. Hjerpe, Y. Andersson-Sköld, E. Glaas, K. André, and L. Simonsson. 2012. Cities' capacity to manage climate vulnerability: Experiences from participatory vulnerability assessments in the lower Göta Älv catchment, Sweden. *Local Environment* 17(6–7): 735–750.
27. Norén, V. 2016. When water becomes a threat—Risk assessment and risk management plans for floods and drinking water in Swedish practice. Ph.D thesis 1416, Uppsala University, Sweden.

28. Nyberg, L., M. Evers, M. Dahlström, and A. Pettersson. 2014. Sustainability aspects of water regulation and flood risk reduction in Lake Vänern. *Aquatic Ecosystem Health and Management* 17(4): 331–340.
29. Patterson, L., and M.W. Doyle. 2009. Assessing effectiveness of flood policy application through spatiotemporal monitoring of socioeconomic exposure. *Journal of the American Water Resources Association* 45(2): 237–252.
30. Plate, E.J. 2007. Flood risk management for setting priorities in decision making. In *Extreme hydrological events: New concepts for security*, ed. O.F. Vasiliev, P.H.A.J.M. van Gelder, E.J. Plate, and M.V. Bolgov, 21–44. The Netherlands: Springer.
31. Poussin, J.K., P. Bubeck, J.C.-H. Aerts, and P.J. Ward. 2012. Potential of semi-structural and non-structural adaptation strategies to reduce future flood risk: case study for the Meuse. *Natural Hazards and Earth System Sciences* 12: 3455–3471.
32. Renn, O. 2005. *Risk governance: Towards an integrative approach*. Geneva: International Risk Governance Council.
33. Roberts, N.J., F. Nadim, and B. Kalsnes. 2009. Quantification of vulnerability to natural hazards. *Georisk: Assessment and Management of Risk for Engineered Systems and Geohazards* 3(3): 164–173.
34. Sarewitz, D., R. Pielke Jr., and M. Keykhah. 2003. Vulnerability and risk: Some thoughts from a political and policy perspective. *Risk Analysis* 23(4): 805–810.
35. Schuster, R.L., and L.M. Highland. 2007. The third Hans Cloos lecture. Urban landslides: Socioeconomic impacts and overview of mitigative strategies. *Bulletin of Engineering Geology and the Environment* 66(1): 1–27.
36. Sharma, U., A. Scolobig, and A. Patt. 2012. The effects of decentralization on the production and use of risk assessment: Insights from landslide management in India and Italy. *Natural Hazards* 64(2): 1357–1371.
37. Singh, A.K. 2010. Landslide management: Concept and philosophy. *Disaster Prevention and Management: An International Journal* 19(1): 119–134.
38. Srivastava, R., and L. Laurian. 2006. Natural hazard mitigation in local comprehensive plans: The case of flood, wildfire and drought planning in Arizona. *Disaster Prevention and Management: An International Journal* 15(3): 461–483.
39. Suh, J., Y. Choi, T.D. Roh, H.J. Lee, and H.D. Park. 2011. National-scale assessment of landslide susceptibility to rank the vulnerability to failure of rock-cut slopes along expressways in Korea. *Environmental Earth Sciences* 63(3): 619–632.
40. UNISDR (United Nations International Strategy for Disaster Reduction). 2005. *Hyogo framework for action 2005–2015: Building the resilience of nations and communities to disasters*. Geneva: UNISDR.
41. UNISDR (United Nations International Strategy for Disaster Reduction). 2015. *Sendai framework for disaster risk reduction 2015–2030*. Geneva: UNISDR.
42. Volchko, Y., J. Norrman, L. Rosén, M. Bergknut, T. Söderqvist, T. Norberg, S. Josefsson, K. Wiberg, and M. Tysklind. 2014. Using soil function evaluation in multi criteria decision analysis for sustainability appraisal of remediation alternatives. *Science of the Total Environment* 485–486: 785–791.
43. Zeng, J., Z.Y. Zhu, J.L. Zhang, T.P. Ouyang, S.F. Qiu, Y. Zou, and T. Zeng. 2012. Social vulnerability assessment of natural hazards on county-scale using high spatial resolution satellite imagery: A case study in the Luogang district of Guangzhou, South China. *Environmental Earth Sciences* 65(1): 173–182.

CHAPTER 12

Study and Analysis of Chennai Flood 2015 Using GIS and Multicriteria Technique

Muthusamy Seenirajan[1], Muthusamy Natarajan[2], Ramasamy Thangaraj[3], Murugesan Bagyaraj[4]

[1]*Civil Engineering, Excel Engineering College, Namakkal, India*

[2]*Civil Engineering, Karpagam University, Coimbatore, India*

[3]*Civil Engineering, Nehru Institute of Technology, Coimbatore, India*

[4]*Gis Analyst Aadhi Boomi Mining Envirotech pvt. Ltd, Salem, India*

ABSTRACT

Terrain characteristics of the land and meteorological properties of the region are the main natural factors for flood. The recent flood in Chennai was unexpected and not triggered by the above factors. Sometimes floods occur when the watershed size is considerably small which leads to the over flow of water inland may due to the encroachment and the urban development of the city. Temporarily used backwater effects in sewers and local drainage channels and creation of unsanitary conditions may cause flooding. Chennai flood was basically claimed to occur due to improper drainage system and underlying strata which was found to be landfill over the ponds and lakes. The Coouam River which flows through the centre of main city was found silting due to the improper drainage facilities and encroachment by the local peoples who causes flood. For the analysis of potentially affected areas Geographical Information System (GIS) integrated with Multicriteria Decision Analysis (MCDA) were employed. Ranking and displaying the potentially risky areas, the spatial Multicriteria analysis was used. It has been revealed that all most all the area's having populations are likely to be exposed to flood hazard. At the end of study, a map of flood risk areas was generated and studied with a view to assisting decision makers on the consequences posed by the disaster.

Keywords: Geographical Information System, Multi-Criteria Decision Analysis, Flood Risk, Flood Potential Map

1. INTRODUCTION

South India experienced heavy rains during the month of November and December, 2015 due to the formation of depression over southwest Bay of Bengal and owing to a strong El Niño. Heavy widespread incessant rainfall activity led to waterlogging and flood like condition in parts of Karnataka, coastal Andhra Pradesh and Tamil Nadu. Tamil Nadu was one of the worst affected states where rainfall in all but two of the 34 districts in the state has exceeded the normally observed levels, while five districts have been pounded by over twice the average rainfall received between October and the end of November, 2015. Chennai in particular was one of the worst hit districts in the state which received almost 1200 mm of rain in November, which is nearly 300 per cent above the normal observation of 407.4 mm. While rainfall in the city broke the nearly century-old record for November, the record for the highest rainfall received in a single day in December was also rewritten. On December 1 - 2, 2015, the Indian city of Chennai received more rainfall in 24 hours than it had seen on any day since 1901. The previous record for rainfall in a single day in December dates back to 1901 when the city received 261.6 mm of rainfall in a span of 24 hours. On December 1, 2015, the city received around 290 mm of rain in a single day causing catastrophic flood disaster. The deluge followed a month of persistent monsoon rains that were already well above normal for the Indian state of Tamil Nadu.

Tamil Nadu, especially Chennai city was terribly hit by the flood during November-December 2015 which claimed more than 400 lives and caused enormous damages, both domestically and economically. Mainly the districts nearby the coast of Chennai, Kanchipuram and Tiruvallur were the most affected. This has given a challenge to the scientific community in developing a comprehensive understanding of the event. Floods are probably the most recurring, wide spread, disastrous and frequent natural hazards of the world [1] . There has been a great confusion and many questions were raised in the minds of people due to the immediate havoc caused by the flood in Chennai. The possible causes for the flood have been identified like heavy rainfall, improper drainage system to drain out the runoff water, soil character, watershed of the river etc, In view of potential extreme climate events, we assert that there is an urgent need to have high resolution data in order to deepen research about the risk in the Chennai city [2] .

But the cause for the Chennai flood is quite different than other usual floods. It was found out that Chennai flood was due to the high intensity rainfall caused by the unfavourable atmospheric condition, rainfall distributed spatially and temporally, quantity of flow occurred in the rivers namely Kosasthalaiyar river, Cooum river and Adyar river.

The operation of the two reservoirs in the upstream of the city were the Chembarambakkam reservoir and the Poondi reservoir, The amount of flow generated in the city is due to rainfall over the city alone, manner of the storm water drainage system responds, areas where the city were inundated, change of topographical features in the city over the years exacerbate the flooding. By global warming and due to the environmental changes the seasonal rainfall fails and causes extreme climatical changes results in cyclone effects causes flood.

Flood is one of the common hydrological phenomena which is to a large extent unpredictable and uncontrollable [3] . More than 1/3rd of the world's land area is flood prone affecting some 82% of the world's population [4] . As per UNDP [5] approximately 170,000 deaths were associated with floods worldwide between 1980 and 2000 [6] .

1.1. Area Selected for Study

Chennai, capital of Tamil Nadu, one of the south most states of India had been selected for the case study on flood. Basically the two coastal districts of Chennai, namely Kanchipuram and Tiruvallur are badly affected by the November-December 2016 flood. These two districts are located at the northeast part of Chennai city (Figure 1). Kanchipuram's latitudinal location is between 12°20'N to 13°00'N and longitudinal location is 79°40'E to 80°20'E. Kanchipuram district occupies the total area of 4433 Sq. km., and Tiruvallur at 80.0088°E and 13.2544°N which is located near by the Chennai city. Terrain characteristics of land and meteorological properties were not considered to be the main factors for causing flood disaster in Chennai. There have been tricky and unexpected reasons that caused flood in Chennai.

These causes will be discussed in detail below. The normal temperature of ranges from 26 to 35 celsius in Kanchipuram district and it receives an average rainfall of 1200 mm. Similarly, the average normal rainfall of Tiruvallur district is 1104 mm.

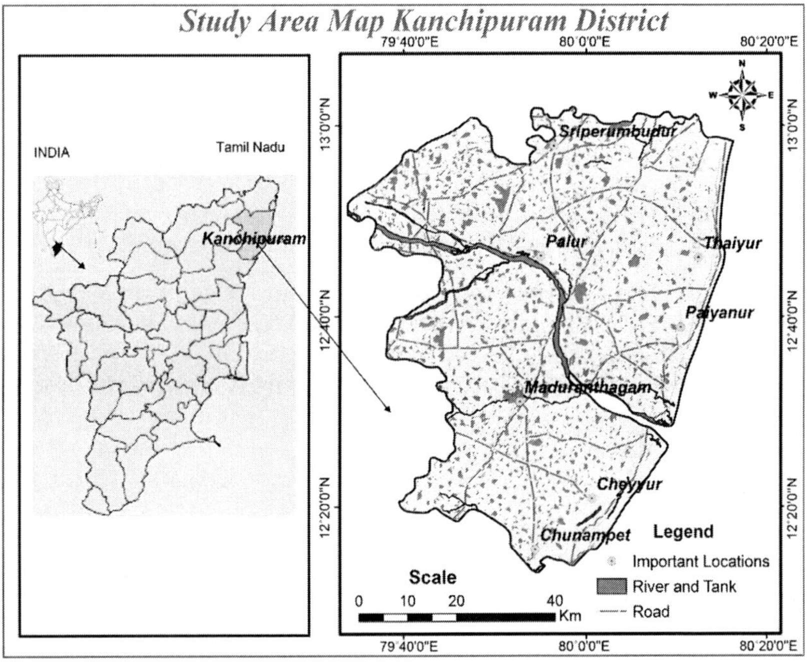

Figure 1. The study area map.

1.2. Materials and Methods

For present work Geo-coded Indian Remote Sensing (Cartosate-1 SRTM) satellite image of April 2005 and topo sheets at 1:50,000 scales collected from Survey of India, used to prepared contour and Drainage Map. Global positioning system (GPS) was used for carrying out field surveys were conducted in Kanchiipuram District flood affected areas. Geological map collected from GSI, Soil map prepared from Soil survey of India and the area rainfall collected in Public Work Department (PWD) Chennai

All the data collect and prepare different thematic map using GIS software finally integrated all maps and give the output for flood hazard zone. The detailed methodology flow chart given in Figure 2.

1.3. Data Source

The principle supporting the data for this study was provided by the NASA which was satellite-based estimation showing rainfall over the south-eastern India on December 1 - 2, accumulating in 30 minutes interval. The integrated Multi-satellite retrievals for GPM (IMERG), a product of Global Precipitation Measurement mission is used to fetch the rainfall precipitation data. On the maps (Figure 3) below represent rainfall total approaching 400 millimetres (16 inches) during the 48 hours period. These are the two regions in Tamil Nadu which received the heaviest rainfall and experienced the maximum damages due to November-December 2015 flood. According to HAL pierce, a scientist on the GPM team at NASA, areas just off the south-eastern coast received the maximum total rainfall that exceeded 500 mm (20 inches).

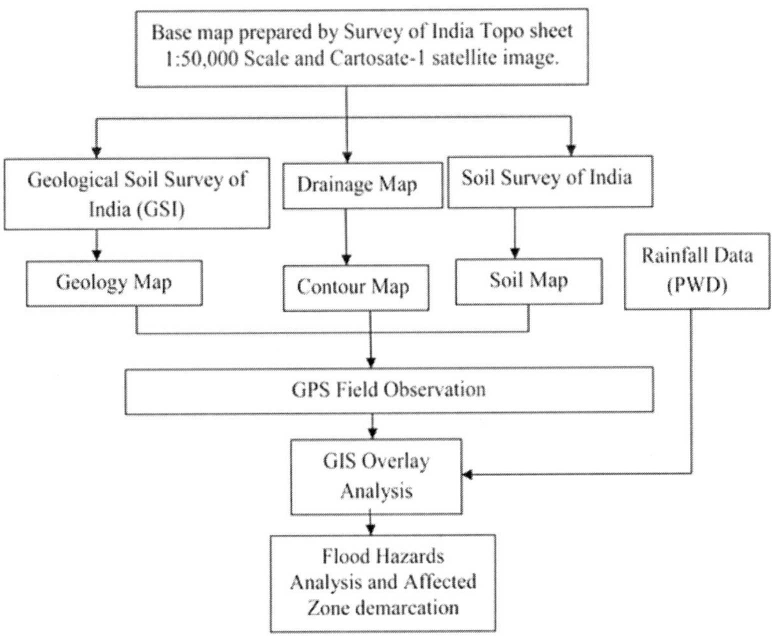

Figure 2. Detailed methodology flow chart.

1.3.1. Geographical and Meteorological Reasons

Between the month of October and December of each year, an enormous area of south India, including Tamil Nadu, receives up to 30percent of its annual rainfall from the northeast monsoon. The northeast monsoon is due to the annual gradual retreat of monsoonal rains from north-eastern India. Unlike during the northeast monsoon, rainfall during the monsoon is sporadic, but the rainfall during November-December 2015 was typically more than the normal rainfall by 90 percent.

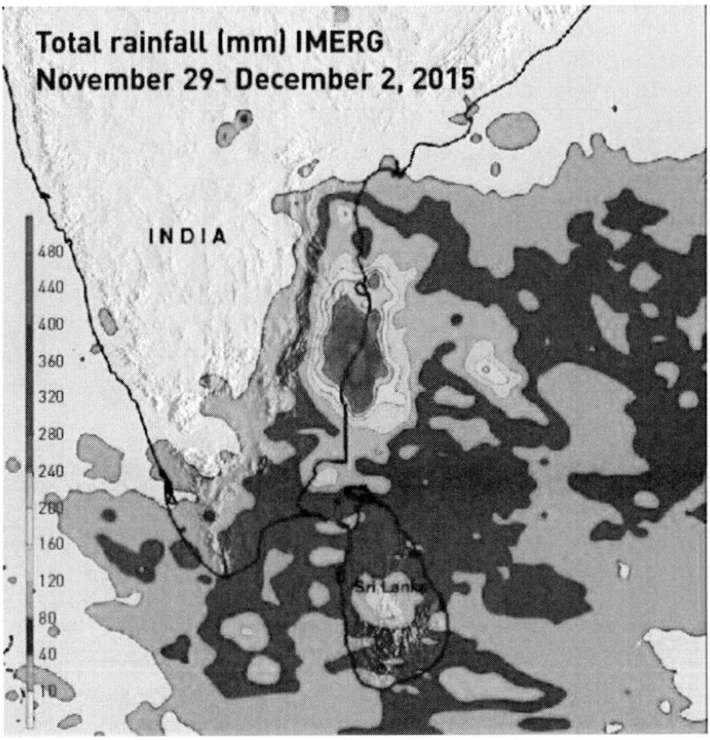

Figure 3. Accumulated rainfalls between November 29 to December 2 over Chennai (Sources NASA).

The coastal districts of Andra Pradesh usually bear a burnt of heavy rains that occur during the northeast monsoon along with Puducherry and eastern Tamil Nadu which becomes prone to flooding with numbers of swelling rivers and wetlands. Between the years from 1943 to 2005 Chennai city had experienced five major floods alone causing particularly severe damage in the year of 1943 and 2005. In addition, illegally and unplanned urban development has caused many wetlands and natural sinks being built over artificially. Along with that, poorly designed drainage systems and aging civil infrastructures, has resulted in an increase frequency of flooding.

1.3.2. Annual Rainfall

Chennai receives the average annual rainfall of about 140 cm (55 in). Starting from mid-October to mid-December the city receives the heaviest rainfall from the northeast monsoon winds. Sporadically the city experiences cyclones formed in the Bay of Bengal. The highest annual rainfall recorded is 257 cm (101 in) in 2005. The highest rainfall of 72.4 mm was recorded in Tiruvelveli-Tamil Nadu.

Tamil Nadu received an average annual rainfall of 1304.1 mm during the year 2005-06. The rainfall was excess in 26 districts and normal in 4 districts as shown by the rainfall records during 2005 and 2006 as compared to the normal annual rainfall. However, the highest recorded rainfall of 2005-06 in Chennai had been beaten by the recent 2015 flood.

GIS system navigated by the NASA was used to view over the south-eastern part of India, Chennai. According to the record, continuing from Tuesday till Wednesday more quantity of moisture started to accumulate over the region of Chennai. These have a serene for heavy precipitation and high chances of flood.

In our work an automated procedure was used to identify the flood plain extent and to determine its depths. But this automated procedure is different from the mentioned above in the way that water levels are obtained. We do not create a water surface level in a raster format and then compare that water surface level with the DTM to delineate flood plains. Rather, the floodplains are directly delineated from the high-resolution, digital terrain model (DTM), LIDAR 20-ft digital elevation model (DEM) was used as simplified, but practical approach, which was developed based on the actual mechanism of flooding.

1.4. Drainage System

The City is drained by 2 rivers the Adyar and the Cooum Rivers, besides a number of major and minor drains through Buckingham Canal into Sea via Ennore Creek and Kovalam Creek. Cooum runs through the heart of the city whereas Adyar wends its way through the southern part of the city before entering into the sea.

Though the river Adyar can be traced to a point near Guduvancheri village, it assumes the appearance of a stream only after it receives the surplus water from the Chembarambakkam tank as wells as the drainage of the areas in the south-west of Chennai. Cooum River starts from Kesavaram Anicut in Kesavaram village built across Kortaliyar River.

The surplus from Cooum tank joins this course at about 8 kms. lower down and this point is actually the head of Cooum River which is located at 48 kms. west of Chennai. The drainage map of the study area is shown in Figure 4.

As the water level crossed the normal limits in residential areas of Kanchipuram and Tiruvallur districts, the drainage system measurably failed to pass the water. (Figure 4) shows the drainage system of Chennai.

The drainage system was blocked due to excessive dumping of garbage and as well as the failure of administration to ensure periodic desilting. Hence water couldn't find way to flow. The failure of drainage system in Chennai and other parts of Tamil Nadu, especially Kanchipuram and Tiruvallur, made the situation worst. Besides that, encroachment was seen on Cooum River, Adyar River and Buckingham Canal, which serves as the main rain water drain for the city. These encroachments were not slump dwellings but concrete directly affecting flow of canal.

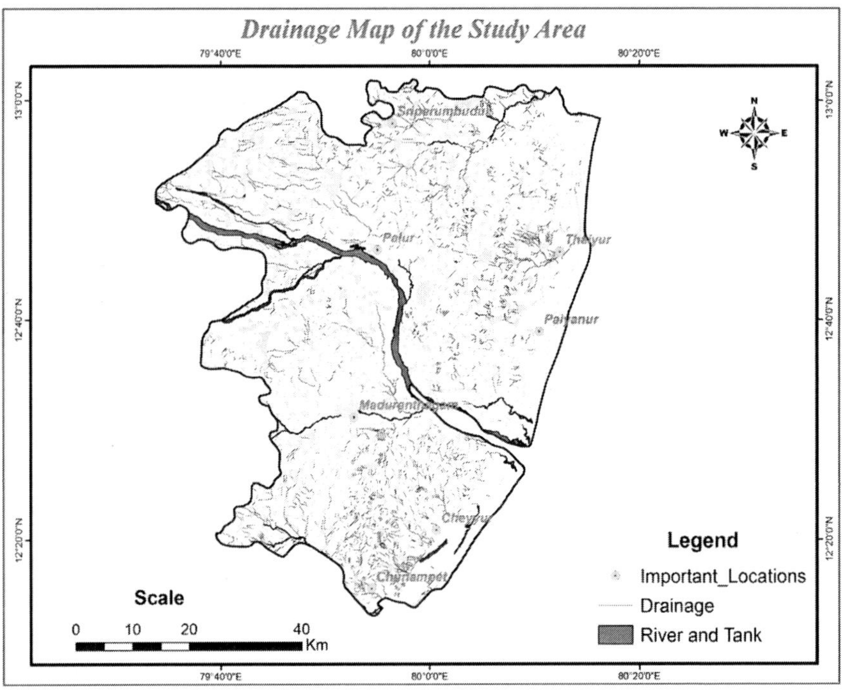

Figure 4. Drainage map of the study area.

In addition to that, the lack of wetland, which acts as a sponge, soaking of rainwater, played vital role in floods. According to a report of leading daily, over 5500 hectares of wetlands in those two districts have been evolved into commercial lands and only 10 percent of the original wetlands remained. Hence, rainwater runoff has nowhere to go and settles instead on to road, causing flooding. This severe flood disaster was caused due to mismanagement and violation of protocols during urban planning.

1.5. Types of Soils

Chennai's soil is mostly clay, sandstone and shale. Areas found along the coasts and the river banks are sandy in nature and in these areas, run water percolates quickly thoroughly the soil. Few parts of Chennai also comprises hard rock surface. The ground water table in Chennai is 4 - 5 m below the ground surface.

Among many soil test apparatus few have been employed to test the soil in Chennai, especially soil moisture test, triaxial soil testing instruments are the few among many which determined the types of soils available in Chennai. The soil map shown in Figure 5.

1.6. Geology of the Study Area

The geological formations are beach sands of quaternary and recent period, Cuddalore sandstone of Mio-pliocene age, shals and sandstone of Upper Gondwannas and charnockits of Archaean era. The detailed geology information shown in Figure 6.

1.7. Slope and Size of Watershed

There are numbers of watershed in Chennai alone. Some of the well known water shed are Velachery, Adyar, Virugambakkam and few other canals. Watersheds range from 10.5 km to 50.99 km and it bears a slope of 1 in 20 which indicates that if the volume of water rises suddenly in the river basin then there will be high chances of flood occurrence due to overflow.

1.8. Spatial Analysis

Rainfall is a continuous phenomenon that affects the whole region, not just the locations of the weather stations. Based on the observed rainfall at the monitoring stations and their locations, we can interpolate and deduce the approximate rainfall across the whole region. We use the Interpolate Points tool from the GIS's spatial analysis service for this study.

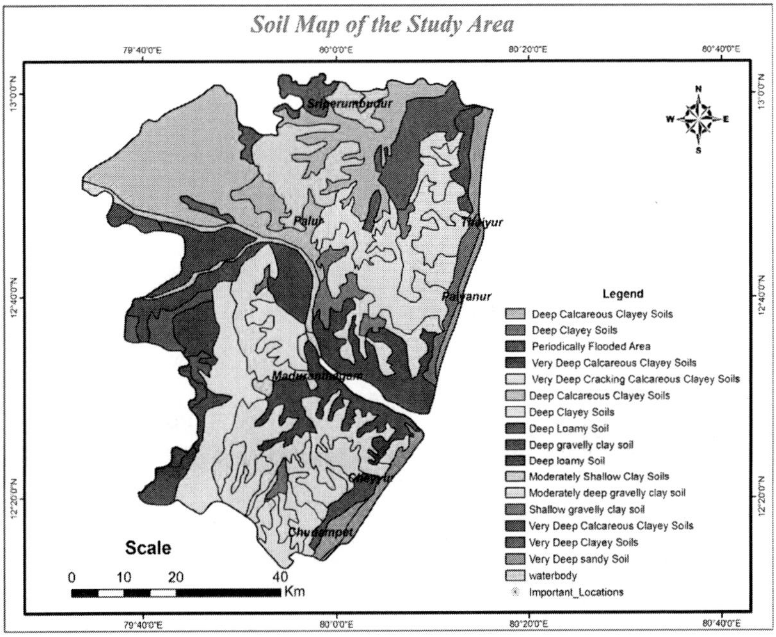

Figure 5. Soil map of the study area.

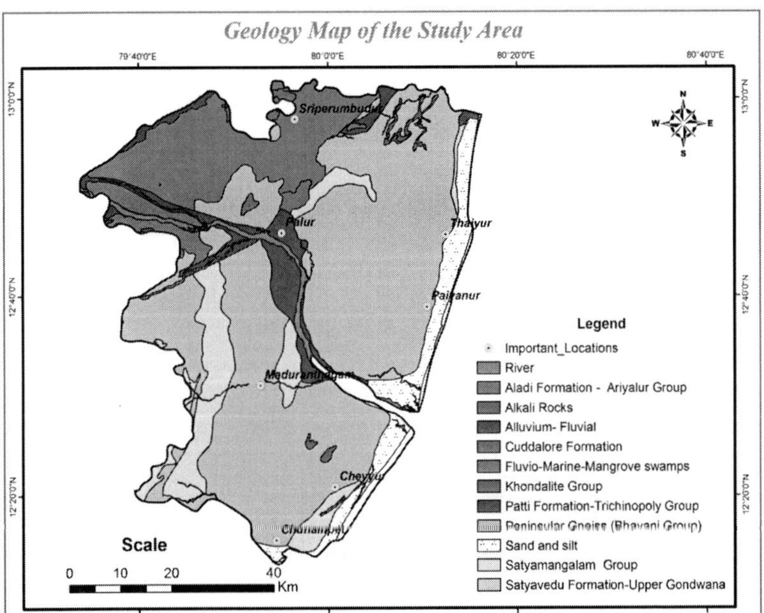

Figure 6. Geology map of the study area.

1.8.1. Pair Wise Comparison Method for Chennai Flood

This method is usually used for the comparison of two criteria at a time. Conversion of subjective assessments of relative importance into linear sets of weights is taken into consideration in this method. This method estimates the weight of the following criteria:

C1 = Geology

C2 = Rainfall (precipitation);

C3 = Drainage network;

C4 = Soil type.

The square pair-wise comparison matrix is presented in Table 1. Each factor was weighted according to the estimated significance for the flood potential project which was to generate the criterion values for each evaluation unit. Table 2 shows the normalized matrix. At the same time, the individual observed record, which never agreed perfectly with the degree of consistency achieved in the ratings were randomly generated. The random indices for the matrices are listed in Table 3. According to the rule thumb, CR less than or equal to 0.1 indicates an acceptable reciprocal matrix, whereas a ratio over 0.1 symbolises that the matrix should be revised.

Table 1. Pair wise comparison matrix for flood risk parameters.

	C1	C2	C3	C4
C1	1.1	2.3	2.1	2.5
C2	0.5	1	2.3	0.5
C3	0.5	0.5	1	0.5
C4	0.5	2	2.1	1.7
Total	2.6	5.8	7.5	5.2

Table 2. Normalize matrix.

	C1	C2	C3	C4	C5
C1	0.4	O.36	0.29	0.5	0.387337662
C2	0.2	0.18	0.29	0.125	0.198133117
C3	0.2	0.09	0.14	0.125	0.139691558
C4	0.2	0.36	0.29	0.25	0.274837662
Total	1	1	1	1	1

Table 3. Random indices for matrices of various sizes (n).

N	R1
1	0.00
2	0.oo
3	0.59
4	0.97
5	1.13
6	1.25
7	1.34
8	1.41
9	1.20
10	1.50
11	1.53
12	1.49
13	1.57
14	1.60
15	1.65

2. RESULT AND DISCUSSION

2.1. Multi Criteria Analysis

Applied and integrated with the spatial data, the causative factors of a phenomenon under concern of Multicriteria analysis, is described. In this study, the risky areas were first produced by numerically overlaying soil, drainage network, geology, rainfall layers and size of the watershed. The selection of these criteria was based on the expert's opinion and availability of data. Boolean overlay was employed to carry out this overlay. Logical operators such as intersection and union combined all criteria for analysis.

In the second phase, ranking method was used, where every criterion under consideration was ranked in the order of the decision maker's preference. Each factor was weighed according to the estimated significance for causing flooding. Factor of rank 1 is the least important and factor of rank 8 is of the most importance. In the third phase, determination of the weight of each criterion was done by pair wise comparison method which was developed to create flood hazard map.

2.2. Flood Risk Map

Evaluation of each unit based on ranking method was generated using the criterion maps combined by logical operations and criterion values. Figure 7 shows the flood risk map created based on GIS and Multicriteria method. Using pair wise comparison the normalized criterion weights were calculated as 0.198, 0.387, 0.275 and 0.14 respectively for basin slope, annual rainfall, drainage network of the river basin and soil type. The study carried out based on this method showed a consistency ratio (CR) value of 0.0, which fell much below the threshold value of 0.1 which indicates a high level of consistency. Therefore, the weights are acceptable.

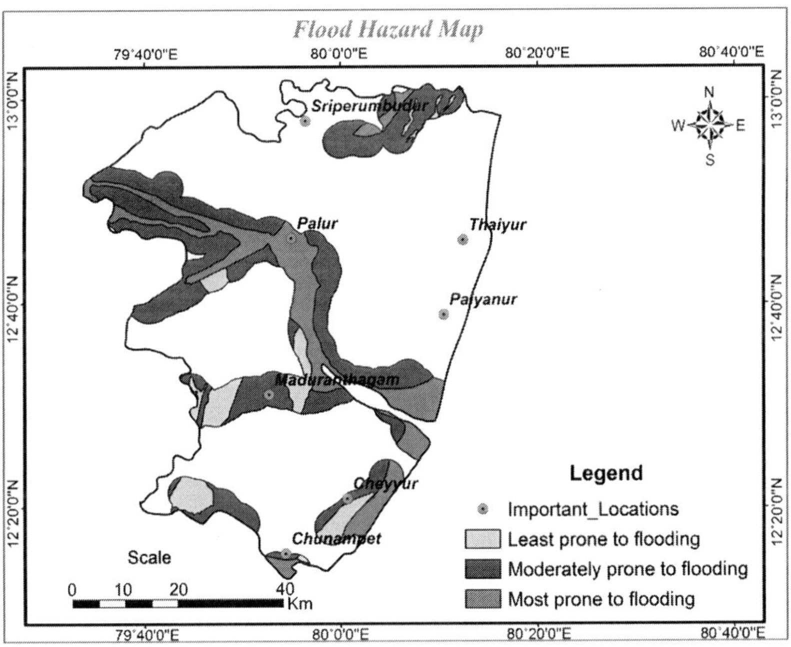

Figure 7. Flood Hazard map.

2.3. Flood Map Validation

The flood potential map which resulted from multi criteria analysis was compared with the original flood map of 2005/2006 obtained from DID in Chennai for the purpose of validation. The original flood potential map (Figure 7) which was classified based on flood vulnerability:

1) 4 for the most prone to flooding;
2) 3 for the moderately prone to flooding;
3) 2 for the least prone to flooding;

All the features which had been designed in grid code values of feature map from the raster pixel values are show in the map below. Correspondingly individual polygon has one grid value (i.e., 1, 2, and 3) based on the cell value at the same location in the raster. This step helped to design four polygon features with each polygon having different flood suitability level. The flood classes which was obtained was converted into external shale file (*.shp) and the external image was compared with the original flood map.

The final extracted flood map which contained the data and features of original and potential flood maps had been analysed and affected villages list out Table 4. The above respective three classes indicate the most prone to flooding area, moderately prone to flooding and least prone to flooding. The final flooding hazard area with village overlay map shown in Figure 7 and Figure 8.

3. CONCLUSIONS

From the field experience and also interaction with local population it has been observed that the urbanization and encroachments of river banks, marshy, low-lying areas especially Adyar River has aggravated the flooding problem. The narrow and constrained river with no flood plains left could not carry the discharges and water simply occupied the adjoining low-lying areas.

A study carried out by Chandan et al. (2014) on analysis of land use change pattern in past 4 decades shows that the total urban area has been increased by more than 20 times mainly from the conversion of grazing, agricultural and open areas to urban impervious surface. They have observed that Vegetation cover has dramatically decreased from 70.47% in 1991 to 35.53% in 2013, whereas the non-vegetation i.e. built up, paved areas etc. have increased 29.53% in 1991 to 64.47% in 2013.

Management of reservoirs, revival of wetlands which act as a "sponge" to absorb the excess water and resizing of the storm water drains keeping in account the present and projected population should be taken as mitigative measures.

Table 4. Most affected zones/places by Chennai flood.

Sl.NO	Affected places/zones
1	Thiruppukuzhi
2	Melambi
3	Putheri
4	Kancheepuram (M)
5	Nattapettai (CT)
6	Erivoy
7	Thenambakkam (CT)
8	Valluvapakkam
9	Palayaseevaram
10	Thangi
11	Palur
12	Kilottivakkam
13	Puliyambakkam
14	Angambakkam
15	Pinayur
16	Villiambakkam
17	Kambarajapuram
18	Seethananjeri
19	Kurumanjeri
20	Sathananjeri
21	Elayanarvelur

Based on 2015 floods experience flood hazard plain zoning needs to be done for future expansion of the city and assessing the threat to habitation living in various municipal zones for better preparedness.

4. RECOMMENDATIONS

After the flood the sense of fear had entered into the mind of people. This forced the experts and scientists to reason out the causes and factors of the flood. After that the suitable and valid remedies should be fetched. Some of them are recommended below:

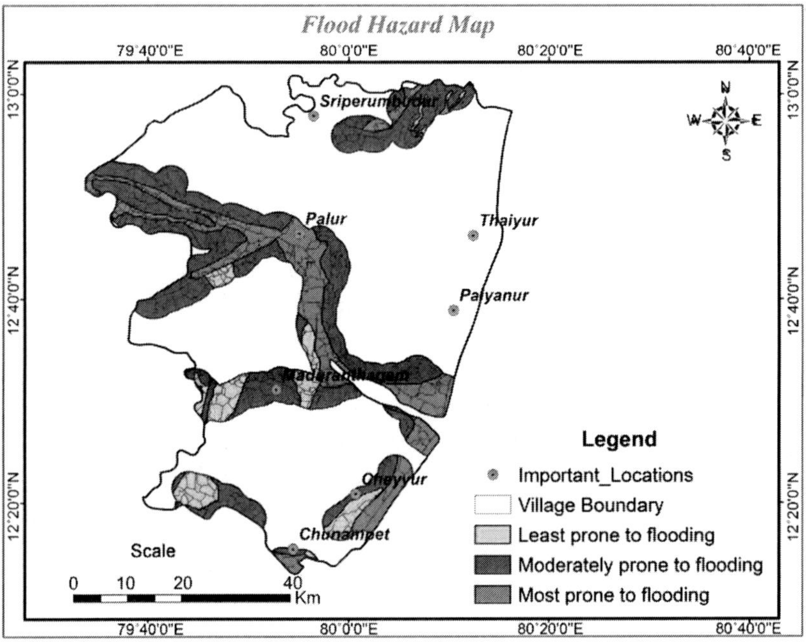

Figure 8. Flood Hazard map with village overlay.

1) Better understanding of the weather

It is always better to understand the weather condition in advance and the data collected should be of great accuracy so that there won't be any havoc created afterwards. By doing so, people can act on the disastrous action easily.

2) Water wiring of cities

Every small water bodies should be taken into consideration along with the large and well known rivers, reservoirs and canals. Only y doing so we would not missed out any small bit of information and thus we could edify the problems.

3) Maintenance of watershed and drainage systems

Watershed should be maintained very well. Although it seems of less important, watershed plays a vital role in controlling the overflow of water. Similarly, drainage system should be checked and kept clear of obstacles. All the channels of drains should be connected properly and the outlet should be properly maintained.

4) Improvements of civil infrastructure and wetlands

Most of the underlying ground soil of Chennai is the refill of ponds and lakes. These soils act as a poor filter for the rainwater. Thus, all the rain water gets collected on the surface which causes floods. This can be avoided by improving the soil permeability and building proper civil infrastructure will allow the water to pass through proper channels.

5) Human activities

One of the factors which contributes to flood or any other natural disasters is human's selfish activities. Burning of harmful gases and fuels, raising the numbers of industries, clearing the forest cover for constructions purposes etc. have dramatically changed the climatic conditions over the decades. Therefore, human needs to reduce their wants and try to live in harmony using the nature wisely.

REFERENCES

1. Sathish, S., Nagendra, H.N. and Ravi, G. (2012) Application of Remote Sensing and GIS for Flood Risk Analysis: A Case Study of Krishna and Tungabadra River Valley. Internal Journal of Social science & Interdisciplinary Research, 1, 50-61.
2. Idelbert Dagbegnon Behanzin (2015) GIS-Based Mapping of Flood Vulnerability and Risk in the Benin Niger River Valley. International journal of Geomatics and Geosciences, 6, 1653-1669.
3. Dhar, O.N. and Nandargi, S. (2003) Hydrometeorological Aspects of Floods. Natural Hazards, 28, 1-33.
4. Dilley, M., Chen, R.S., Deichmann, U., Lerner-Lam, A.L., Arnold, M., Agwe, J., Buys, P., Kjekstad, O., Lyon, B. and Yetman, G. (2005) Natural Disaster Hotspots: A Global Risk Analysis. Washington, D.C: International Bank for Reconstruction and Development/the World Bank and Columbia University. Washington, DC.
5. U.N.D.P. (2004) United Nations Development Programme, Bureau for Crisis Prevention and Recovery. New York, USA. p. 146.
6. Machado-Mosquera, S. and Ahmad, S. (2007) Flood Hazard Assessment of Atrato River in Colombia. Water Resources Management, 21, 591-609.

CHAPTER 13

The Sustainable Procedure Framework for Disaster Risk Management: Illustrated by the Case of the EU Floods Directive in Sweden

Beatrice Hedelin[1]

[1] Department of Environmental and Life Sciences, Centre for Climate and Safety, Karlstad University, 651 88 Karlstad, Sweden

ABSTRACT

How can the concrete meaning of the ambiguous and theoretical concept of sustainable development (SD) be defined and implemented, without losing sight of its fundamental principles? This study introduces a theoretical framework that supports studies of SD implementation in the context of strategic disaster risk management, by defining what SD implies with regard to planning procedures. The framework is based on the procedural SD principles of participation and integration. It was originally developed for, and has shown great value in, the field of water resource management. In-depth interviews with senior risk management researchers indicate that the framework is also applicable to and valuable for disaster risk management studies. To illustrate the application of the framework, a study of the EU Floods Directive in Sweden is summarized with the framework as the basis for the analysis.

Keywords: Disaster risk management, EU Floods, Directive Planning, Sustainable development, Sustainable Procedure, Framework Sweden

1. INTRODUCTION

Sustainable development (SD) implies integration of social, ecological, and economic concerns (Robinson 2004; Ginson 2006; Levin et al. 2013). This can be viewed as the "substantive" dimension of SD (Robinson 2004). In order to accomplish this integration, however, considering the "procedural" dimension of SD is an equally important imperative[1] (Robinson 2004). This procedural dimension of SD, and its connection to the field of disaster risk management (DRM), is the focus of this article, specifically with respect to preventive and strategic DRM, such as in regional flood risk planning and municipal land use planning.[2] A sustainable procedure can be described as a political conversation about desirable futures informed by scientific knowledge from a broad range of integrated disciplines (Bradshaw and Bekoff 2001; Clarke and Dickson 2003; McMichael et al. 2003). Such a procedure also needs to be informed by the knowledge and perspectives of the actors who are affected in different ways by the plan or decision to be made (Pretty 2003; Ostrom 2009; Vaidya and Mayer 2014).

In the last 10 years the importance of integrative and participatory planning procedures has become recognized in the field of DRM (Pelling 2007; Evers 2008; Ikeda et al. 2008; Vinet 2008; Gamper and Turcanu 2009; Wisner 2011; Renn 2015). In natural resources management (NRM), where similar considerations have been well established, a theoretical framework that describes what an integrative and participatory procedure implies for strategic planning procedures has been developed (Hedelin 2007). The framework takes the SD principles of participation and integration as its point of departure. It has been successfully applied and has delivered practical recommendations for how to develop planning procedures and legislation in relation to SD (Hedelin 2008b; Hedelin and Lindh 2008).

The idea of this study is to assess, and possibly adjust, the NRM framework in relation to the field of strategic DRM, with a view to making it applicable to DRM studies. While there are already a number of important efforts to establish theoretical basis and frameworks for linking DRM and SD—see for example Becker (2014) and Nyberg et al. (2014)—the NRM framework would contribute to the DRM field in two main ways: its ability to systematically and explicitly relate DRM practices to the theoretical and ambiguous SD concept, and its procedural approach to SD. These points are explained in more detail in Sect. 2.

The theoretical and general character of the NRM framework, and the fact that it concerns procedure, suggest transferability to the DRM field. But important differences between the fields could affect the applicability of the framework. One difference is the character of the bio-geophysical processes of natural hazards, which are both more uncertain and faster than most natural resources processes—for example, compare a flooding event with eutrophication or the long-term depletion of an aquifer. Another difference is related to participation. Disasters may engage the public more than natural resource issues because the consequences of disasters and response measures often affect the public more directly. Compare the flooded basement of a private house with the slow deterioration of an ecosystem. Another difference is that, in contrast to the NRM field, very few nongovernmental organizations (NGOs) are dedicated to disaster risk issues. These circumstances have consequences for how to effectively design planning procedures.

This study aims to contribute to the research on sustainable DRM procedures by introducing the sustainable procedures framework, focusing on participatory and integrative procedures. As a first step, the objective here is to assess whether the NRM framework contains any inconsistencies in relation to strategic DRM procedures, and, if inconsistencies can be identified, to adjust the framework in relation to the DRM field. Finally, to illustrate the application of the framework, the implementation of the EU Floods Directive in Sweden is used as a case study and analyzed on the basis of the framework.

2. THE SUSTAINABLE PROCEDURE FRAMEWORK

The Sustainable Procedure Framework and its key ideas are summarized here (Table 1; for a detailed explanation of the framework and how it was developed, see Hedelin 2007). The framework was originally developed for natural resource management, such as river basin management and municipal land use planning. It is intended for research-based analysis and development of planning procedures of strategic character at national, regional, and municipal levels.

The framework can be seen as a response to the need to establish ways to explicitly and systematically relate practical planning procedures to the concept of SD. The difficulty of relating a specific practice to the abstract and theoretical concept of SD is well recognized (Robinson 2004). The framework's approach is to use two SD-principles—participation and integration—as the points of departure.

Table 1. Overview of the sustainable procedure framework (for explanation of the criteria, see Hedelin 2007)

Sustainability principle	Structural component	Criteria Sustainable planning procedures must include, support, or promote:
Integration	…across disciplines	A Integration of knowledge from all relevant disciplines
		B Handling of different views of knowledge (for example, positivist, relativist)
		C Handling of different kinds of uncertainty
	…across values	D Identification of the most relevant values in relation to the current issue
		E Rational argumentation based on the identified values, by relating them to alternative choices in the planning process
Participation	…contributing to the process	F Inclusion of knowledge owned by relevant actors
		G Inclusion of the ideological orientations represented by relevant actors
		H Participation in the most critical phase(s) of the process
	…generating commitment, legitimacy, or acceptance	I A procedure for defining the actors who should be involved
		J Handling of power asymmetries
		K Procedures that ensure that ideological orientations are not suppressed
		L Learning

These are well-established procedural principles of SD[3]and by far the most cited SD procedural principles recognized in the NRM field (see for example Gregersen et al. 2007; Sawhney et al. 2007; Henriksen et al. 2009). The principles still say little about how to design a planning process, however. Therefore, a set of criteria has been derived that describe necessary constituents of an integrative and participatory procedure. While the criteria do not describe a specific optimal design for a NRM process (specific design is context dependent), they say more than the SD principles about what a sustainable NRM procedure implies. The criteria are derived from the review and synthesis[4] of theoretical and empirical literature on what integration and participation implies in relation to strategic planning procedures. The scope of the relevant literature is wide and includes multiple levels of complexity. As a consequence, the criteria of the framework are still relatively general in character. A high level of generality is required in order to keep the framework applicable to a wide range of NRM cases.

One important contribution of the framework is its interdisciplinary character. The framework establishes a structure of integrated and composed scientific knowledge, systematized in relation to the concept of SD. In earlier studies the use of the framework has produced practical recommendations for how to make the studied planning cases better in relation to SD (Hedelin 2008b; Hedelin and Lindh 2008). In addition to analysis, another way to use the framework is as a baseline for transdisciplinary development of planning procedures.[5]

In short, criteria A to E stem from the concept of integration and are influenced by research on integrated planning and management (Born and Sonzogni 1995; Bellamy et al. 1999; Margerum 1999; Sneddon et al. 2002). They are structured based on the idea that integration can be obtained across disciplines (A–C) and across values (D and E) (Jepson 2001). Criterion A says that disciplinary knowledge must be integrated in the planning process (Lubchenco 1998; Ludwig et al. 2001; McMichael et al. 2003). To be able to do so, paying close attention to the issue of interaction between persons and knowledge fields that have different knowledge views is essential (B) (Ludwig et al. 2001), and the issue of uncertainty needs to be handled in a systematic way (C)—the research on adaptive management has much to contribute here (see for example, Holling 1978 and Folke et al. 2005). Criteria D and E say that values need to be integrated in the planning process. A first step is to identify the most relevant values[6] (D) so that these can be explicitly related to during the course of the process (E) (Klosterman 1983; Dahlgren and Khakee 1990; Rayner 1999; Söderbaum 2000).

Criteria F–L are closely linked to participation and influenced by participatory and collaborative planning, and by deliberative democracy

(Dryzek 2000; Grote and Gbikpi 2002; Healey 2006). The criteria are structured according to the main aims related to participation—increasing the quality of decisions, and generating the necessary commitment, legitimacy, or acceptance (Hemmati 2002). Criteria F and G say that the knowledge owned and the ideological orientations[7] represented by the concerned actors are vital for making well-informed plans and decisions (Olsson and Folke 2001; Hemmati 2002). Criterion H says that the participatory efforts need to be connected to the most critical phases of the process (Palerm 1999; Diduck and Sinclair 2002). In general, concerned actors (other than those directly responsible for the formal process) come in too late in the process, making it difficult for them to influence the fundamental direction of the decisions instead of details (Asplund and Hilding-Rydevik 1996).

Criterion I says that one must have a systematic approach for identifying the actors that should be involved (Dryzek 1997; Lidskog 2005). Criteria J and K state that taking into account issues of power asymmetries is key to establishing a rational and democratic process (Flyvbjerg 1998; Wiklund 2002). Also, learning (L) is a key aspect in relation to democracy as well, because those involved need to have sufficient understanding of the complex issues at hand (Dryzek 1997; Dahl 1998; Lidskog 2005).

The criteria are linked in various ways—for example, the integration of knowledge and values (A–E) is dependent on the involvement of concerned actors' knowledge and ideological orientations (F and G), and in order to meet criterion L (learning), almost all of the other criteria have to be fulfilled.

3. METHOD

The study design is based on the idea that if no inconsistencies in the NRM framework can be identified in relation to strategic DRM procedures it can be seen as methodologically valid to apply and evaluate the framework in the context of DRM procedures.

To identify inconsistencies 12 senior researchers of integrated or participatory approaches to DRM were interviewed (hereafter referred to as respondents; Table 2).[8] Respondent-driven sampling was used to select respondents. The 12 interviews, ranging from 1 to 2 h, were conducted by phone or by face to face meetings, and recorded digitally.

Table 2. Researchers interviewed for the purpose of assessing the Sustainable Procedure Framework's applicability to the DRM field

Discipline, organization, title, and research interests
Geographer, Bonn University and Karlstad University Ph.D., Professor, Water resources and flood risk management, decision support systems, participation
Chemist, Karlstad University Ph.D., Professor, Risk management, systems theory
Physical geographer, Karlstad University and Swedish Civil Contingencies Agency Ph.D., senior researcher, Risk management, learning from accidents
Psychologist, Swedish Defence University Ph.D., Professor, Behavioural science in relation to risks, crisis management
Physical planner, Swedish Civil Contingencies Agency Licentiate, senior expert, Learning from accidents, disaster risk analysis
Hydrologist, Karlstad University Ph.D., senior researcher, Risk management
Chemist (atmospheric), Swedish Geotechnical Institute and Gothenburg University Ph.D., senior researcher, Risk assessments and risk evaluation and the connection to sustainable development
Political scientist, national economist, political economy, Linköping University Ph.D., senior researcher, Climate policy, economic sustainability criteria for water management, participation, climate vulnerability, and adaptation, socioeconomic aspects, climate negotiations
Political scientist, Uppsala University Ph.D., senior researcher, Public policy, public management, crisis management, learning from and for risk management
Political scientist, Gothenburg University Ph.D., Professor, Water and flood risk management, participation and democracy in multilayer governance
Sociologist, Mid Sweden University Ph.D., senior researcher, Organizational theory, collaboration for management of risks and crisis
Political scientist, Uppsala University Ph.D., Professor, Evaluation, public policy, collaborative governance

The background, objectives, and method of the study were explained in the invitation e-mail. The respondents were asked to read the article explaining the NRM framework (Hedelin 2007) before the interview. During the interviews the meaning of the framework and its criteria were clarified, criterion by criterion. This enabled the respondents to qualitatively assess the framework with respect to DRM procedures using their scientific knowledge and perspectives of DRM procedures as the basis for the assessment. During the interviews the question of inconsistency was brought up repeatedly: In light of your knowledge and perspective of strategic DRM procedures, are there aspects of the framework as a whole or in parts that are inconsistent with such procedures? Could you point out a necessary issue, aspect, or principle that is currently lacking/that needs to be reformulated/that needs to be excluded, in order to make the framework consistent with DRM procedures?

The respondents were also asked specifically about suggestions made by preceding respondents (adding, removing, or reformulating issues/aspects/principles). This was done to assess, develop, and integrate the suggestions during the process (thereby applying a flexible research design; see Robson 2002, pp. 163–200). Data collection was considered complete when no new arguments that were assessed to have an impact on the study result came up during several succeeding interviews.[9] As a final step, the respondents were invited to comment on the main results from the interview process by e-mail.

4. RESULTS

The interviews did not identify any inconsistencies in the framework as a whole, or in parts. None of the respondents could point out a principle, issue, or aspect that needed to be included, excluded, or changed. The main reason provided was that the criteria are general in character, and that differences between the NRM and DRM fields would be apparent only in their application or specific interpretation in practice. Another reason presented was that the framework concerns procedure (not substance).

During the first interviews the question whether the framework lacked any principles for sustainable DRM procedures was specifically posed, citing the precautionary principle as an example. The question was dropped after seven interviews because clear arguments for not including the precautionary principle had been expressed by the respondents. The main argument made was that in DRM, deciding the level of acceptable risk is often the very focus of the procedure and much related to prioritizing among values (criterion E). Stating by principle that risk should preferably be avoided is inapplicable in most DRM cases. The precautionary principle, the respondents argued, springs from the management of complex environmental issues and is therefore more important in relation to such issues.

Related to the issue of SD principles, and specifically to the principle of integration (integrating sectors, administrations, geographical scales, and so on), the first respondent suggested that the framework might be lacking contextual issues such as who is in charge of the procedure, its relationship to formal legislation, roles, mandates, and relationships between the involved actors. The succeeding interviewees were asked for their assessment in relation to this suggestion. Most respondents argued that the extension of the scope of the criteria would be valuable, while some argued that such an expansion would not make the framework better. None of the respondents argued, however, that the lack of such contextual issues would make the framework inconsistent with respect to DRM procedures.

Another suggestion (made by the fourth respondent) that found broad support among the subsequent respondents was to make the framework easier to use for DRM practitioners. The theoretical and abstract style of the framework, they argued, makes the framework suitable as a research tool rather than a practitioners' tool.

A number of additional suggestions for how to describe individual criteria were also made. These suggestions were all intended to clarify the meaning of individual criteria and did not concern issues of consistency with DRM procedures. One respondent suggested changing "handling" to "management" in the definition of criteria B, C, and J. Another respondent suggested changing the term "ideological orientation" to "perspective" (criterion G). Another suggestion was to divide the criterion of uncertainty (C) into two or more criteria that would identify different kinds of uncertainties. None of these suggestions, however, had broad support from the succeeding respondents.

The respondents showed a positive attitude to the framework, for example, arguing that it covered important aspects: "I think all of these [criteria] are important" and "I think that it is really good that you have included [aspects of] power in this [...] I am very positive about the criteria." One respondent, however, was concerned about the normative character of the framework—the fact that it defines what a sustainable procedure "should" support or promote. The same respondent asked for an explanation of the relationship between procedure and state by raising the question of what guarantee the procedure, as defined by the criteria, will provide for a sustainable state.

5. DISCUSSION

The objective of this study was to assess whether the NRM framework contains any inconsistencies in relation to strategic DRM procedures, and, if inconsistencies are identified, to adjust the framework to the DRM field. Based on the interviews, no inconsistencies exist in the framework, neither as a whole nor in parts. Accordingly, there is no need to adjust the framework at this stage. Further application of the framework in the field of DRM will provide more insights into the potential need for adjustment or further development.

The interviews laid the basis for constructive discussions about the framework—related to its fundamental ideas and principles and to possible ways of developing it further. One respondent pointed out the normative character of the framework. The normative character of the framework is a consequence of the normative character of the SD concept (Leach et al. 2010). The SD concept holds that ecological, social, and economic values must be integrated, that we must see to the needs of future generations, and that on a long-term basis we must keep within the boundaries of nature, and so on. But since it also holds to the principle of participation, it follows that those who are most affected by a decision must be consciously engaged in making that decision. A sustainable state can never be defined based on scientific standards alone. However, the fact that those affected by a decision are involved does not safeguard that the decision taken is sustainable. Power imbalances, lack of scientific knowledge input, among others are factors that hamper a participatory procedure. It is here that the normative character of the procedural framework introduced finds its obvious support; power asymmetries must be handled (criterion J), the procedure must support stakeholder learning (criterion L), and so on.

This is connected to the second fundamental issue raised by one of the respondents—whether a procedure according to the framework criteria guarantees a sustainable state. While the argument that a sustainable procedure always brings a sustainable state needs to be proven, procedures that do not meet the procedural SD principles have no place in sustainable development. The framework's approach is to define SD for procedures to provide a tool for studying procedures. Other criteria, emanating from state SD principles, need to be applied when studying decisions and plans (outcomes of procedures). Deliberative processes, as described by the framework, are thought to support long-term and common interests by means of the discursive mechanism, which allow individual arguments to interact and transform rather than aggregating unchallenged individual interests.[10] Therefore, a procedure as described by the framework can be seen to have large potential to deliver a sustainable state.

Related to the procedure-state issue is the issue of how the framework supports integration of DRM and NRM. The substance (state) of a decision or plan depends heavily on the process by which it is developed. Since integration is one of the two main conceptual components of the framework, the application of the framework will support integration of NRM and DRM. For example, the framework states that all relevant values of the issue at hand need to be integrated (criteria D, E, G). This means that a process of, for example, water resources management needs to involve systematic handling of flood risk. Lack of such considerations in a studied planning process will generate case specific knowledge and recommendations for how to include values threatened by floods in the specific case. And vice versa, when a DRM process is studied, the framework will support identification of potential barriers and possibilities to make the process better in terms of integrating the main environmental/natural resource values concerned by the plan or decision.

In relation to the issue of SD principles, the result does not suggest that any additional SD principles need to be included in the framework. Several of the respondents agreed, however, that inclusion of contextual issues such as the integration between organizations, legislations, sectors, and geographical scales, would be valuable ways to develop the framework further. The following Swedish case will provide a good example of this, illustrating how relationships between administrative levels and policy fields are key factors for establishing integrated flood risk management. Some of the framework's criteria provide important support for studying such issues (mainly A, B, D, F, G, and I), but integration with respect to organizational issues are currently not the focus of the framework. Inspired by the study results, a study has been carried out for developing the framework further in this respect (Hedelin 2015b).

Based on the interviews, another idea for further developing the framework was to develop it into a practitioners' planning support. Development work would be needed to adjust the framework to meet the needs of practitioners, for example, as a toolbox, handbook, or guide. In such a project, as well as in a project aimed at developing a specific planning procedure based on the framework, transdisciplinary research approaches could be useful. Designing forms for involvement of actors, tools for handling of data, and methods for structuring a rational discourse that explicitly relates to the main values involved, requires a thorough understanding of the planning context in addition to the scientific knowledge synthesized by the framework (see van Herk et al. 2011 and Ribarova et al. 2011 for recent examples of participatory and transdisciplinary research in the fields of flood risk and water resource management). One important use of the framework is to provide a scientific point of departure for transdisciplinary research, aimed at the operationalization of SD for strategic planning and decision-making procedures.

Forthcoming work related to the framework can be engaged in both developing the framework further and in developing specific planning procedures and practical support for planners. The perhaps most obvious use of the framework is for the analysis of planning procedures. The Swedish case, presented below, provides an example of the type of results such assessments can provide.

6. ILLUSTRATION OF THE SUSTAINABLE PROCEDURE FRAMEWORK: THE CASE OF THE EU FLOODS DIRECTIVE IN SWEDEN

During the autumn of 2012 the framework was used as a basis for the analysis of the ongoing procedures of implementing the EU Floods Directive in Sweden (see Hedelin 2015a for a full account of the study). A summary of the case study is presented here to illustrate the kind of results the framework can provide.

6.1 The EU Floods Directive

All EU member states are working on implementing the EU Floods Directive (EU 2007) that aims to reduce and manage the risks associated with floods. One of the directive's primary ideas is that an integrated management approach is needed to successfully handle flood risks. Many formulations of the Floods Directive (FD) support this: the requests to coordinate actions within river basins, the broadly formulated objective to protect human health, the environment, cultural heritage, and economic activities, and the request to coordinate the FD and the Water Framework Directive (WFD) (EU 2000). Another fundamental idea behind the FD relates to participation—the active involvement of the actors concerned with flood risk management: "Member States shall encourage active involvement of interested parties" (EU 2007, Article 10:2). In order to render these fundamental ideas operational, the FD prescribes three main steps: preliminary flood risk assessments in all water districts (step 1); the development of flood hazard maps and flood risk maps for the areas identified (steps 2a and 2b); and the establishment of flood risk management plans (step 3). The work will be carried out in 6-year cycles—the first ended in December 2015 with the completion of the first generation of risk management plans.

The main actors for the handling of flood risks in Sweden are the local municipalities (around 300 in number). The municipalities are directly responsible for land use planning, flood risk planning, and crisis management. But the Flood Risk Ordinance (Government of Sweden 2009)—which is how the FD has been written into Swedish legislation—does not place any new formal obligations on the municipalities.

At the regional level, the county administrative boards (21 in number) have mostly a supporting and supervisory role in risk management. The implementation of the FD has added two important tasks to the county administrative boards (CAB)—the development of flood risk maps and flood risk management plans (steps 2b and 3). At the national level, the key actor is the Swedish Civil Contingencies Agency (MSB) who is the main national authority on risks and crises. The MSB leads and coordinates the national FD work, represents Sweden at the European level, and carries out the preliminary flood risk assessments (step 1) and the development of flood hazard maps (step 2a). At the central governmental level the Ministry of Defence is responsible for FD issues (Figure 1).

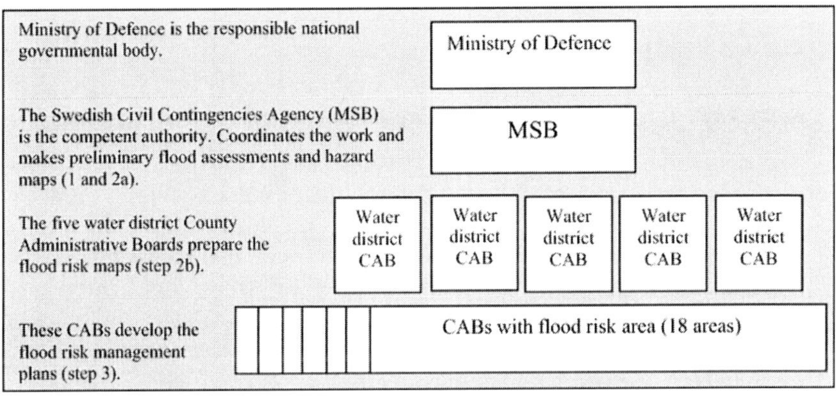

Figure 1. Outline of the formal organizational system for the Floods Directive in Sweden

6.2 Case Study Method

A full account of the method used to apply the Sustainable Procedure Framework to the analysis of the procedures of implementing the EU Floods Directive in Sweden is provided by Hedelin (2015a). The FD study was based on interviews with the key persons responsible for designing and leading the work of the FD in Sweden at the national and regional levels:

- the leader of the early working group for the implementation of the FD in Sweden who investigated the regulatory and organizational settings;
- the person who handles FD-related issues at the Ministry of Defence;
- the two project leaders at the national competent authority (MSB);
- four (of the five) key persons at the CABs responsible for the flood risk maps in the five super-regional water districts. (The fifth person

had only just started in his position at the time during which the interviews were conducted.)

The eight interviews were undertaken in the fall of 2012, each lasting 1–2 h. The framework provided structure for the interviews, and a basis for phrasing follow-up questions to the main interview question—"how is the FD work carried out (or will it be carried out)?" Directed content analysis was used and the framework was used as the theoretical basis for categorization (Shannon and Hsieh 2005).

6.3 Case Study Results

6.3.1 (A) Integration of knowledge from all relevant disciplines

The respondents all agreed that a broad range of knowledge is required in the management of flood risks, including geology, hydrology, environmental protection, GIS, modelling, technical infrastructure, flood prevention measures, crisis management, land use planning, law, and cultural heritage. Many referred to the four areas highlighted in the directive itself—health, environment, cultural heritage, and economic activities (EU 2007, Article 1)—as effectively steering the need for knowledge integration and data collection.

Actions supporting the integration of knowledge are taken at national and regional levels. The MSB has established a national reference group including those authorities that have formal responsibilities connected to the FD, and also seeks to encourage knowledge and information exchange in relation to the CABs, by arranging meetings, producing "dialogue letters," and maintaining a homepage. The CAB respondents, however, primarily perceive these efforts as serving the interests of top-down information provision. All water district CABs collaborate, or plan to collaborate, internally with different departments and externally with other CABs.

6.3.2 (B) Handling of different views of knowledge (for example, positivist, relativist)

The CAB collaborations include persons with different knowledge backgrounds from different departments. These collaborations are perceived to work very well. One respondent noted: "We have raised the understanding of each other's tasks and perspectives and we work well together."

Three of the CAB respondents, however, viewed the collaboration between the county and the national level as being somewhat complicated, relating perhaps to aspects of academic background and professional culture. One noted: "There are different schools involved [...] I guess it's a functional personal thing that makes it a little hard to get things going sometimes." No specific strategy currently exists for the avoidance of potential problems relating to knowledge views and academic backgrounds.

6.3.3 (C) Handling of different kinds of uncertainty

On the question of which kinds of uncertainties are important in their work with the FD, the consequence of alternative scenarios and the areal distribution of water during floods are the primary uncertainties raised by the respondents. The work involves the promotion of systematic ways to handle these uncertainties, for example, by developing differentiated scenarios (50-, 100-, and 200-year floods), and by creating a new national digital elevation model.

Another kind of uncertainty concerns the preconditions of the work, such as the content and formats of maps, the time plan for deliverables (draft reports and maps), financing, and the formal status of the flood risk management plans. Lower level organizations wait for guidelines from higher levels (CAB–MSB–EU), and a sense of frustration is expressed: "And that, I feel, has permeated the whole FD implementation. These uncertainties become a hindrance" and "I'm feeling held back right now." Project plans exist and are being developed within the organizations involved, but no shared, overall project plan exists.

6.3.4 (D) Identification of the most relevant values in relation to the current issue

The FD categories—human health, environment, cultural heritage, and economic activities—function as a broad basis with respect to value identification. More detailed instructions for the selection of values to be presented on the risk maps will be provided by the EU. Both the MSB and the CABs plan to use a wider set of values (data) than those required.

6.3.5 (E) Rational argumentation based on the identified values, by relating them to alternative choices in the planning process

The preliminary assessment of flood risks (step 1) is an important step in the handling of alternatives and in relating the planning process to values. GIS analysis using flood scenarios and the four value categories were carried out by the MSB, resulting in 18 identified population centers affected by potential flooding. The values could be weighted to generate alternative selections of areas.

Another important step in the prioritization of values is the establishment of flood risk management plans (step 3). The MSB is of the view that the plans will primarily include measures that are municipally planned, based in part on the new risk and hazard maps provided. The CAB respondents, however, have a more expert-based view of the plans where the measures are more considered as a logical outcome of the produced maps.

6.3.6 (F) Inclusion of knowledge owned by relevant actors

Besides the responsible authorities, the municipalities are the main actors in terms of knowledge. The MSB sees the municipalities as the key users of the assessments and risk/hazard maps. As such, knowledge of their needs is crucial to rendering the material as useful as possible for them. Open meetings bringing together the municipalities and the CABs are planned, where different municipal departments will need to participate. The MSB also sees the municipalities as the main contributors to the flood risk management plans. The CABs tend to focus on the flood risk maps. If needed, the municipalities will be asked for complementary data. For the flood risk management plans, the general picture provided is that the CABs will develop preliminary plans that will then be discussed with the municipalities concerned.

The general view is that non-municipal actors (for example, river regulators, private citizens, fishing organizations) can become involved indirectly in the work via municipal processes for the handling of risks (for example, in land use planning) regulated by other legislations that prescribe consultation.

In addition, all respondents refer to the formal consultation process prescribed by the FD, which provides an opportunity to comment on draft plans. One respondent highlighted the possibility of focusing on river groups[11] in relation to the inclusion of knowledge.

6.3.7 (G) Inclusion of the ideological orientations represented by relevant actors

During the process of preliminary flood risk assessment a number of expert authorities were consulted. A workshop was conducted with the national reference group where the method including weighting for selection of flood risk areas was explained and discussed. None of the CABs were involved at this stage.

According to the MSB, stakeholders will be indirectly involved in the FD work, through formalized municipal processes, such as land use planning and risk and vulnerability assessments. But the issue of the ideological orientations of those local actors involved in the process has not been fully developed.

6.3.8 (H) Participation in the most critical phase(s) of the process

Thus far, a number of expert authorities have been engaged in the process while municipal involvement has mainly concerned the flow of information from higher levels. The key phase for municipal involvement is seen to be step 3 (flood risk management plans). According to the MSB, the municipalities' planned measures for reducing flood risk will constitute the basic data for the establishment of the plans. The CAB respondents plan to consult the municipalities after having established plans based on the flood hazard and flood risk maps (see also E and F).

The flood risk management plans will be subject to formal public consultation and will be made available at the MSB's homepage. The MSB has the view that stakeholders and the general public will primarily become involved indirectly, via the normal municipal processes for the handling of risks regulated by other legislative documents that prescribe a process of public consultation.

6.3.9 (I) A procedure for defining the actors that should be involved; (J) Handling of power asymmetries; (K) Procedures that ensure that ideological orientations are not suppressed; and (L) Learning

Concepts like participation and democracy are not generally used by the respondents. The Ministry of Defence and the leader of the early working group for the FD implementation describe democracy as an issue of low relevance in relation to the FD. One respondent at the CAB level noted that: "The issue of democracy and anchorage is not included as an important issue in this work. [...] I don't know why, but that is how I perceive it."

Much in line with these approaches to democracy, there are no systematic activities to identify those actors that are most affected by the decisions undertaken, and no concrete plans to involve local actors other than the municipalities. Issues of handling power imbalances or stakeholder learning are not covered.

6.4 Case Conclusions and Recommendations

The results (Table 3) show that there is strong potential in the current work in relation to the criteria concerning integration across disciplines and values (A–E). Still, the work can be developed further, especially with regard to knowledge exchange between the national and the regional levels. Frustration concerning uncertainties connected to administrative issues, and divergent views of roles and mandates (for example, formal status of risk management plans) show that an extended exchange between levels is urgent.

Table 3. Overview of the results from the evaluation of the EU Floods Directive in Sweden

Criteria	Summary of results
A Integration of disciplinary knowledge	A broad spectrum of knowledge is seen as important and is also represented in the work. Horizontal knowledge integration is facilitated while vertical integration needs to be further developed
B Handling of different views of knowledge	Differences in academic and professional backgrounds are seen as both valuable and as a possible difficulty in terms of learning and communication. No strategy exists for avoiding potential problems connected to academic backgrounds and knowledge views
C Handling of different kinds of uncertainty	Systematic ways of handling uncertainties related to floods are part of the work, for example through differentiated scenarios. Systematic ways of handling institutional uncertainties are less developed
D Identification of the most relevant values	A broad set of values are identified based on the value categories highlighted in the FD (health, environment, cultural heritage, and economy). See also G
E Rational argumentation based on values and alternatives	A method supporting rational argumentation in relation to alternatives and values is used in the preliminary flood risk assessment. The risk maps being produced can also facilitate such argumentation
F Inclusion of relevant actors' knowledge	Knowledge from national and regional authorities with formal responsibilities is included. Local knowledge is not yet included, but there are plans to involve the municipalities concerned
G Inclusion of ideological orientations	Ideological orientations from the key authorities are included. It is uncertain to what degree municipal perspectives will be included
H Participation in the most critical phase(s)	The municipalities concerned, who are the main actors concerned, will be involved only indirectly, as providers of basic data for risk maps and for input to the plans. The management plans will be subject to traditional public consultation
I A procedure for defining the actors that should be involved	The local actors who are the most directly affected by the decisions under way have not been systematically identified
J Handling of power asymmetries	There is no explicit approach for handling power asymmetries
K Safeguarding ideological orientations	There is no explicit approach to ensure that ideological orientations are not suppressed
L Learning	There is no strategy for stakeholder learning, but the hazard and risk maps can support such learning

The establishment of a national-regional working group would be a first step. Another would be to develop a shared process plan, as the means for a transparent, systematic, and continuous conversation and learning about how the work can be carried out.

Furthermore, the results clearly show that issues of participation and democracy are still waiting to be developed by the FD administration. In relation to criteria F–H (participation contributing to the process) the work is characterized by the knowledge and perspectives owned by experts rather than by the provision of input of knowledge and perspectives from those at the local level. Another key objective here is anchorage and democracy (criteria I–L). In order to attain the advantages of a participatory planning approach, the FD administration needs to develop their work much further in this respect. One strategy is to involve the appropriate actors directly in the tasks prescribed by the FD. It would be useful to engage the river groups when designing participatory structures. Another strategy would be to support and encourage the municipalities to involve their citizens and other stakeholders in the existing municipal processes for the handling of flood risks. Handbooks and other information activities are possible tools for operationalizing such a strategy.

7. CONCLUSION: THE APPLICABILITY OF THE SUSTAINABLE PROCEDURE FRAMEWORK TO DRM PROCEDURES

This study did not identify any theoretical inconsistencies in the Sustainable Procedure Framework in relation to strategic DRM procedures. It can therefore be seen as methodologically valid to apply and test the framework in relation to such procedures. The Swedish case illustrated the use of the framework and exemplified the kind of practicable recommendations that such studies can provide. Future use of the framework will produce more knowledge, not only of the studied DRM procedures, but also about the usefulness and possible further development needs of the framework. The basis is now laid, however, for application of the Sustainable Procedure Framework to a wide range of cases of strategic DRM.

FOOTNOTES

1 The substantive and the procedural dimensions of SD are inherently connected. See the Sect. 5 for details on the relationship between SD procedure and substance (state).

2 Management of crisis situations, or management and design of specific preventive measures, were not part of this study.

3 Well-established fundamental principles of SD have been identified in an extensive literature review (Hedelin 2008a). The principles identified have main bearing on either procedure or substance. The substantive principles are: use of resources within nature's boundaries, and, a fair distribution of resources in space and time. In addition to participation and integration, the procedural principles identified are: the principles of polluter pays, precaution, and a strategic and long-term approach. The latter is to some degree also included in the framework because of the character of its intended area of application (strategic, long-term planning), which constituted a point of departure for the development of the framework criteria.

4 The production of a synthesis entails the integration of the relevant areas of the literature into a new whole (Kirkevold 1996). Rather differently from a traditional review, a synthesis claims to present connections that have not previously been made (Kirkevold 1996).

5 A transdisciplinary research approach addresses societal problems through interdisciplinary research as well as collaboration between researchers and other actors. Its aim is to enable mutual learning processes between science and society (Jahn et al. 2012).

6 The most suitable method for identifying those values is dependent on context.

7 Value systems or ethics are other terms used with a similar meaning (Söderbaum 2000).

8 Interdisciplinary and scientific knowledge are used as the basis for the assessment of the framework because of the interdisciplinary and deductive character of the framework. If the framework can be established as a valid theoretical tool, it can be applied and assessed in relation to DRM practice and practitioners' knowledge. Detailed arguments for using a group of experts as a basis for understanding complex issues that integrates multiple perspectives can be found in the Delphi literature (Novakowski and Wellar 2008).

9 The adopted methodology supports a knowledge-based selection of respondents and interview process. The study design does not rule out,

however, that another group of respondents, or a considerably larger number of respondents, could have an influence on the study result.

10 See literature on deliberative democracy and collaborative planning for explanation of the deliberative processes and the discursive mechanism, for example Dryzek (2000).

11 River groups are local, river system-based stakeholder groups mainly composed of regulation companies, power companies, municipalities, and CABs (convener). Their main task is to spread knowledge among, and coordinate the involved actors, especially in times of high flows.

REFERENCES

1. Asplund, E., and T. Hilding-Rydevik. 1996. Knowledge, environment and the future. Plan-EIA—case studies in two municipalities. Stockholm: Department of Regional Planning, Royal Institute of Technology (in Swedish).
2. Becker, P. 2014. *Sustainability science—Managing risk and resilience for sustainable development.* Amsterdam: Elsevier.
3. Bellamy, J.A., G.T. McDonald, G.J. Syme, and J.E. Butterworth. 1999. Evaluating integrated resource management. *Society & Natural Resources* 12(4): 337–353.
4. Born, S.M., and W.C. Sonzogni. 1995. Integrated environmental management: Strengthening the conceptualization. *Environmental Management* 19(2): 167–181.
5. Bradshaw, G.A., and M. Bekoff. 2001. Ecology and social responsibility: The re-embodiment of science. *TRENDS in Ecology and Evolution* 16(8): 460–465.
6. Clarke, W.C., and N.M. Dickson. 2003. Sustainability science: The emerging research paradigm. *Proceedings of the National Academy of Sciences* 100: 8059–8061.
7. Dahlgren, L., and A. Khakee. 1990. *Values in societal planning.* Stockholm: Byggforskningsrådet (in Swedish).
8. Dahl, R.A. 1998. *On democracy.* New Haven: Yale University Press.
9. Diduck, A., and A.J. Sinclair. 2002. Public involvement in environmental assessment: The case of the non-participant. *Environmental Management* 29(4): 578–588.
10. Dryzek, J.S. 1997. *The politics of the Earth: Environmental discourse.* Oxford: Oxford University Press.
11. Dryzek, J.S. 2000. *Deliberative democracy and beyond: Liberals, critics, contestations.* Oxford: Oxford University Press.
12. EU (European Union). 2000. Directive 2000/60/EC of the European Parliament and of the Council of 23 October 2000 establishing a framework for community action in the field of water policy. *Official Journal of the European Communities* 43: 1–72.
13. EU (European Union). 2007. Directive 2007/60/EC of the European Parliament and of the Council of 23 October 2007 on the assessment and management of flood risks. *Official Journal of the European Union* 50: 27–34.
14. Evers, M. 2008. *Decision support systems for integrated river basin management. Requirements for appropriate tools and structures for a comprehensive planning approach.* Ph.D. thesis. Germany: Leibniz University.
15. Flyvbjerg, B. 1998. *Rationality and power: Democracy in practice.* Chicago and London: University of Chicago Press.
16. Folke, C., T. Hahn, O. Olsson, and J. Norberg. 2005. Adaptive governance of social-ecological systems. *Annual Review of Environmental Resources* 30: 441–473.
17. Gamper, C.D., and C. Turcanu. 2009. Can public participation help managing risks from natural hazards? *Safety Science* 47(4): 522–528.

18. Ginson, R.B. 2006. Beyond the pillars: Sustainability assessment as a framework for effective integration of social economic and ecological considerations in significant decision-making. *Journal of Environmental Assessment Policy and Management* 8(3): 259–280.
19. Government of Sweden. 2009. Flood risk ordinance (SFS 2009:956). Stockholm: Swedish Government (in Swedish).
20. Gregersen, H.M., P.F. Folliott, and K.N. Brooks. 2007. *Integrated watershed management: Connecting people to their land and water.* Cambridge: CABI.
21. Grote, R., and B. Gbikpi (eds.). 2002. *Participatory governance: Political and social implications.* Wiesbaden, Germany: VS Verlag für Sozialwissenschaften.
22. Healey, P. 2006. *Collaborative planning: Shaping places in fragmented societies.* Macmillan: Palgrave.
23. Hedelin, B. 2007. Criteria for the assessment of sustainable water management. *Environmental Management* 39(2): 151–163.
24. Hedelin, B. 2008a. *Planning for sustainable use of water.* Ph.D. thesis. Sweden: Karlstad University.
25. Hedelin, B. 2008b. Criteria for the assessment of processes for sustainable river basin management and their congruence with the EU water framework directive. *European Environment* 18: 228–242.
26. Hedelin, B. 2015a. The EU floods directive in Sweden: Opportunities for integrated and participatory flood risk planning. *Journal of Flood Risk Management.* doi:10.1111/jfr3.12162.
27. Hedelin, B. 2015b. Further development of a sustainable procedure framework for strategic natural resources and disaster risk management. *Journal of Natural Resources Policy Research* 7(4): 247–266.
28. Hedelin, B., and M. Lindh. 2008. Implementing the EU water framework directive – Prospects for sustainable water planning in Sweden. *European Environment* 18(6): 327–344.
29. Henriksen, H.J., J.C. Refsgaard, A.L. Hojberg, N. Ferrand, P. Gijsbers, and H. Scholten. 2009. Harmonised principles for public participation in quality assurance of integrated water resources modelling. *Water Resources Management* 23(12): 2539–2554.
30. Hemmati, M. 2002. *Multi-stakeholder processes for governance and sustainability: Beyond deadlock and conflict.* London: Earthscan.
31. Holling, C.S. 1978. *Adaptive environmental management and assessment.* Chichester: Wiley.
32. Ikeda, S., T. Sato, and T. Fukuzono. 2008. Towards an integrated management framework for emerging disaster risks in Japan. *Natural Hazards* 44(2): 267–280.
33. Jahn, T., M. Bergmann, and F. Keil. 2012. Transdisciplinarity: Between mainstreaming and marginalization. *Ecological Economics* 79: 1–10.
34. Jepson, E.J. 2001. Sustainability and planning: Diverse concepts and close associations. *Journal of Planning Literature* 15(4): 499–510.
35. Kirkevold, M. 1996. Res – A way to strengthen the integration of nursing research. In *Vision, Knowledge and Work,* ed. T. Bjerkreim, J. Mathisen, and R. Nord, 100–118. Oslo: University Press (in Norwegian).
36. Klosterman, R.E. 1983. Fact and value in planning. *Journal of the American Planning Association* 49(2): 216–225.
37. Leach, M., I. Scoones, and A. Stirling. 2010. *Dynamic sustainabilities.* London and Washington, DC: Earthscan.
38. Levin, S., T. Xepapadeas, A.S. Crépin, J. Norberg, A. de Zeeuw, C. Folke, T. Hughes, K. Arrow, S. Barrett, G. Daily, P. Ehrlich, N. Kautsky, K.G. Mäler, S. Polasky, M. Troell, R.J. Vincent, and B. Walker. 2013. Social-ecological systems as complex adaptive systems: Modeling and policy implications. *Environment and Development Economics* 18(2): 111–132.
39. Lidskog, R. 2005. Siting conflicts—democratic perspectives and political implications. *Journal of Risk Research* 8(3): 187–206.
40. Lubchenco, J. 1998. Entering the century of the environment: A new social contract for science. *Science* 279(5350): 491–497.

41. Ludwig, D., M. Mangel, and B. Haddad. 2001. Ecology, conservation, and public policy. *Annual Review of Ecological Systems* 32: 481–517.
42. Margerum, R.D. 1999. Integrated environmental management: The foundations for successful practice. *Environmental Management* 24(2): 151–166.
43. McMichael, A.J., C.D. Butler, and C. Folke. 2003. New visions for addressing sustainability. *Science* 302(5652): 1919–1920.
44. Novakowski, N., and B. Wellar. 2008. Using the Delphi technique in normative planning research: Methodological design considerations. *Environment and Planning A* 40(6): 1485–1500.
45. Nyberg, L., M. Evers, M. Dahlström, and A. Pettersson. 2014. Sustainability aspects of water regulation and flood risk reduction in Lake Vänern. *Journal of Aquatic Ecosystem Health and Society* 17(4): 331–340.
46. Olsson, P., and C. Folke. 2001. Local ecological knowledge and institutional dynamics for ecosystem management: A study of Lake Racken Watershed, Sweden. *Ecosystems* 4(2): 85–104.
47. Ostrom, E. 2009. A general framework for analyzing sustainability of social-ecogical systems. *Science* 325(5939): 419–422.
48. Palerm, J.R. 1999. Public participation in EIA in Hungary: Analysis through three case studies. *Environmental Impact Assessment Review* 19: 201–220.
49. Pelling, M. 2007. Learning from others: The scope and challenges for participatory disaster risk assessment. *Disasters* 31(4): 373–385.
50. Pretty, J. 2003. Social capital and the collective management of resources. *Science* 302(5652): 1912–1914.
51. Rayner, S. 1999. Mapping institutional diversity for implementing the Lisbon principles. *Ecological Economics* 31: 259–274.
52. Renn, O. 2015. Stakeholder and public involvement in risk governance. *International Journal of Disaster Risk Science* 6(1): 8–20.
53. Ribarova, I., P. Kalinkov, D. Assimacopoulos, P. Jeffrey, D. Inman, K.A. Daniell, L.S. Vamvakeridou-Lyroudia, T. Melin, K. Tarnaki, and N. Ferrand. 2011. Research-supported participatory planning for water stress mitigation. *Journal of Environmental Planning and Management* 54(2): 283–300.
54. Robinson, J. 2004. Squaring the circle? Some thoughts on the idea of sustainable development. *Ecological Economics* 48: 369–384.
55. Robson, C. 2002. *Real world research*, 2nd edn. Oxford: Blackwell Publishers.
56. Sawhney, P., M. Kobayashi, M. Takahashi, P.N. King, and H. Mori. 2007. Participation of civil society in management of natural resources. *International Review for Environmental Strategies* 7(1): 117–131.
57. Shannon, S.E., and H.F. Hsieh. 2005. Three approaches to qualitative content analysis. *Qualitative Health Research* 15(9): 1277–1288.
58. Sneddon, C., L. Harris, R. Dimitrov, and U. Özesmi. 2002. Contested waters: Conflict, scale and sustainability in aquatic socio-ecological systems. *Society & Natural Resources* 15(8): 663–675.
59. Söderbaum, P. 2000. *Ecological economics, a political economics approach to environment and development*. London: Earthscan.
60. van Herk, S., C. Zevenbergen, J. Rijke, and R. Ashley. 2011. Collaborative research to support transition towards integrating flood risk management in urban development. *Journal of Flood Risk management* 4(4): 306–317.
61. Vaidya, A., and A.L. Mayer. 2014. Use of the participatory approach to develop sustainability assessments for natural resource management. *International Journal of Sustainable Development and World Ecology* 21(4): 369–379.
62. Vinet, F. 2008. From hazard reduction to integrated risk management: Toward adaptive flood prevention in Europe. In *Flood recovery, innovation and response*, ed. D. Proverbs, C.A. Brebbia, and E. Penning-Roswell, 113–122. Southampton, UK: WIT Press.

63. Wiklund, H. 2002. Arenas for democratic deliberation: Decision-making in an infrastructure project in Sweden. Ph.D. thesis. Sweden: Jönköping International Business School.
64. Wisner, B. 2011. Are we there yet? Reflections on integrated disaster risk management after ten years. *Journal of Integrated Disaster Risk Management* 1(1): 1–14.

CHAPTER 14

Coastal Flooding in the Solent: An Integrated Analysis of Defences and Inundation

Matthew P. Wadey [1,*], Robert J. Nicholls [1] and Craig Hutton [2]

[1] *Faculty of Engineering and the Environment, University of Southampton, Southampton SO17 1BJ, UK*

[2] *Geodata Institute, University of Southampton, Southampton SO17 1BJ, UK*

ABSTRACT

This paper demonstrates a methodology for integrating existing models for the rapid simulation of coastal flood events across a large and varied case study area on the UK south coast. Following validation against observations from real coastal floods, synthetic events driven by realistic waves and water levels and the full range of failure mechanisms were modelled for a range of loadings to generate peak flood water depths and an overview of impacts across this spectrum of possible floods. Overtopping is relatively important compared to breaching as coastal floodplains are small. This modelling system supports multiple potential applications, such as planning flood warnings, coastal defence upgrade, and land use, including under sea-level rise. The concepts drawn from this study are transferable to similar coastal regions.

Keywords: coastal flooding; regional inundation modelling; defence failure; flood risk assessment; sea-level rise

1. INTRODUCTION

Floods are an important long term risk to society. While society generally adapts to them, e.g., [1,2], this is usually accomplished in response to real events, which can include significant losses and even deaths. In Europe, developments in coastal flood risk management accelerated following the 1953 North Sea floods which killed 1836 people in the Netherlands, 307 in the UK and 17 in Belgium [3,4,5] and floods on the German Bight in 1962 when more than 300 people lost their lives [6,7]. In comparison with understanding at that time, coastal planners now benefit from a better understanding of the nature and degrees of exposure to flooding due to advances in coastal modelling e.g., [8]. Numerical simulation of coastal inundation is well demonstrated [9,10,11,12], although uncertainties, practical constraints, and the conceptual understanding of coastal flood systems can be limiting [13]; with poor flood incident management still resulting in widespread loss of life and other significant damages [14]. Hurricane Katrina demonstrated the catastrophic effects of defence failures in low-lying and populated coastal areas [15], while flooding in Bangkok has highlighted the impacts of flooding upon a large populous city, *cf.* [16]. The consequences of un-developed flood management measures were tragically apparent on the French Atlantic coast in February 2010 when more than 40 people died following failure and outflanking of coastal defences during the well forecast storm Xynthia [17]. In the hours before the event, warnings were available, but not specific to coastal flooding consequences, leaving authorities incapable of planning evacuation [18].

This study is differentiated from most flood risk research by focusing upon analysis of inundation over a large area where coastal floodplains are relatively small, and water flowing over defences (overflow and wave overtopping) may be more significant compared to breach (a reduction in the effective crest height). The latter failure mechanism is a particular threat for low-lying and expansive floodplains such as those on the east UK and Belgian, Dutch and German North Sea coasts and has received considerable attention [19]. Sea-level rise is increasing the probability of extreme events in most coastal regions e.g., [20,21,22], hence it is essential that methods are available to identify the likelihood and characteristics of flooding. In Europe, the EU Flood Directive (2007/60/EC) has determined that a framework is required to assess flood risk, including emphasis upon the frequency, magnitude and consequences of flooding. Few flood maps contain information on consequences [23], although flood risk assessment methodologies [24,25,26,27] are available to provide valuable insights to vulnerable defences and floodplain areas. Some countries benefit from national-scale assessments which make optimal use of the information available, e.g., [28], although places remain where flood dangers are unknown, or where spatial and temporal resolutions of broader assessments provide an abstraction of the reality of single coastal flood events.

High resolution datasets and numerical tools are currently available for many coastal areas, although are not routinely integrated; particularly where risks are not obvious or recently experienced This paper describes a method for such integration that can support coastal management measures such as forecasting and warning, defence planning, and land-use zonation. This is intended to provide outputs that are easily communicated to the public and flood managers. The Solent is selected as a case study; a shoreline on the south coast of England which comprises a large estuarine system and wave exposed open coast. Defended and undefended floodplains coexist, whilst flood consequences and risks are not well understood, and sea-level rise is increasing flood risk [20]. The aims of this paper are to:

1) Describe a coastal flood modelling methodology, where flood simulations account for real flood defence characteristics and a full range of defence failure mechanisms (ranging from simple overflow or outflanking, through to full breaching).
2) Simulate synthetic coastal floods within the Solent, to indicate flood event impacts under present and 21st century sea levels across this range of failure mechanisms.
3) Compare the flood consequences for different failure mechanisms in these floods.

A background to the Solent will now be described, followed by description of the methodology, results, discussion and conclusions.

2. THE SOLENT CASE-STUDY REGION AND INITIAL ASSESSMENT OF COASTAL FLOOD RISK

The Solent case-study is within the Environment Agency's Southeast Region. Alongside the South Downs, the Solent is a sub-region for which defences and flood warnings are managed; and contains two shoreline management plan (SMP) [29] areas (the mainland, and the Isle of Wight). SMP studies provide a framework for dealing with flooding and erosion by dividing sections of shoreline on the basis of coastal processes [30] and related defence systems. Figure 1 introduces the location of the Solent case study area, the areas broadly at risk of coastal flooding, and six 'sub-regions' for analysis described later in this paper.

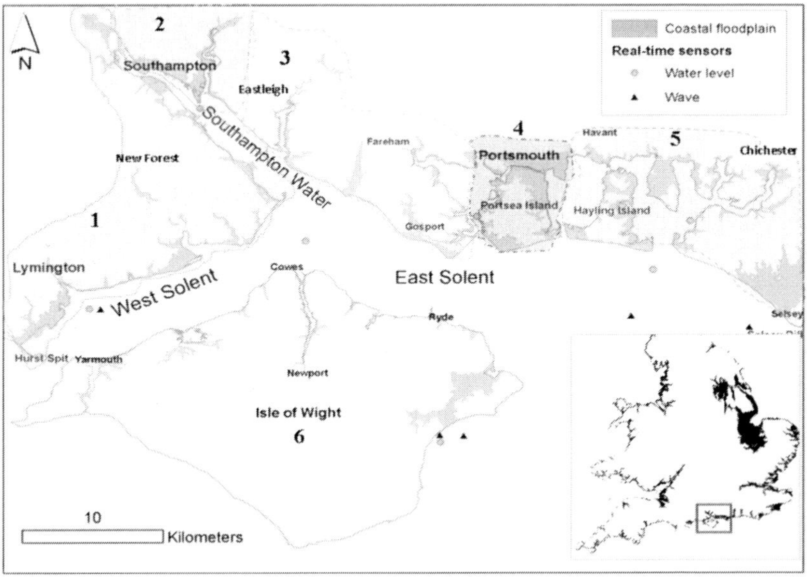

Figure 1. The Solent case-study, locations of wave and water level recorders, and the approximate 1 in 1000 year coastal floodplain (refer to Section 3.2). The base map (lower right) is an Environment Agency 1 in 1000 year indicative coastal floodplain for England and Wales [31]. Sub-regions used to describe flood modelling results later in the paper are: (**1**). The New Forest; (**2**). The city of Southampton; (**3**). The boroughs of Eastleigh, Fareham and Gosport; (**4**). The city of Portsmouth; (**5**). The boroughs of Havant and Chichester; (**6**). The Isle of Wight.

Much of the Solent is sheltered from south-westerly Atlantic waves by the Isle of Wight, and a managed shingle barrier at the western end known as Hurst Spit [32]. Storm surges in the Solent mainly occur as a result of low pressure systems that move from the Atlantic eastward over southern England [33] whilst smaller surges also occur as a result of large North Sea storm surge events transmitted into the English Channel through the Dover Strait [34]. Tidal residuals rarely exceed 1 m; with only a 0.33 m difference between a 1 in 10 and 1 in 1000 year water-level [35]. Coastal floods in the Solent during the 20th and early 21st centuries have been frequent but usually involving small water depths, and with no recorded loss of human life [36].

The form and extent of the estuary has largely been controlled by changes in sea level until progressive development and land reclamation began in the 18th century. This was most significant around Southampton Water and Portsmouth; the two cities having grown together to form an urban area containing over one million inhabitants. The tidal range increases from approximately 2 m at Hurst to 5 m at Selsey. The region is internationally renowned for its complex tides [37]. Double high waters occur and are particularly pronounced during large spring tides, and at mid-flood tide the tide is constant for about an hour (Figure 2). Meteorological induced sea level effects on the UK south coast are generally less severe than on the east and west coasts; although surge events [20,38] and Atlantic swells [39] have been associated with coastal flooding. A national assessment of flood risk [40] identified that the South coast would experience some of the largest increases in flood risk during the 21st century. It has been estimated that 24,138 properties in the Solent (excluding the Isle of Wight) are exposed to a 1 in 200 year coastal flood [41] with the largest concentration of risk in Portsmouth which contains more than 15,000 of these properties alone [41,42]. After London and Hull, Portsmouth may contain the greatest coastal flood risk for any city in the UK [43]. It is estimated that 25 percent of the city's coastal defences provide less than a 1 in 200 year level of protection (the indicative standard of protection for urban coastal areas in England and Wales) [44].

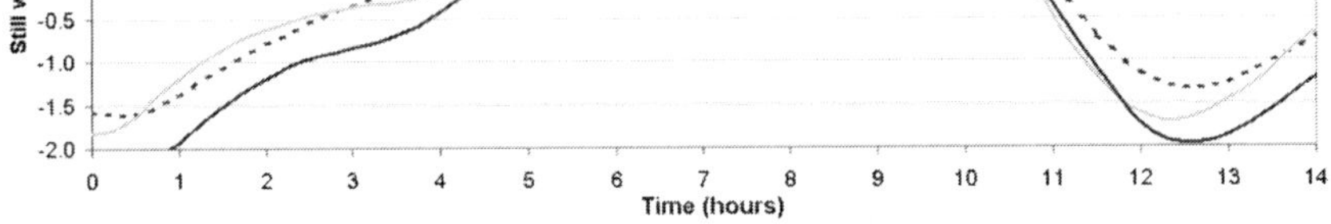

Figure 2. Water level time-series; recorded at three locations across the Solent (refer to Figure 1) during a typical spring tide.

Shown in Figure 2 are typical semi-diurnal lunar spring tidal cycles for three locations in the Solent. In Figure 3 are water levels recorded during a recent storm surge event. The Solent's defences are frequently subjected to hydraulic action, which due to the double high tides can be prolonged during storms. This effect is most prevalent in the central and eastern area of the region. The double high-water is significant at Southampton (which has almost no coastal flood defences). Portsmouth contains floodplains that are situated much lower (relative to extreme water levels) than Southampton, although there are flood walls and/or beaches around the city. Defence design and floodplain characteristics are highly variable across the remainder of the region, as shown in this paper.

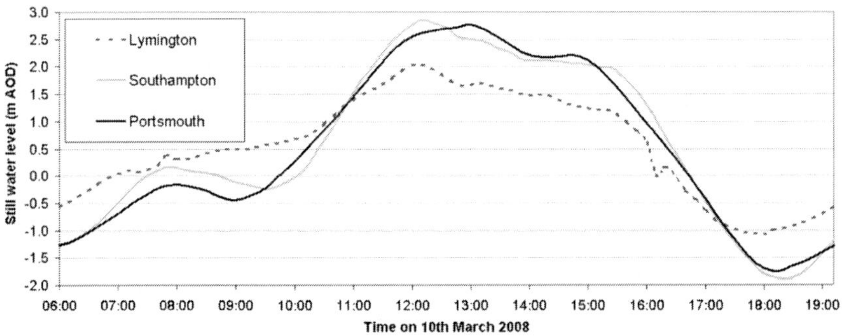

Figure 3. Water level time-series recorded during a storm surge event on the 10th March 2008.

3. METHODS

The modelling approach is summarised in Figure 4. Regional estimates of exposure to coastal flooding in the Solent were generated by combining planar water levels with a digital elevation model (DEM). This approximates coastal flood outlines and depths without considering defences or the dynamics of flooding. The second method simulates flooding more realistically by inclusion of hydraulic connectivity and mass conservancy; by coupling flood defence analysis and storm-tide water level time-series with a suitable numerical inundation model. Data layers were used to help prepare the model and view outputs, using geographical information systems (GIS). These included land-use maps, property datasets (containing commercial, residential, and unoccupied buildings), and aerial photography.

3.1. Regional Topography, Water Level and Defence Datasets

The DEM used for this work was mostly constructed from Light Detection and Ranging (LiDAR) surveys, processed in the format of a digital terrain model (DTM) (artefacts such as buildings removed). At the upper reaches of several tidal rivers the DTM was extended using a 10 m resolution topographic dataset derived from photogrammetry [45].

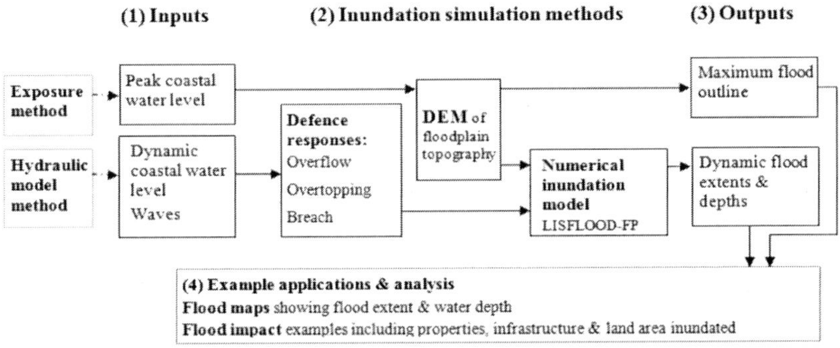

Figure 4. Method and model overview showing (**1**) inputs; (**2**) inundation simulation approaches; (**3**) outputs; and (**4**) example applications.

The LiDAR data was collected in 2007 and 2008 at 1 m and 2 m spatial resolution. Accompanying metadata indicates survey points are within ±0.15 m vertical accuracy [46], although an absolute error of ±0.30 m is often quoted [47,48]. For the hydraulic modelling method, shoreline flood defences were masked from the DEM, which was re-sampled from its original 1 m and 2 m resolution to 50 m cells; using a bilinear interpolation technique recommended for floodplain modelling [49,50]. This grid cell size allowed computational efficiency, and was commensurate with the available flood defence survey data (which was re-inserted onto the DEM). Large areas of the region comprise rural floodplain, and this resolution has been considered suitable in previous studies e.g., [51]; although benefits of finer floodplain grid cell sizes are acknowledged (Section 5.3).

Still water level return periods in the Solent have been periodically updated, and older methods for estimation of extreme still water level probabilities [52,53,54] are noted to significantly over-estimate return levels on the UK south coast [55]. For example, the 1 in 200 year water level definition shows variations of more than 0.1 m between those published in 2007 [56] and recent analysis of tide data at Southampton and Portsmouth [35]; one reason that quantification of property exposed to the 1 in 200 year flood in this work will vary in comparison to previous estimates mentioned in Section 2 of this paper. The latest available still water level return periods are from a national study which applied the Skew Surge Joint Probability Method around the UK coast [57] (e.g., Table 1). These latest levels were used in this work, and prescribed spatially to simulations of flood events by methods used for the Environment Agency's flood zone mapping and a previous flood risk assessment in the Solent [44] (see Figure 5). In relation to Portsmouth this equates to extreme water levels that are approximately 0.8 m less at the western end of the region (Hurst) and to 0.5 more at the eastern extent (Selsey Bill).

Table 1. Return still water levels at Portsmouth in 2008 [57], mean high water springs, and the still water level recorded during a recent storm surge. The water level datum referred to (mAOD) is approximately mean sea level.

Return period (years)	1	10	20	50	100	200	1000	MHWS	10th Match 2008
Elevation (mAOD)	2.56	2.81	2.88	2.98	3.05	3.12	3.28	1.97	2.77

Regional sea-level rise in the past decade has been approximately 1.7 mm per year [58] hence the most recently derived return periods do not require significant adjustment.

Wave data was compiled from different sources, to enable wave loading scenarios to be combined with regional sea level extremes. Significant wave height return periods were available at measured sites (Figure 1) [59]; and fetch-limited conditions were calculated in all areas [60]. Records of the largest swell wave heights and periods that have occurred since records began [46] were also noted for exposed coastline east of Portsea Island.

Defence datasets describing condition and structural elements (such as roughness, berms, and crest heights) were available across the region. Information was also gained from the DEM, aerial photography, photos of defences, literature and site visits. Crest heights were mostly from real-time kinematic (RTK) GPS surveys with a vertical accuracy of ±0.03 m [46]. Recent condition grade scores (between 1 for excellent, and 5 for poor) were supplied by local authorities, allocated by coastal engineers to defences in England and Wales during routine inspections, and which can be mapped to defence performance under loading [61,62].

3.2. The Planar Water Level Method

Regionally uniform synthetic coastal events were modelled across a range covering normal tidal levels to extremes which included mid-range estimates of 21st century sea-level rise [63]. The planar water level method defined maximum flooding for given levels of still water (and allowance for 2 m wave run-up to define properties exposed to flooding on the open coast of Hayling Island). This identified 99 hydraulically discrete and independent flood 'compartments' covering 120 km^{2}of land.

These compartments were separate on the basis of a 1 in 1000 year regional planar water level plus 0.3 m allowance for vertical data errors.

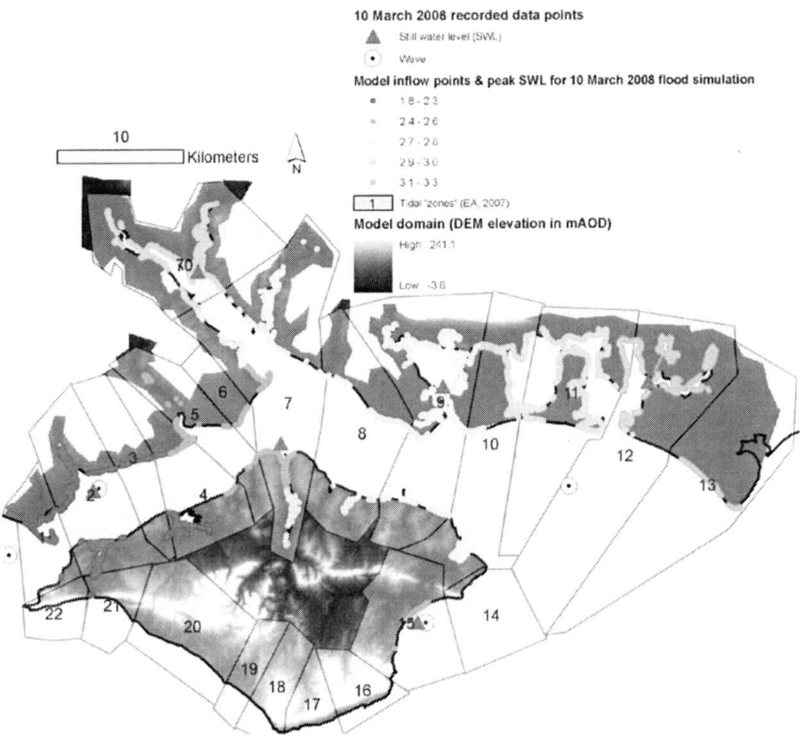

Figure 5. The computational domain and boundary still water levels for the 10th March 2008 regional flood simulation. For this example water levels at inflow points were spatially (linearly) interpolated between data recorders. For simulations of synthetic flood events (Section 4) the zones shown were used to allocate shared still water level characteristics to the model boundary (refer to Section 3.1). Water level time-series for the hydraulic modelling were based upon a Lymington storm-tide curve for zones **1–6**; Southampton Water for zones 7 and 70; and Portsmouth for zones **7–15** (refer to Figure 3).

3.3. Hydraulic Modelling

Where inundation from defence failures was simulated, peak flow was calculated at the boundary of the model. This was spread to adjacent cells by solving of a continuity and momentum equation in each direction using the inertial version of the raster-based inundation model LISFLOOD-FP [64]. LISFLOOD-FP originated as a simple storage cell model solving an analytical approximation of the 2D shallow water equations [65,66]. Use of high spatial resolution and good quality topographic data has allowed similar results to full 2D codes (for sub-critical gradually varied flows only), and LISFLOOD-FP has been previously been used for modelling coastal floods [9,12,51,67].

At higher grid resolutions the requirement for small time-steps to allow stable and realistic simulations of flow [66,68] prompted incorporation of an inertial formulation of the two dimensional shallow water equations into LISFLOOD-FP, which has been noted to significantly increase computational efficiency [64,69]. Flow between neighbouring floodplain cells is calculated as in Equation (1):

$$q^{t+\Delta t} = \frac{q^t - gh^t_{flow}\Delta t \dfrac{\Delta(h^t + z)}{\Delta x}}{\left(1 + gh^t_{flow}\Delta t n^2 \left|q^t\right| / (h^t_{flow})^{10/3}\right)} \quad (1)$$

where q is the flow per unit width, h_{flow} is the depth of water available to flow, Δt is the model time step, Δx is the model grid resolution, z is the bed elevation, h is the water depth, and n is the Manning's friction coefficient. Water depths are updated using Equation (2):

$$h^{t+\Delta t}_{i,j} = h^t_{ij} + \Delta t \frac{Q^t_{xi-1j} - Q^t_{xij} + Q^t_{yij-1} - Q^t_{yij}}{\Delta x^2} \quad (2)$$

where Q is the discharge, h_{ij} is the depth of water in the cell (i,j). The stability criterion for this numerical model is given by the Courant–Freidrichs–Levy condition for shallow water flows such that the stable model time step, Δt, is a function of the grid resolution, Δx, and the maximum water depth, h, within the domain:

$$\Delta t_{max} = \alpha \frac{\Delta x}{\sqrt{gh_t}} \quad (3)$$

where α varies between 0.2 and 0.7. The reader is referred to Bates *et al.* (2010) [64] for a fuller explanation of this model.

Importantly, when coupled to the defence response models this model is rapid enough for real-time regional simulations up to the peak water level, driven by applying water level time-series at the model boundary.

Almost 5000 potential inflow locations were defined along 246 km of intertidal shoreline comprising point vectors with data attached (including defence crest height floodplain height at that location, and parameters for overtopping and breach analysis). These were placed in cells on the seaward edge of the 50 m resolution DEM, covering all flood compartments. A line of points was also placed in cells immediately inland of the defences for wave overtopping failures to be simulated. Water level time series at these locations were based upon tide-surge water level time-series (see Figure 3), adapted according to defence responses if necessary. Failure of defence systems is quite rare, although outflanking, overflow, overtopping and breach were considered.

Scope for calibration by varying surface friction values was limited by a lack of accurate flood observations, although boundary conditions (defence failure, waves, water level, *etc.*) dominated flood inundation modelling results. A uniform representative composite value of n = 0.035 was used, based on the range of surface roughness values found across the region. Some allowance was made for DEM error by excluding analysis of flooding in cells where water depths were less than 0.05 m [70]. As shown in Table 2, flood simulations over a relatively large area could be run quickly, even using a standard PC.

Table 2. Summary of the model scale and run-time (for a standard 2.5 GHz desktop PC).

Model	Domain size		DEM resolution (m) and file size (MB)		Number of inflow points	Time-step (seconds)	Run-time to simulate 14 hours of real-time flooding (minutes)
	Km^2	columns, rows					
Solent	105.7	1227,846	50	3.87	4909	4.2–6.3	15
City of Portsmouth	15.2	150, 168	50	0.01	595	5.4–8	<1

3.3.1. Overflow

Overflow failure is when the still water level exceeds the crest height of a defence (or of the land if there is no defence). For flood simulations comprising no waves an estimate of overflow during a flood event was calculated by applying still the storm-tide water level time-series to inflow points lying seaward of defences in the DEM. When wave conditions were included, the original peak still water level was routed through boundary cells, although with water level time-series adjusted for the duration that the still water level could theoretically exceed the defence crest height. The latter method is most relevant to shorelines where defence crests are at a similar elevation to the ground level of the surrounding floodplain; or where seepage, gaps, and alongshore variations in crest heights (for any 50 m defence section) may allow a greater inflow.

3.3.2. Overtopping

Flooding due to overtopping is when the combined effects of water levels and waves results in water entering a floodplain at a faster rate than it can drain away. This mode of failure can occur when the still water level is below the height of the defence crest or the floodplain. To generate boundary conditions for the inundation model, the mean overtopping discharge, percentage of overtopping waves, and run-up were calculated at stages of the tidally varying freeboard using empirical methods from the EurOtop Manual [71]. For sloping structures the formula used was that by Owen (1980) [72]; TAW (2002) [73] for beaches; and for vertical and composite sea walls more iterative formulae (to discriminate between pulsating, impulsive and broken wave conditions, according to the standard EurOtop guidance). A bathymetry model was constructed using nearshore sounding data [46] to approximate the regionally variable effects of wave shoaling upon significant wave heights. The duration of overtopping events was delineated by time intervals where mean overtopping rates or run-up exceeded the known capacity of defences. If for any inflow point an overtopping event was simulated; the onset and termination of floodplain discharge was defined by exceedance of the run-up two percent value (at sloping defences) or a mean overtopping volume discharge of greater than 20 l/s/m. When defence crest freeboard was negative and the floodplains below still water level; the still water level time series was simply routed through the failed boundary cell (although restricted to the calculated overtopping duration). For most floodplains the peak (or equilibrium) water level in the overtopped boundary cell was limited to the seaward still water level. This limit was higher for floodplains above still water levels, based upon prior analysis of run-up and overtopping volumes in the relevant flood compartments. The results in this paper assumed a peak significant wave height (H_S) and mean period (T_m) to occur for the duration of an overtopping event; although this could be adjusted to account for more dynamic conditions if required. At each inflow location this was converted to a vertical change in water level expected to occur in the lee of defences during a single tidal cycle; and added to the DEM floodplain level in the boundary cell, forming the peak level in a modified time series of water levels.

Analysis of overflow and overtopping was applied to over 224 km of the region's shoreline, including beaches, embankments and vertical sea walls. It was found that approximately 116 km of these defences rise far enough above the floodplain (whilst preventing seawater from reaching hinterland) for breaching to be relevant.

3.3.3. Breach Analysis

Breaching, although rare, is considered because this type of defence failure is recognised to greatly increase the consequences of a flood event, especially for large flood plains [19]. Analysis was allocated to defences where breaching is relevant, allowing for the identification of possible regional 'weak spots' and information for the modeller to route water through boundary cells if chosen. A simplified reliability analysis generated fragility curves for three selected failure modes. This was facilitated by software developed under FLOOD*site* Task 7 [74], which enables a Level III reliability approach (*i.e.*, Monte-Carlo simulation), (see [27] for a more thorough description). Three breach failure mechanisms were selected from a much wider inventory [75]. These were: (1) analysis of damage initiated by overflow and/or overtopping [76]; (2) failure caused by water internally eroding defences (piping) [77]; and (3) critical erosion of shingle beaches in response to wave and water level loadings [78]. The former two failure mechanisms were analysed using condition grade scores and structural descriptors to imply defence strength and stability [79]. Topographic profile surveys were used for analysis of shingle beaches (which are important natural and managed defences in many areas of the Solent). Methods exist to calculate progression of breaching, and subsequent release of water onto the floodplain but any deterministic prediction remains highly uncertain [80,81]. Hence the modelling results in this paper take a simplified approach with breaches assumed to be broken down to floodplain level and water flowing through for the entire tidal cycle. Section 3.4.2 (this paper) describes inundation simulations where specific breach locations were approximated for a validation case study; whereas results in Section 4 describe flood simulations where breaching scenarios consider all shoreline defences to have been removed (and wave height reduction coefficients removed where foreshore usually alleviates overtopping). This allowed a simplified comparison with overtopping scenarios, and a more realistic worst-case scenario than the planar-water level results. There is some discussion of breach probabilities during an extreme flood event simulation in Section 4.2.

3.4. Model Validation

A model is said to be validated if its accuracy and predictive capability in the validation period have been proven to lie within acceptable limits or errors for a particular purpose [68]. However, for many coastal areas information to validate inundation models is not available. Flood simulations were compared with observations gained from two actual floods: (1) a regional event on 10th March 2008, which comprised widespread although mostly minor flooding across the Solent; and (2) an event in December 1989 where inundation resulted from overtopping and sea wall breaching between Hurst Spit and Lymington.

3.4.1. 10th March 2008 Regional Flood Event

On the 10th March 2008 an approximately 1 m storm surge and large waves resulted in widespread flooding across the Solent and other coastal areas in the English Channel [82]. The region's cities escaped major inundation, although an example of a near miss in Southampton is shown in Figure 6. Observed flood extents were in the range 0.01–0.03 km^2 with peak flood water depths mostly less than 1 m. An exception was more severe flooding in eastern extent of the region at Selsey (located at the largest and most easterly floodplain shown in Figure 1), where the only known breach of defences occurred in the region that day due to waves flattening an 800 m section of shingle barrier beach [83]. At the time, the peak still water levels were approximated as a 1 in 20 year extreme at Portsmouth and a 1 in 50 year event at Southampton [20]. Waves and water level observations from the event were available from real-time data recorders used for regional coastal monitoring (Figure 1) and information collated by the Environment Agency [84]. These were spatially interpolated between recording stations and applied to the model boundaries (Figure 5). Outputs were compared with observed flooding in GIS. Observations were interpreted from photos, descriptions and reports, e.g., [84] and close examination of a high resolution version of the floodplain DEM which enabled flood extents and depths to be (approximately) verified and reconstructed.

Previous coastal flood modelling studies [9,11] have compared modelled and observed coastal flood extents using a fit measure (F_A) which comprises the intersection and union of predicted (E_p) and observed (E_0) flooded pixels. A value of zero corresponds to no agreement, a value of 1 to perfect agreement:

$$F_A = \frac{E_p \cap E_0}{E_p \cup E_0} \tag{4}$$

For each measurement of F_A it was noted whether the model under or over-predicted flooding. Over-prediction or under-prediction by the model was considered negligible unless more than one full flooded cell lay outside of the observed extent. Hence in the case of a good model fit ($F_A > 0.75$) not all values qualified as over or under-predictions. Limitations to this exercise are acknowledged; such as the use of simplified boundary inputs, the relatively large pixel size (50 m) of the model, uncertainty in the timing of the observed flood data, and interference from other flood sources (during the real event). Flood depth was also noted, using photos and the DEM. The uncertainties from this method meant that if the modelled flood depth was within 30 cm of the observed depth (at any location), this was regarded as acceptable. The results of this exercise are summarised in Table 3.

Figure 6. High water at Southampton docks on 10th March 2008 [85].

Table 3. Summary of the 10th March 2008 model validation.

Test type	Criteria	Result
Detection of observed floods	The model indicated flooding at 29 out of 30 locations where floods were reported (the location missed was on a tidal river where additional sources of floodwater may have occurred). At 27 locations, some description of flooding was available to allow comparison with the model; which is summarised below.	
Flood extent fit scores	Mean F_A value (for the 27 compared locations)	0.65
	over-predictions	6
	under-predictions	8
	neither definitive over or under-prediction	13
	Number of locations with a good fit ($F_A > 0.75$)	**12**
	2 were slight over-predictions, 1 was a slight under-prediction	
	Number of locations with a moderate fit ($F_A = 0.50$ to 0.75)	7
	2 over-predictions, 3 under-predictions	
	Number of locations with a poor fit ($F_A < 0.50$)	**8**
	2 were over-predictions, 4 were under-predictions	
Flood depth	All modelled depths were within 30 cm of observed flood depths.	

The mean F_A value was reduced considerably by over-prediction of flood extent at two small urban floodplains, where the F_A values were 0.10 and 0.32. The former of these values was at a location where waves overtopped a shingle beach at the southeast corner of Hayling Island, the latter was overtopping of sea walls at Old Portsmouth (on the southwest corner of Portsea Island). These over-predicted floods comprised water depths of less than 20 cm. It is possible that these were reasonable predictions of overtopped volumes although in reality ponding of overtopped water may be lessened by floodplain drainage and the presence of surface features such as buildings (neither of which were included in the model). Reducing wave shoaling coefficients in the overtopping model (from approximately 0.75 to 0.60) improved these predictions; although the original values were retained for further modelling on the assumption that drainage may be limited in extreme events whilst shoaling may be expected to change with foreshore and nearshore sediment dynamics during storms. Furthermore, these variables are likely to be negligible in comparison to significant defence failures.

Specific locations of the properties affected by this event were not available, although a post-event report [84] suggested that 30 properties in the region were flooded. Viewing photographs taken at 24 of the flooded locations suggested that at least 120 buildings could actually have been in contact with flood waters (see Figure 7). There is uncertainty over how thoroughly floods were reported, whilst these numbers exclude flooding and evacuation of a holiday park at Selsey [86]. The modelled flood outline of this event contained more than 2000 properties, more than half of these accounted for within the shallow flooded cells in the aforementioned over-predicted flood extent on southeast Hayling Island. Filtering these results by the number of properties in grid cells with flood water deeper than 1 m decreased this estimate of inundated properties to 162; and to only 38 with water deeper than 1.5 m. It is probable that flood waters threatened many more properties than reported, but walls, fences, raised floor levels, and flood prevention measures such as sandbagging prevented significant flooding to interiors of buildings.

3.4.2. 17th December 1989 Flood Event (Breach Failure)

A historic flood was used to evaluate the model's ability to simulate larger floods resulting from significant defence breaching. Such an event occurred on the 17th December 1989 at a mostly rural flood compartment between Hurst Spit and the town of Lymington, on the west Solent. Rapid onset and widespread coastal flooding resulted in environmental damage and severe inundation of 10 properties in the rural area of the floodplain; with overtopping also affecting 50 properties in Lymington [88]. Measurements of wind speed, peak water level, breach locations, and defence heights are available for the event [88] and were used to model flooding over a single tidal cycle. Variations of boundary inputs were used to run several flood simulations, and these were compared to observed flood extents producing F_A values between 0.7 and 0.8. However, these F_A values do not take into account the two notable discrepancies between the present day DEM and the reported flooding of December 1989 (as noted in Figure 8). The landfill area is dynamic and flooding cannot be assessed, while in the vicinity of Milford on Sea, the breaching of Hurst Spit may have sealed a channel, protecting Milford-on-Sea from flooding. It is possible that other areas of the floodplain surface may also have altered between the time of the event and collection of the recent LiDAR data; although the old defence data (which is less detailed than the present-day dataset), nature of breaching, and wave data provides greater uncertainty to the model inputs for this validation exercise. As with the 10th March 2008 validation case-study, filtering properties within grid cells of certain depths provided a closer comparison with reported flooding. Counting only properties where modelled flood water depths exceeded 0.5 m indicated 104 properties to be inundated; depths of greater than 1 m indicated 47 properties to be flooded (60 properties were reported to have actually experienced flood damage [88]).

Figure 7. Flood waters surround properties in Emsworth, Borough of Havant near high water on 10th March 2008, [87].

The fragility curve methodology and recent defence data was partly verified by this event. Information that indicated the condition and height of the sea wall in 1989 [88] produced breach failure probabilities ranging between 0.8–1.00 when subjected to the 17th December 1989 loadings (*i.e.*, the reliability analysis indicating definitive exceedance of the limit state for several sections). The same loadings applied to the present-day sea wall data indicated both minimal inundation from overtopping, and no definitive defence section failures, although breach probabilities were between 0.6 and 0.9 for some sections. This site did not flood on the 10th March 2008, even though the water levels exceeded those of the 17th December 1989.

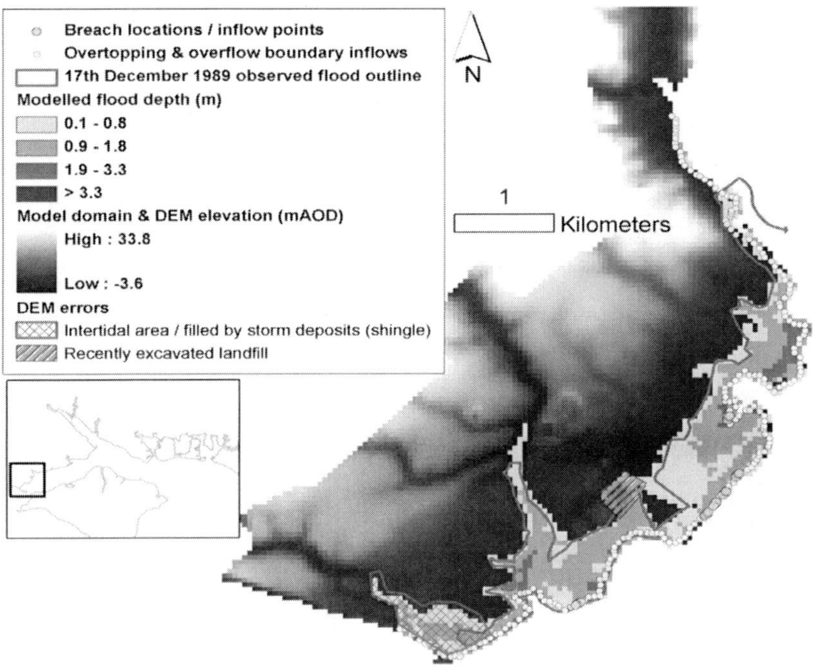

Figure 8. The 17th December 1989 flood simulation and observed flood outline. A boundary water level of 2.1 mAOD recorded at Lymington [88] was applied across the entire site for this simulation. This may be a relevant approximation since the breach of Hurst Spit that day e.g., [89] would be likely to have altered local hydrodynamics [90,91] (whereas present-day event flood simulations assume Hurst Spit stays in place, and that tidal range/extreme water levels increase eastwards across the site). Over-washed material from the storm may also have blocked the intertidal area shown (at Milford on Sea) complicating comparison between the observed and modelled floods.

4. RESULTS

Performance measures verified that the main components of the model (DEM, defence data, *etc*) are sufficiently accurate for modelling hypothetical coastal flood events. This section firstly focuses upon simulation of a 1 in 200 year regional coastal still water level flood (indicative of the standard of protection for urban coastal defences in England and Wales and definition of Environment Agency flood zone maps). Secondly, inundation across a range of coastal flood simulations is described. This indicates model sensitivity and uncertainty related to boundary water levels, and provides a regional overview of potential flood event impacts at present and higher sea levels. In this paper, the inundation simulations driven by defence failures are in a deterministic form, rather than comprising a sampling-based flood system analysis e.g., [26]. Inundation scenarios modelled for each still water level include: (a) the planar water level flood outline; (b) a worst-case hydraulically modelled flood that includes breach of all defences; and (c) three categories of wave loading simulated with the hydraulic model although *without* breaching (these include no waves, an annual extreme, and the maximum possible wave conditions) hence accounting for overflow, overtopping and outflanking failures only (for one tidal cycle). The still water levels (and definition of flood event extremity) applied to the model boundaries were based upon the extreme water level analysis described in section 3.1 and Figure 5. The water level time-series applied to the inflow points simply comprised shifting storm-tide curves (Figure 3) to the relevant peak still water level.

4.1. Simulation of an Extreme Present-Day Flood Event

Table 4 summarises the results of a 1 in 200 year water level flood. As expected e.g., [9], hydraulic modelling (even with extreme wave overtopping and breach scenarios) estimated less flooding than calculated by the planar water level method. The flood simulation using an annual wave extreme suggests that more than 10,000 properties could be impacted by this water level event. The model validation case studies indicated that waters exceeding 1 m depth could be a more useful indicator of the number of buildings that experience flooded interiors. This criterion suggests that for the synthetically modelled floods, increasing wave loadings from the annual to maximum extreme (and with no breaching) almost doubles the number of properties that would be significantly inundated. Inclusion of breaching (of all the region's defences) in the hydraulic flood simulation method almost trebles the effects from waves alone (for the count of properties inundated to greater than 1 m depth).

Table 4. Modelling results for a 1 in 200 year water level coastal flood in the Solent (refer to Figure 1 for locations).

REGION (see Figure 1)	Location	Exposure (numbers in brackets is land area inundated to >1m depth)		Properties flooded by hydraulic simulations Total (number in brackets are properties inundated to >1 m depth)			
		Floodplain area (km^2)	Properties	Breach of all defences and max. waves	Max. waves	1 in 1 yr waves	No waves
1	New Forest	14.1 (6.6)	736 (134)	639 (147)	591 (112)	430 (55)	329 (34)
2	City of Southampton	2.0 (0.1)	2,148 (249)	978 (21)	973 (21)	919 (21)	715 (10)
3	Eastleigh, Fareham & Gosport	7.2 (2.4)	1,789 (106)	1,395 (109)	1,321 (93)	1,163 (91)	564 (52)
4	City of Portsmouth	13.8 (4.9)	14,055 (3,483)	10,922 (2,734)	6,437 (459)	4,660 (110)	1,923 (8)
5	Havant & Chichester	29.8 (13.2)	3,781 (1,657)	3,168 (613)	2,995 (448)	2,602 (329)	1,033 (90)
6	Isle of Wight (all)	11.4 (5.6)	617 (155)	505 (114)	377 (67)	354 (67)	255 (25)
Total	**Solent**	78.3 (32.8)	23,126 (5,784)	17,607 (3,728)	12,694 (1,200)	10,128 (673)	4,816 (219)

In the city of Portsmouth, a substantial amount of the region's property is threatened by flooding under all loading scenarios. However the city's flood defences greatly reduce the number of properties exposed to more than 1 m flood depth; and the extreme wave and overtopping and breaching scenarios are required to significantly increase the count of buildings likely to be severely flooded. The reliability analysis indicated that loads associated with the 1in 200 water level (combined with waves) would generate low probabilities of breach for most of the city's defence structures. The probability of failure from the fragility curves was less than 20 percent for defences protecting two of the largest clusters of property in the central south and mainland areas, although failure probabilities are higher for sea walls on the east coast of Portsea Island. In the city of Southampton the added effect of waves with still water level expands the flooded area significantly, despite the fetch-limited nature of the shoreline. However, only 21 properties in the city are inundated to greater than 1 m depth. Hence, at present sea level Southampton is likely to be subject to disruptive impacts from coastal flooding, rather than significant risk to life and property.

Note that 531 properties were flooded to greater than 2 m depth in the case of an extreme breaching scenario (and approximately 130 for maximum and annual wave scenarios). The majority of these are in Portsmouth (311 properties) under the breach scenario; whilst most are in Chichester and Havant (102 properties) when applying only the maximum waves scenario (to generate extreme overtopping-dominated defence failures). There are also substantial areas of floodplain where flood depths are between 1 m and 2 m, which can be particularly dangerous when floodwaters are fast moving [92]. The extraction of this information from the model indicates pockets of particularly high damage potential and risk to life. Inclusion of widespread breaching is shown to increase regional flooding impacts substantially (Table 4), although only a relatively small length of coastline actually appears at high risk of this kind of failure. Locations with the highest probability of breaching are mostly in rural areas on tidal rivers and in the harbours. Simulating inundation through defence sections for which the failure probability exceeded 90 percent (along approximately 35 km of defence line) is not shown here, although generated a similar inundation extent to that of the extreme wave overtopping scenario. Slightly greater floodplain water depths resulted from inclusion of these breaches, although many of these defences are likely to have already significantly overtopped prior to breaching. Such failures would inevitably increase the consequences of a flood event, although the overall impacts would be dependent upon factors such the timing of breach initiation and growth, and repair measures.

4.2. Coastal Flood Simulations Across a Range of Loadings

Flood simulation results across a range of still water level are shown in Figure 9a, Figure 10a and Figure 11a (all flood simulations assumed the present defences). Also shown for each measure of inundation, is the proportion of flooding caused by non-breach defence failures in comparison to inundation simulations where all defences were removed (Figure 9b, Figure 10b and Figure 11b). These results indicate that substantial land areas and numbers of property are threatened with flooding from frequently occurring water levels, with defences playing a crucial role in the magnitude of these consequences. The number of properties in the modelled flood outlines increases steadily with increasing boundary water levels. The difference in the total number of flooded properties and land area when comparing the hydraulically simulated floods and the planar water level (exposure) floodplain indicates that much of the wider floodplain may not be hydraulically well-connected over a single tidal cycle. At the same time, the effects of overtopping can be significant and these results show that such floods can almost match the planar water level estimates for properties flooded to greater than 1 m depth (Figure 10a). These results show that for some individual flood compartments the number of properties inundated by the non-breach defence failure mechanisms increases most distinctly within the (approximately) 10 cm band of water levels either side of the 1 in 200 year extreme (because the protective capacity of many coastal defences is exceeded in this range of loadings). Using the 1 m depth criteria shows that for the Solent as a whole, there a particularly large increase in the count of flooded properties beyond the 1 in 200 year still water level.

Figure 9b and Figure 11b show a general increase (in response to increasing still water levels) in the percentage of inundation that arises from overflow/overtopping failures in relation to breach. This is expected as defences become increasingly overwhelmed beyond their protective capacity due to rising still water level boundary conditions allowing larger and prolonged discharges during flood events. However, for the 1 m depth criteria (Figure 10b) the potential for breach to exert the greatest flood damages across the Solent is most apparent for small water levels (where breach is unlikely and significant overtopping not possible), and a wide range of present-day extremes. Beyond present-day still water level extremes, the gap between breaching and overtopping impacts is greatly diminished.

UKCP09 [63] global mean sea level projections derived from the IPCC Fourth Assessment Report [93] give an estimated range (5th to 95th percentile) for sea level increase of 18–59 cm between present day (assuming a 1980–1999 baseline) and 2090–2099. For simplicity the effects of a medium UKC09 scenario (of approximately 0.5 m sea-level rise) upon a 1 in 200 year water level flood event are shown in Figure 9, Figure 10, Figure 11. This approximated scenario of 21st century sea-level rise substantially increases the amount of property affected by at an extreme water level coastal flood event in the Solent. For example, in the case of a 1 in 200 year water level flood combined with an annual wave extreme, the total number of buildings in the regionally modelled flood outline increases from 10,128 to almost 24,000 (approximately 140 percent change), although there is a more dramatic 600 percent increase if considering >1 m floods and probably provides a better indication of actual impacts assuming no adaptation (Figure 10a). Even without the effects of waves, a 1 in 200 year flood event in addition to 0.5 m sea-level rise (and no defence improvements) allows nearly 4400 properties to be inundated to greater than 1 m depth.

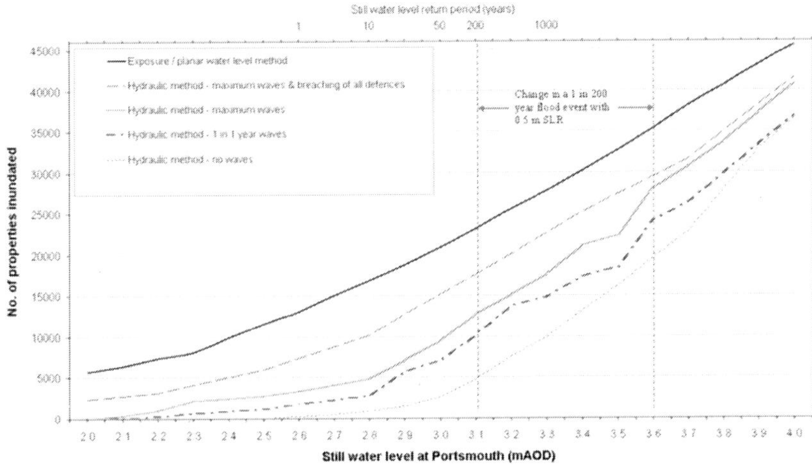

(a)

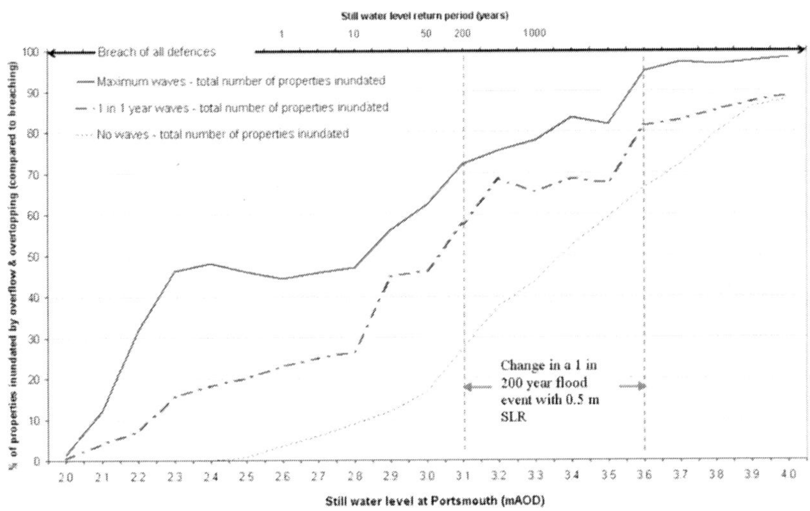

(b)

Figure 9. (**a**) Raw inundation modelling results for properties within modelled coastal flood outlines across the Solent–water level return period refers to annual probability of that water level for 2011; (**b**) Percentage of properties inundated by overflow and overtopping defence failures (compared to breach of all defences).

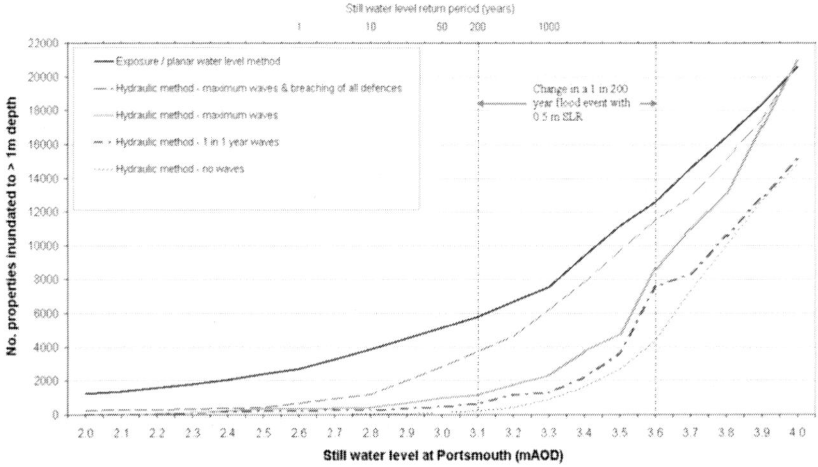

(a)

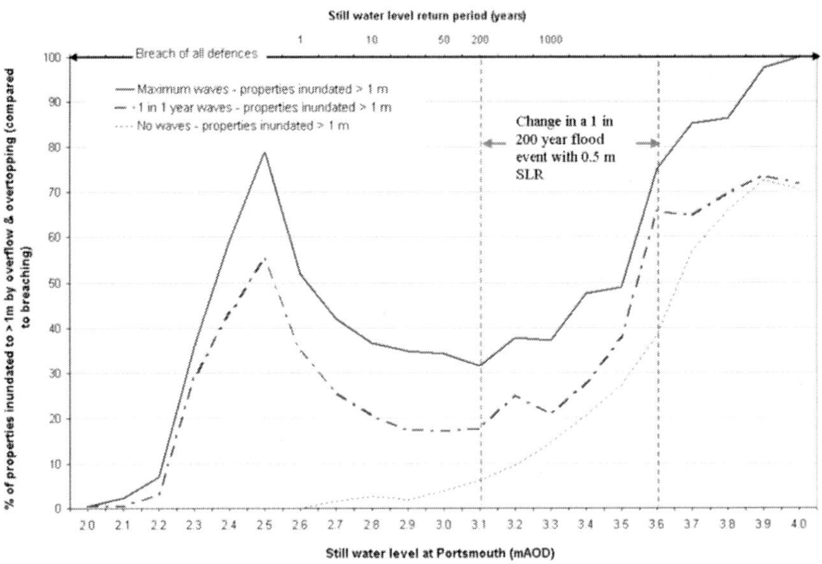

(b)

Figure 10. (**a**) The number of properties in the Solent inundated to a water depth of greater than 1 m; (**b**) Percentage of properties inundated to greater than 1 m water depth by overflow and overtopping defence failures (compared to breach of all defences).

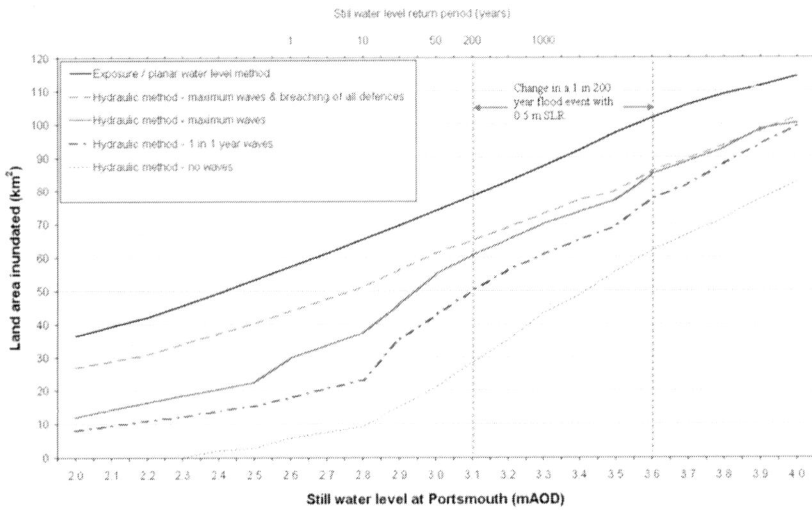

(a)

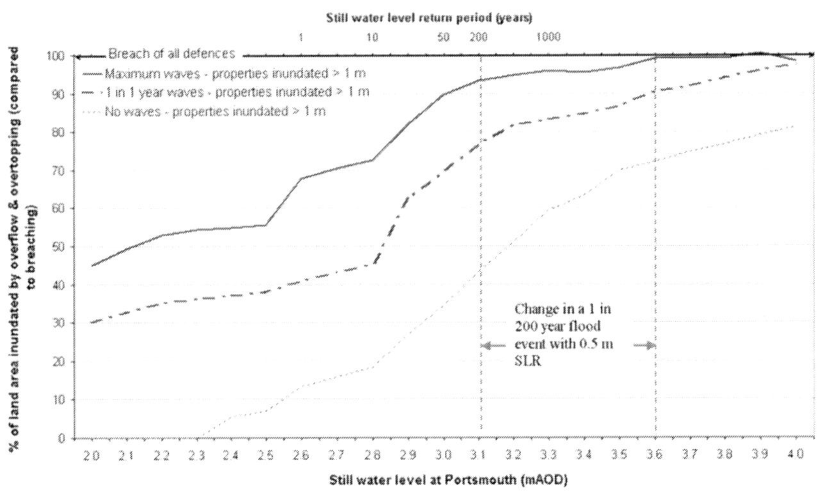

(b)

Figure 11. (**a**) Coastal flood modelling results across the Solent by land area; (**b**) Percentage of land area inundated by overflow and overtopping defence failures (compared to breach of all defences).

5. DISCUSSION

5.1. Application of the Methodology

The case study results illustrate how the described methodologies provide a regional overview of the impacts of coastal flood events; and also allow a comparison between the consequences of different types of defence failure. In the region of interest, overtopping is more important than in many previously investigated sites, such as eastern England. This shows that investigating flooding across a range of defence failure mechanisms provides useful insights.

Estimates of coastal flood exposure by the planar water level method were subject to substantial uncertainty, stemming from vertical accuracy of the DEM, compilation of property datasets, and definition of what comprised inundation (*i.e.*, whether the interior or garden of the property is affected). Based upon ±0.15 m variation in the vertical LiDAR accuracy, the regional exposure estimate (generated by the count of buildings in the flood outline) could vary upwards by more than 30 percent, and downwards by approximately 10 percent.

The reason for this asymmetry in uncertainty was because the modelled flood outline could be used to select point vectors representing property locations. The upper absolute exposure estimate was different because it also comprised adding a 20 m buffer (an approximation for the spatial footprint of properties within the dataset). Similar principles are applicable to the hydraulic modelling outputs, although accounting for defence failures, mass conservancy and hydraulic connectivity of flood spreading outweigh this effect.

The hydraulic modelling method performed well, using high quality defence and floodplain data that exists in the Solent. The greatest uncertainties in boundary condition processes and resultant inundation was for wave overtopping onto floodplains that are above the peak still water level. In these situations the irregular nature of waves transmitted over a structure and interactions with drainage may be particularly complex to model in a simplified way.

The flood modelling method was computationally fast and differentiated from many existing regional approaches by the high level of spatial detail in schematisation of defences and floodplain, with wave overtopping and outflanking flood mechanisms included in all inundation simulations.

5.2. Flood Impacts in the Solent

The 1 in 200 year flood simulation demonstrates how the impacts of an extreme present day flood event could be significantly greater than impacts seen on the 10th March 2008. Under equivalent wave loadings, the approximately 30 cm increase in the boundary water levels quadruples the amount of property likely to be affected by inundation (which would increase further with larger waves). The effect of more severe and widespread defence failures (than those seen on the 10th March 2008) would also involve more prolonged , widespread and faster flowing floodplain water; with the 1 in 200 year water level potentially causing sea water to inundate between 30–65 km^2 land. Flood simulations identified significant flooding in populated areas outside of the building dataset; notably at the largest holiday park in the region at Selsey which can accommodate up to 12,500 people. Areas of this site were inundated to depths between 1–2 m in both the annual and extreme overtopping scenarios for the 1 in 200 year simulations. Statistics for people per property and occupancy rates from the 2001 UK Census suggest that on average at least two people may occupy each building in the dataset; hence the population impacted by the large flood events in the region is likely to be substantial and will grow significantly with 21st century sea-level rise (although the defences at Selsey are currently being upgraded). Without changes to defences there is a large increase in the amount of property threatened by flooding when applying the 0.5 m sea-level rise example. Hydraulic modelling of the worst possible 1 in 200 year coastal flood at present-day sea level suggested that approximately 500 properties lay in areas where flood waters may reach 2 m or deeper; this number trebling with the 0.5 m sea-level rise example. Without breach (at present day sea levels) this is far less, although with 0.5 m sea-level rise this level of flooding occurs when a 1 in 200 year event combines with only an annual extremity of wave scenario (without breach failures included). Due to this, and the increasing impacts of overflow and overtopping failures with sea-level rise (shown in Figure 9b, Figure 10b and Figure 11b) the importance of future defence improvements is apparent, or a significant increase in flood consequences is inevitable.

Of the ninety-nine flood compartments in the Solent (Section 3.2), twelve are (at present day sea levels) likely to experience significantly greater flood event impacts due to breaching failures (in comparison to severe overtopping events). These compartments are quite large, and account for 40 percent of the land area that is at risk of flooding in the Solent. However, it is apparent that other failure mechanisms are currently significant to flood impacts of both frequently occurring and extreme events in this region.

5.3. Directions for Further Research

This methodology would be complimented if used alongside more spatially descriptive data of regional coastal flood sources, and receptors. The former would be possible by applying nearshore hydrodynamic modelling and the latter by supplementing existing datasets with spatial distributions of population, vulnerability and key infrastructure. The defence and floodplain datasets in the region could be used for higher resolution modelling.

Inundation modelling using a coarsened, interpolated DEM has been observed in other studies to overestimate flood extent and underestimate flood depth [49], and modelling of urban inundation benefits significantly from finer discretisation of space [49,70]. This may only be feasible or offer significant benefits at some locations in the region; the most obvious choices being the region's two cities (Portsmouth and Southampton), and low-lying flood compartments where breaching is a threat.

The fast run-time of the regional model and easy integration with GIS could have other applications. For example, within operational real-time coastal flood forecasting systems [94] the modelling of flood pathways (defence failures and inundation mapping) could be better integrated to benefit real-time flood management [31,95,96,97], e.g., by enabling more efficient and strategic flood warning dissemination, improved visual information (to communicate flood event outcomes), and more quantitative appraisal of warning system performance.

The development of practical methods for the application of wave overtopping to the boundaries of coastal inundation models is an important area of future research, for which limited guidance is currently available. Uncertainty analysis is also recognised as crucial to numerical modelling studies [98] and could be more formally integrated for a more comprehensive analysis of present and future flood impacts e.g., [51]. Better calibration is recommended; with compilation of flood data when available to improve models and their interpretation.

The 10th March 2008 data (Section 3.4.1) indicated lack of damage to property to flooding below certain water depths, although this also demonstrated the difficulty of using counts of flooded properties to validate models. However, this information probably helped to generate a more realistic interpretation of flood impacts from the synthetic flood event simulations.

6. CONCLUSIONS

A method for regional analysis of coastal flooding has been demonstrated, using the Solent, UK as a case study. It integrates a number of existing approaches for modelling defence response and floodplain inundation. This allowed individual, synthetic and real, coastal flood events to be simulated, providing a simple evaluation of regional coastal flood impacts. By modelling across different water levels, an overview of the uncertainty relating to boundary condition variations is available. This also showed how overtopping and overflow mechanisms contribute a large present day flood threat across the Solent. Breaching has the potential to generate significant impacts across a range of present day still water levels; although across the range of still water levels expected with 21st century sea-level rise the effects of overflow and overtopping are similar to the effects of broad-scale breaching.

This research offers benefits for the understanding of coastal flooding in a region that has undergone less analysis than the east and west coasts of the UK. This research was also produced in recognition that there is an increasing need to establish methods which improve the understanding of coastal flood events. The run-time for flood event simulations is rapid, allowing multiple uses; such as (1) flood warning; (2) defence prioritisation; (3) floodplain mapping and land-use planning; and (4) operational coastal management. Furthermore outputs are simple and flexible enough to communicate flood impacts to non-scientific audiences. It is recommended that these approaches be further explored, particularly for Portsmouth and Southampton where the majority of coastal population and infrastructure reside, as well as areas where breaching is a threat. Risks and management approaches could change differently for Portsmouth and Southampton over the coming century. For the former, risk is already high and investments in defences already exist with the need for upgrade recognised. Southampton is at present mostly undefended, therefore adaptation will represent a bigger change from current practice, although the needs for a response are increasingly recognised. Globally there are many coasts such as the Solent where surges are not large and flood risk is currently not a major issue; but sea-level rise will significantly exacerbate flood risks. The transferability of this modelling to other coastal regions is in principle quite simple. However, flood predictions were sensitive to small variations in crest height; whilst high resolution and recent datasets (e.g., for defences, floodplains, and extreme sea level analysis) is not available everywhere.

ACKNOWLEDGMENTS

This research was part of a Ph.D funded by the Engineering and Physical Sciences Research Council (EEPSRC) and the 'SemSorGrid4Env' (SG4E) EU research project (http://www.semsorgrid4env.eu/home.jsp). Data was provided by the Channel Coastal Observatory (http: www.channelcoast.org), the Environment Agency Southern Region and the Havant Borough Council/Portsmouth City Council Coastal Defence Partnership. Paul Bates, Jeff Neal and Tim Fewtrell from the University of Bristol provided LISFLOOD-FP and expert guidance.

REFERENCES

1. Van Koningsveld, M.; Mulder, J.P.M.; Stive, M.J.F.; van der Valk, L.; van der Weck, A.W. Living with sea-level rise and climate change: A case study of the netherlands. J. Coast. Res. 2008, 24, 367–379.
2. Dawson, R.J.; Ball, T.; Werritty, J.; Werritty, A.; Hall, J.W.; Roche, N. Assessing the effectiveness of non-structural flood management measures in the thames estuary under conditions of socio-economic and environmental change. Glob. Environ. Change 2011, 21, 628–646.
3. Gerritsen, H. What happened in 1953? The big flood in the netherlands in retrospect. Philos. Trans. R. Soc. A Math. Phys. Eng. Sci. 2005, 363, 1271–1291.
4. Baxter, P.J. The east coast big flood, 31 January–1 February 1953: A summary of the human disaster. Philos. Trans. R. Soc. A Math. Phys. Eng. Sci. 2005, 363, 1293–1312.
5. McRobie, A.; Spencer, T.; Gerritsen, H. The big flood: North sea storm surge. Philos. Trans. R. Soc. A Math. Phys. Eng. Sci. 2005, 363, 1263–1270.
6. Bütow, H. Die große flut in Hamburg: Eine Chronik der Katastrophe vom Februar 1962; Verlag: Hansestadt, German, 1963.
7. Von Storch, H.; Woth, K.; Gönnert, G. Storm surges—The case of Hamburg, Germany. ESSP OSC Panel Session on "GEC, Natural Disasters, and Their Implications for Human Security in Coastal Urban Areas", Available online: http://www.safecoast.nl/editor/databank/File/hamburg-storms.pdf (accessed on 26 January 2012).
8. Battjes, J.; Gerritsen, H. Coastal modelling for flood defence. Philos. Trans. R. Soc. Lond. Ser. A Math. Phys. Eng. Sci.2002, 360, 1461–1475.
9. Bates, P.D.; Dawson, R.J.; Hall, J.W.; Horritt, M.S.; Nicholls, R.J.; Wicks, J.; Mohamed Ahmed Ali Mohamed, H. Simplified two-dimensional numerical modelling of coastal flooding and example applications. Coast. Eng. 2005, 52, 793–810.
10. Brown, J.D.; Spencer, T.; Moeller, I. Modeling storm surge flooding of an urban area with particular reference to modeling uncertainties: A case study of Canvey Island, United Kingdom. Water Resour. Res. 2007, 43, W06402.
11. Gallien, T.W.; Schubert, J.E.; Sanders, B.F. Predicting tidal flooding of urbanized embayments: A modeling framework and data requirements. Coast. Eng. 2011, 58, 567–577.
12. Smith, R.A.E.; Bates, P.D.; Hayes, C. Evaluation of a coastal flood inundation model using hard and soft data. Environ. Model. Softw. 2012, 30, 35–46.
13. Narayan, S.; Hanson, S.; Nicholls, R.J.; Clarke, D.; Willems, P.; Ntegaka, V.; Monbaliu, J. A holistic assessment of coastal flood systems using system diagrams and the source-pathway-receptor (spr) concept. J. Nat. Hazards Earth Syst. Sci. 2012. in press.

14. Dawson, R.; Peppe, R.; Wang, M. An agent-based model for risk-based flood incident management. Nat. Hazards2011, 59, 167–189.
15. RMS. Hurricane Katrina: Profile of a Super Cat, Lessons and Implications for Catastrophe, Risk Management Solutions; Risk Management Solutions; Risk Management Solutions, Inc.: Newark, CA, USA, 2005. Available online: http://www.Rms.Com (accessed on 14 June 2011).
16. Hanson, S.; Nicholls, R.; Ranger, N.; Hallegatte, S.; Corfee-Morlot, J.; Herweijer, C.; Chateau, J. A global ranking of port cities with high exposure to climate extremes. Clim. Change 2011, 104, 89–111.
17. Lumbroso, D.M.; Vinet, F. A comparison of the causes, effects and aftermaths of the coastal flooding of england in 1953 and france in 2010. Nat. Hazards Earth Syst. Sci. 2011, 11, 2321–2333.
18. Kolen, B.; Slomp, R.; van Balen, W.; Terpstra, T.; Bottema, M.; Nieuwenhuis, S. Learning from French Experiences with Storm Xynthia–Damages after a Flood; ©HKV LIJN IN WATER: Waterdienst, The Netherlands, 2010.
19. Muir Wood, R.; Bateman, W. Uncertainties and constraints on breaching and their implications for flood loss estimation. Philos. Trans. R. Soc. A. Math. Phys. Eng. Sci. 2005, 363, 1423–1430.
20. Haigh, I.; Nicholls, R.; Wells, N. Rising sea levels in the English Channel 1900 to 2100. Proc. ICE Maritime Engineer.2011, 164, 81–92.
21. Menéndez, M.; Woodworth, P.L. Changes in extreme high water levels based on a quasi-global tide-gauge data set. J. Geophys. Res. 2010, 115, C10011.
22. Wahl, T.; Jensen, J.; Frank, T.; Haigh, I. Improved estimates of mean sea level changes in the german bight over the last 166 years. Ocean Dyn. 2011, 61, 701–715.
23. De Moel, H.; van Alphen, J.; Aerts, J.C.J.H. Flood maps in Europe—Methods, availability and use. Nat. Hazards Earth Syst. Sci. 2009, 9, 289–301.
24. Reeve, D.E.; Burgess, K.A. A method for the assesment of coastal flood risk. IMA J. Manag. Math. 1993, 5, 197–209.
25. Hall, J.; Dawson, R.J.; Sayers, P.B.; Rosu, C.; Chatterton, J.B.; Deakin, R.A. A methodology for national-scale flood risk assessment. Water Marit. Eng. ICE 2003, 156, 235–247.
26. Dawson, R.; Hall, J. Adaptive importance sampling for risk analysis of complex infrastructure systems. Proc. R. Soc. A Math. Phys. Eng. Sci. 2006, 462, 3343–3362.
27. Gouldby, B.; Sayers, P.; Mulet-Marti, J.; Hassan, M.; Benwell, D. A methodology for regional-scale flood risk assessment. Water Manag. 2008, 161, 169–182.
28. EA. Flooding in England: A National Assessment of Flood Risk; Environment Agency: Rotherham, UK, 2009. Available online: http://publications.Environment-agency.Gov.Uk/pdf/geho0609bqds-e-e.Pdf (accessed on 23 April 2012).
29. Defra. Shoreline Management Plan Guidance Volume 1: Aims and Requirements; DEFRA: London, UK, 2006. Available online: http://www.Defra.Gov.Uk/publications/ files/ pb11726-smpg-vol1-060308.Pdf (accessed on 25 January 2012).
30. Motyka, J.M.; Brampton, A.H. Coastal Management: Mapping of Littoral Cells Ref: SR328; HR Wallingford: Oxfordshire, UK, 1993.
31. DEFRA; Environment Agency. Best Practice in Coastal Flood Forecasting: R&d Technical Report FD2206/TR1; HR Wallingford Report TR 132; DEFRA: London, UK, 2004.
32. Bradbury, A.; Kidd, R. Hurst spit stabilisation scheme: Design and construction of beach recharge. In 33rd MAFF Conference of River and Coastal Engineers, Keele University, Staffordshire, UK, 1–3 July 1998.
33. Haigh, I.; Butcher, P.; Harris, J.M.; Cooper, N.J.; Trip, I. Developing an improved understanding of storm surges in the solent. In Proceedings of the39th DEFRA Flood and Coastal Management Conference, York, UK, 29 June–1 July 2004.
34. Law, C.R. Storm surges in the English Channel. Hydrogr. J. 1975, 2, 30–34.
35. Haigh, I. Extreme Sea Levels in the English Channel. Ph.D Dissertation, University of Southampton, Hampshire, UK, 2009.
36. Ruocco, A.; Nicholls, R.; Haigh, I.; Wadey, M. Reconstructing coastal flood occurrence combining sea level and media sources: A case study of the Solent, UK since 1935. Nat. Hazards 2011, 59, 1773–1796.

37. Pugh, D.T. Tides, surges and mean sea-level. In A Handbook for Engineers and Scientists; Wiley: Chichester, UK; Volume 1987.
38. Wells, N.C.; Baldwin, D.J.; Wang, J.Y.; Collins, M.B. Modelling of extreme storm surge events in the English Channel for the period 14–18 December 1989. Glob. Atmos. Ocean Syst. 2001, 7, 275–294.
39. Mason, T.; Bradbury, A.; Poate, T.; Newman, R. Nearshore wave climate of the English Channel—Evidence for bi-modal seas. In 31st International Conference on Coastal Engineering, Hamburg, Germany, 31 Auguest–5 September 2008; Jane, M.S., Ed.; World Scientific: Singapore; Hamburg, Germany, 2009; pp. 605–616.
40. Evans, E. Foresight: Scientific Summary. Volume ii, Managing Future Risks; Office of Science and Technology: London, UK, 2004.
41. NFDC. Cabinet 06/07/09 Report A: North Solent Shoreline Management Plan; New Forest District Council: Hampshire, UK, 6 July 2009. Available online: http://www.Newforest.Gov.Uk/committeedocs/cab/cdr05054.Pdf (accessed on 15 December 2011).
42. Portsmouth City Council (PCC). Portsmouth City Local Plan, First Review 2001–2011. Supplementary Planning Guidance: Flood Protection. Approved by the Executive Member for Planning, Regeneration, Economic Development and Property 6th February 2004; Portsmouth City Council: Wales, UK, 2004. Available online: http://www.portsmouth.gov.uk/media/Flood_Protection_SPG.pdf (accessed on 26 August 2009).
43. RIBA; ICE. Facing up to Rising Sea-Levels: Retreat? Defend? Attack? The Future of Our Coastal and Estuarine Cities; Royal Institute for British Architects; Institute of Civil Engineers: London, UK, 2009. Available online: http://www.buildingfutures. org.uk./downloads/Facing_Up_To_Rising_Sea_Levels.pdf (accessed on 31 January 2010).
44. Atkins. Strategic Flood Risk Assessment for the Partnership for Urban South Hampshire; Document Number: 5049258/72/DG/048; Atkins limited: London, UK, 2007. Available online: http://www.environment-agency.gov.uk/static/documents/Research/ PUSH_SFRA.pdf (accessed on 24 April 2012).
45. OS. Ordnance Survey Land-Form Profile™ User Guide; V4.0–5/2001© Crown Copyright; Ordnance Survey: Southampton, UK, 2001. Available online: http://www. Ordnance survey.co.uk (Accessed on 6 July 2011).
46. CCO. Strategic Regional Coastal Monitoring Programmes; Channel Coastal Observatory: Southampton, UK, 2011. Available online: http://www.channelcoast.org (accessed on 23 April 2012).
47. Xharde, R.; Long, B.; Forbes, D. Accuracy and limitations of AIrborne Lidar surveys in coastal environment. In Proceedings of theGeoscience and Remote Sensing Symposium, 2006; IGARSS 2006. IEEE International Conference, Denver, CO, USA, 31 July–4 August 2006; pp. 2412–2415, ISBN: 0-7803-9510-7.
48. Webster, T.L.; Forbes, D.L.; MacKinnon, E.; Roberts, D. Flood-risk mapping for storm-surge events and sea-level rise using lidar for southeast new brunswick. Can. J. Remote Sens. 2006, 32, 194–211.
49. Fewtrell, T.J.; Bates, P.D.; Horritt, M.; Hunter, N.M. Evaluating the effect of scale in flood inundation modelling in urban environments. Hydrol. Process. 2008, 22, 5107–5118.
50. De Mers, M.N. Fundamentals of Geographic Information Systems, 4th ed; John Wiley Sons Inc.: New York, NY, USA, 2009.
51. Purvis, M.J.; Bates, P.D.; Hayes, C.M. A probabilistic methodology to estimate future coastal flood risk due to sea level rise. Coast. Eng. 2008, 55, 1062–1073.
52. Dixon, M.J.; Tawn, J.A. Extreme Sea-Levels at the UK a-Class Sites: Site-by-Site Analysis; Pol Internal Document Number 65; Proudman Oceanographic Laboratory: Liverpool, UK, 1994. Available online: http://www.pol.ac.uk/ntslf/pdf/id65.pdf (accessed on 23 April 2012).
53. Dixon, M.J.; Tawn, J.A. Extreme Sea-Levels at the UK a-Class Sites: Optimal Site Bysite Analysis and Spatial Analyses for the East Coast; Pol Internal Document Number 72; Proudman Oceanographic Laboratory: Liverpool, UK, 1995; p. 298.
54. Dixon, M.J.; Tawn, J.A. Estimates of Extreme Sea Conditions—Final Report, Spatial Analysis for the UK Coast; Pol Internal Document Number 112; Proudman Oceanographic Laboratory: Liverpool, UK, 1997; p. 217.

55. Haigh, I.D.; Nicholls, R.; Wells, N. A comparison of the main methods for estimating probabilities of extreme still water levels. Coast. Eng. 2010, 57, 838–849.
56. Flood Risk Mapping and Data Management Team. Environment Agency Extreme Water Levels: Bournemouth to Littlehampton; EA: Rotherham, UK, 2007.
57. EA. Coastal Flood Boundary Conditions for UK Mainland and Islands; Project: Sc060064/tr2: Design sea Levels; EA: Bristol, UK, 2011; p. 142.
58. Haigh, I. Quantification of sea-level rise at the Southampton and Portsmouth tide gauges. Personal Communication.
59. NFDC. North Solent Shoreline Management Plan, Appendix c Baseline Process Understanding, c.1 Assessment of Shoreline Dynamics; New Forest District Council: Hampshire, UK, 2010. Available online: http://www.newforest.gov.uk/ media/ adobe/ o/s/C1_Coastal_Processes _Literature_Review_%28final_draft%29.pdf (accessed on 30 June 2011).
60. CERC. Coastal Engineering Research Center 1984: Shore Protection Manual; Department of the Army, Waterways Experiment Station, Corps of Engineers, Coastal Engineering Research Center, Vicksburg, United States Government Printing Office: Washington, DC, USA, 1984; p. 652.
61. Dawson, R.; Hall, J. Probabilistic condition characterisation of coastal structures using imprecise information. In Proceedings of the 28th International Conference Coastal Engineering Cardiff, UK, 8–12 July 2002; McKee-Smith, J., Ed.; World Scientific: Hackensack, NJ, USA; Cardiff, UK, 2003; pp. 2348–2359.
62. Flikweert, J.; Simm, J. Improving performance targets for flood defence assets. J. Flood Risk Manag. 2008, 1, 201–212.
63. Lowe, J.A.; Howard, T.; Pardaens, A.; Tinker, J.; Holt, J.; Wakelin, S.; Glenn, M.; Leake, J.; Wolf, J.; Horsburgh, K.; et al. UK Climate Projections Science Report: Marine and Coastal Projections; Met Office Hadley Centre: Exeter, UK, 2009; p. 99.
64. Bates, P.D.; Horritt, M.S.; Fewtrell, T.J. A simple inertial formulation of the shallow water equations for efficient two-dimensional flood inundation modelling. J. Hydrol. 2010, 387, 33–45.
65. Bates, P.D.; de Roo, A.P.J. A simple raster-based model for flood inundation simulation. J. Hydrol. 2000, 236, 54–77.
66. Hunter, N.M.; Horritt, M.S.; Bates, P.D.; Wilson, M.D.; Werner, M.G.F. An adaptive time step solution for raster-based storage cell modelling of floodplain inundation. Adv. Water Resour. 2005, 28, 975–991.
67. Dawson, R.J.; Hall, J.W.; Bates, P.D.; Nicholls, R.J. Quantified analysis of the probability of flooding in the thames estuary under imaginable worst-case sea level rise scenarios. Int. J. Water Resour. Dev. 2005, 21, 577–591.
68. Hunter, N.M.; Bates, P.D.; Horritt, M.S.; Wilson, M.D. Simple spatially-distributed models for predicting flood inundation: A review. Geomorphology 2007, 90, 208–225.
69. Fewtrell, T.J.; Duncan, A.; Sampson, C.C.; Neal, J.C.; Bates, P.D. Benchmarking urban flood models of varying complexity and scale using high resolution terrestrial lidar data. Phys. Chem. Earth Parts A/B/C 2011, 36, 281–291.
70. Hunter, N.M.; Bates, P.D.; Neelz, S.; Pender, G.; Villanueva, I.; Wright, N.G.; Liang, D.; Falconer, R.A.; Lin, B.; Waller, S.; et al. Benchmarking 2D hydraulic models for urban flooding. Proc. Inst. Civil Engineers Water Manag. 2008, 161, 13–30.
71. Pullen, T.; Allsop, N.W.H.; Bruce, T.; Kortenhaus, A.; Schuttrumpf, H.; van der Meer, J. Eurotop, Wave Overtopping of Sea Defences and Related Structures—Assessment Manual, 2007. Available online: www.overtopping-manual.com(accessed on 24 April 2012).
72. Owen, M.W. Design of Seawalls Allowing for Wave Overtopping; Report Ex 924; HR Wallingford: Oxfordshire, UK, 1980.
73. Technical Report—Wave Run-Up and Wave Overtopping at Dikes; TAW: Delft, The Netherlands, 2002.
74. Gouldby, B.; Kingston, G.; Wills, M.; Van Gelder, P.; Buijis, F.; Kortenhaus, A. Flood Defence Reliability Calculator: Reliable User Manual; Floodsite Report Number: T07-10-08; January 2008.

75. Allsop, W.; Kortenhaus, A.; Morris, M. Failure Mechanisms for Flood Defence Structures; Floodsite Report Number: T04–06–01; HR Wallingford Ltd.: Oxfordshire, UK, 2007.
76. Vrouwenvelder, A.C.W.M.; Steenbergen, H.M.G.M.; Slijkhuis, K.A.H. Theoretical Manual of Pc-Ring, Part B: Statistical Models (in dutch); No. 98-con-r1431. Delft, The Netherland, 2001.
77. Terzaghi, K.; Peck, R.B. Soil Mechanics in Engineering Practice; John Wiley & Sons: New York, NY, USA, 1967; p. 729.
78. Powell, K.A. Predicting the Short Term Profile Response for Shingle Beaches; Hydraulics Research Report sr219; HR Wallingford: Oxfordshire, UK, 1990.
79. DEFRA; EA, Performance and Reliability of Flood and Coastal Defences; R&d Technical Report Fd2318/tr 2; Buijis, F., Simm, J., Wallis, M., Sayers, P., Eds.; DEFRA: London, UK, 2005.
80. Morris, M.; Hassan, M.; Kortenhaus, A.; Visser, P. Breaching Processes: A State of the Art Review; Floodsite Report Number t06–06–03, Revision Number 5_2_p01; FLOODsite Consortium: Oxfordshire, UK, 2009.
81. Morris, M.; Hanson, G.; Hassan, M. Improving the accuracy of breach modelling: Why are we not progressing faster? J. Flood Risk Manag. 2008, 1, 150–161.
82. Wadey, M.; Nicholls, R.J.; Haigh, I.; Hutton, C. Creating a coastal flood event dataset for the solent region-the 10th March 2008 storm surge event. 2012. in preparation.
83. Cope, S. Details of the breach of the medmerry/selsey shingle barrier during the storm on 10 March 2008. Personal Communication, New Forest District Council/Channel Coastal Observatory: Southampton, UK.
84. EA, Solent & South Downs Area Report 10 March Tidal Flooding Event, Hampshire & Isle of Wight: A report of events, Consequences and Conclusions, Environment Agency, Rotherham, UK, 2010. Version 2.
85. Wadey, M.P. Photograph taken by Matthew Wadey from the Channel Coastal Observatory overlooking Southampton Docks. National Oceanography Centre, Southampton, UK, 2:05PM, 10 March 2008.
86. Evacuations as waves flood park. BBC World News, 10 March 2008. Available online: http://news.bbc.co.uk/1/hi/england/sussex/7287880.stm (accessed on 25 January 2012).
87. Emsworth Residents Association. Flooding in Emsworth, Borough of Havant, during a Storm Surge in March 2008; Emsworth_Flood_2008_1.jpg. 10 March 2008. Available online: http://www.floods.emsworthhants.org.uk/ (accessed on 24 April 2012).
88. NRA, Lymington/Pennington Flood Investigation–Interim Report; Report for National Rivers Authority Southern Region; Ea Report Reference Number: IMP/sch/008/5054, Southern Projects Ltd. Jersey, UK, 1990.
89. Stripling, S.; Bradbury, A.P.; Cope, S.N.; Brampton, A.H. Understanding Barrier Beaches; R&D Technical Report Fd1925/Tr; JOINT Defra/ea Flood and Coastal Erosion Risk Management R&D Programme: London, UK, 2008. Available online: http://sciencesearch.defra.gov.uk (accessed on 24 April 2012).
90. Nicholls, R. The Stability of the Shingle Beaches in the Eastern Half of Christchurch Bay. Ph.D Dissertation, University of Southampton, Hampshire, UK, 1985.
91. Nicholls, R.J.; Webber, N.B. The past, present and future evolution of hurst castle spit, Hampshir. Prog. Oceanogr.1987, 18, 119–137.
92. Jonkman, S.N.; Penning-Rowsell, E. Human instability in flood flows. JAWRA J. Am. Water Resour. Assoc. 2008, 44, 1208–1218.
93. Pachauri, R.K., Reisinger, A., Eds.; Contribution of Working Groups I, II and III to the Fourth Assessment Report of the Intergovernmental Panel on Climate Change; IPCC, Climate Change 2007: Synthesis Report; IPCC: IPCC, Geneva, Switzerland, 2007. Available online: http://www.ipcc.ch/publications_and_data/ar4/syr/en/contents.html (accessed on 24 April 2012).
94. Flather, R.A. Existing operational oceanography. Coast. Eng. 2000, 41, 13–40.
95. Ray, J.; Coulling, E.; Evans, S.Y.; Spink, K. Developing a practical tool for coastal flood forecasting and warning. ICE Coastal Management: Innovative Coastal Zone Management. Sustainable Engineering for a Dynamic Coast, Belfast, UK, 2011.

96. Wadey, M.; Nicholls, R.; Hutton, C. Threat of coastal inundation in the solent: Real-time forecasting. ICE Coastal Management: Innovative Coastal Zone Management. Sustainable Engineering for a Dynamic Coast, Belfast, UK, 2011.
97. DEFRA; EA, Improving Flood Warnings: Final Report: Improving Institutional and Social Responses to Flooding; Science Report: Sc060019-Work Package 1a; Twigger-ross, C., Fernández-bilbao, A., Tapsell, S., Walker, G., Watson, N., Collaborators, C., Rose, P., Dade, H., Deeming, E., Eds.; DEFRA: London, UK, 2009; p. 67, ISBN: 978-1-84911-058-7.
98. Pappenberger, F.; Beven, K.J. Ignorance is bliss: Or seven reasons not to use uncertainty analysis. Water Resour. Res.2006, 42, W05302.

CHAPTER 15

Response and recovery measures for two floods in north China during the nineteenth century: a comparative study

Yu Ye1,2*, Xiuqi Fang1 and Fan Li1

1 School of Geography, Beijing Normal University, 19 Xinjiekou Wai Street, Haidian District, Beijing 100875, China.

2 Key Laboratory of Environment Change and Natural Disaster, Ministry of Education, BNU, Beijing 100875, China

ABSTRACT

Background

The process of human response to natural disasters and its mechanisms as revealed by historical events still has a broad significance for modern society. This study analyzed the disaster relief process and the social response for two floods in China: the Yongding River flood in 1801 and the Yellow River flood in 1841. These two floods reflect the different response processes between the national and provincial capitals during a stage of climate cooling and social transition in the Qing dynasty.

Results

Applying methods of historical documents analysis and qualitatively comparative analysis to the materials such as *Relief Chronicles Authorized by the Emperor in XinYou* and *Flood Description in Bian Liang*, it shows that: (1) In 1801, the central government took on a lead position, from flood surveying to relief processes. However, local government and gentries played an important role in 1841. (2) In 1801, the government successfully undertook a series of relief measures addressing production, housing, food prices, taxes, and water conservancy and administration. In 1841, the response measures were relatively simple, focusing mainly on providing shelter and food for victims. (3) The government carried out long-term disaster prevention measures such as dredging channels after the flood in 1801. In 1841, however, the efforts were focused mainly on emergency rescue. (4) Refugees in the 1801 flood were effectively managed by a centralized authority. In 1841, regulation of the flooding was delayed by corruption and conflicts between officers, leading to an expansion of the disaster's impact.

Conclusions

Above results have led to the conclusion that disaster relief systems and response measures had a significant effect on the consequences of those floods. Various flood relief measures and natural disasters management regimes have implications for contemporary flood hazard mitigation.

Keywords: Floods, The Yongding River, The Yellow River, Coping strategies, Impact of climate change

1. BACKGROUND

The process of human response to natural disasters and its mechanisms as revealed by historical events still has a broad significance for the great challenge of modern society in coping with global climate change. The fifth report of the International Panel of Climate Change (IPCC) shows that global warming has resulted in more frequent extreme climatic events that have effects on agriculture and forest management, human health, and economic and social development (IPCC 2014). Continued global warming would most likely bring severe challenges for regional social and economic development and adaptation ability. Strengthening the research on the impact of extreme disasters on society and human coping strategies is not only an important scientific problem but also urgently needed for the healthy, stable and sustainable development of present and future societies in China.

In recent years, international studies examined the possible impact of abrupt climate change or natural disasters on social development. These studies have proven that the focus should be on both the feedback of the sub-system of society on abrupt climate change and the dynamics of climate change impact on society. For example, the abrupt climatic change that occurred in 2200 B.C. induced considerable land-use degradation and evidently caused the collapse of the Akkadian empire (Weiss et al. 1993). Cooling from A.D. 1560 to 1660 caused successive agro-ecological, socioeconomic, and demographic catastrophes, leading to the General Crisis of the seventeenth century in Europe (David et al. 2011). Widespread episodes of extreme drought due to as much as a 40% reduction in annual precipitation shaped the demise of the Maya civilization (Medina-Elizalde and Rohling 2012). Richard and Wagner (2010) analyzed the relationship between climate change and violent conflict in Europe over the last millennium. These above-mentioned international studies have revealed the relationships between historical extreme disasters and the decline of society, and the evolution of human civilization. However, it is necessary to reveal the process and dynamics of past climate change impacts, especially with regard to the impacts of climatic disasters.

The impact of historical, extreme climatic events and human adaptation has become one of the most important and most frequently researched areas in China, with abundant and extensive historical documents. Currently, this research can be classified into three types. First, case studies on the impact of natural disasters on important historical events; for example, the North Wei Dynasty moved the capital of Pingcheng (present-day Datong in Shanxi Province) to Luoyang in Henan Province during the second cold event of the North and South Dynasty of Weijin, 450–530s (Man et al. 2000). Second, estimates of the impact of climate change that compare historical climate or flood/drought series with a social and economic proxy index series; for example, population (Li 1999), food production (Hao et al. 2003), the scale of agricultural development (Zhang et al. 2005), the movement of the agricultural boundaries and grazing zones (Zou 1995), the prosperity and depression of dynasties (Zhang et al. 2004), and the frequency of disturbance events (Ye et al. 2004). Some national research also analyzes the impact of climate change on social unrest and dynastic transition in ancient China (Zhang et al. 2005; Ge 2011). All of the above research emphasizes the impact of climate change or extreme natural disasters on society. Discussions on impact dynamics and adaptation are still scarce.

In recent years, new progress has been made on this research area in China. Some scholars note that the social background, government decision-making and social response should receive more attention to further understand the impact dynamics, with the exception of the corresponding relationship between extreme climatic events and social events (the events which impact the development of society such as population fluctuation, violent conflict, replacement of dynasties etc.). Many Chinese historical case studies reflect that social responses and governmental decision-making play obvious roles in impacting the process of climate change. These studies control the level of impact, the manner of response to a certain extent, and the final outcome. These factors differ between flourishing and declining periods of society (Xiao et al. 2011, 2012; Ye et al. 2012). The influence of historical climate change on food security and social vulnerability are gradually being recognized. Fang et al. (2014) presented a new theoretical concept based on research on the impact of historical climate change on historical agricultural society.

The turn of the nineteenth century was characterized by a period of little ice ages when the climate was cooling down. For the Qing dynasty in China, the turn of the nineteenth century was also a time of transition from a period of flourishing society to a period of decline society (Fang et al. 2013). This paper analyzes the disaster relief process and the social response for two floods in China, the Yongding River flood in 1801 and the Yellow River flood in 1841 using methods of historical documents analysis and qualitative comparative analysis. These two floods reflect different response processes between the national and provincial capitals during this stage of climate cooling and social transition in the Qing dynasty. This paper explores the functions of governmental decision-making and measures in response to these flood events. Comparative studies of historical climate events and human response processes are helpful to better understand human adaptation mechanisms and to provide a reference for the modern decision making and for reducing social vulnerability.

2. METHODS

2.1 Study area

The two flood events in our study occurred in the Yongding River and Yellow River basins (Figure 1). This study focused on the comparison of impact and response processes between the national and provincial capitals, which are located along the Yongding River and the Yellow River, respectively.

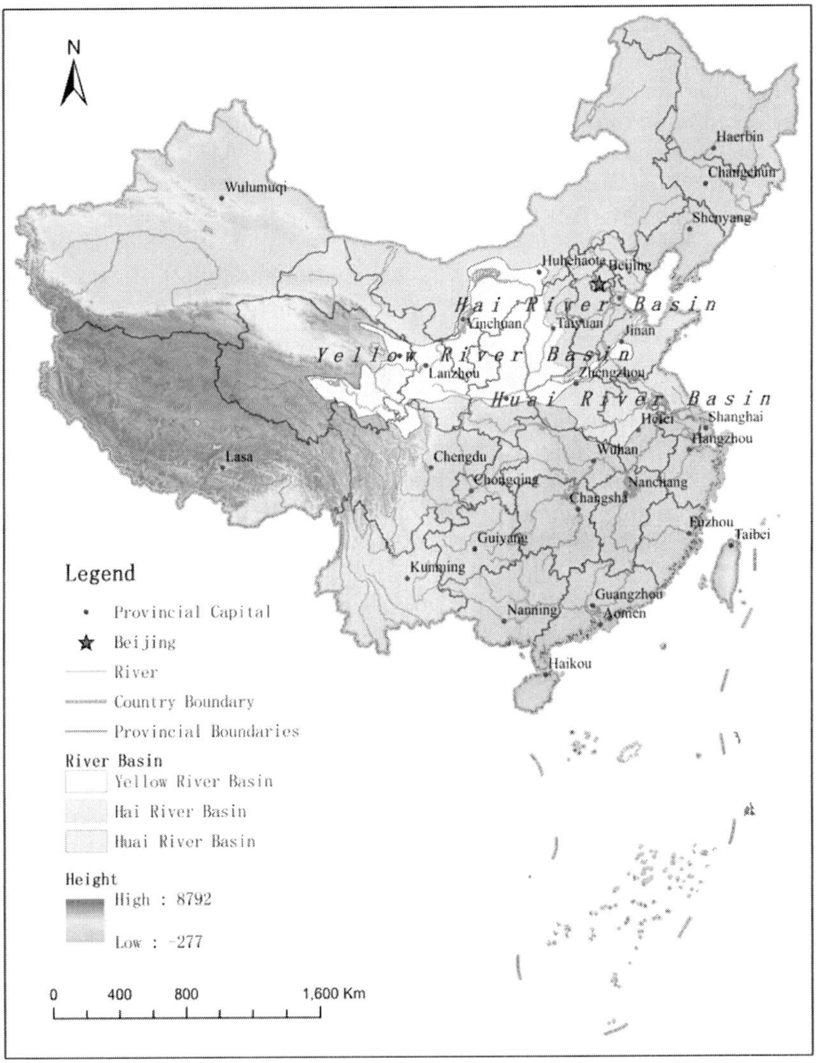

Figure 1. Location of study areas

The Yongding River basin lies between 39°N 112°E and 41°N 118°E. It is the largest river in the Hai River basin and flows through five provinces or autonomous regions: Shanxi, Inner Mongolia, Hebei, Beijing and Tianjin. It is one of the four main rivers that receives more attention with regard to flood prevention in China. In the Yongding River basin, the inter-annual variability of precipitation is apparent and distributed unevenly over time and space. Some areas have frequent heavy rain, which can easily cause natural disasters. Upstream of the Yong-ding River, the slope is steep, and the water is fast flowing. When this water flows downstream to the Beijing plain, the terrain is flat and the river channel so irregular that it was referred to as the Wuding River throughout history; hence, flood disasters often occurred. Since Jin dynasty (1115–1234 A.D.), government flood prevention measures included the building of dikes to avert floods and the construction of a dam to divert water. During the Kangxi Kingdom in Qing dynasty, the initial dike system was developed downstream of Lugou Bridge (Management Office of Yong-ding River in Beijing 2002). However, during large floods, the dam and dikes (levees) constructed to mitigate against flood water inundation were sometimes over topped and breached, especially in the capital city located at the middle and lower reaches of the Yongding River.

The Yellow River flows through nine provinces or autonomous regions: Qinghai, Sichuan, Gansu, Ningxia, Inner Mongolia, Shaanxi, Shanxi, Henan and Shandong. The annual sediment discharge in the Yellow River was once as high as 1.6 billion tons, and the annual sediment content is close to 40 kg/m^3, which is the highest in the world (Ren 2006). Large amounts of sand and less water make the lower reaches of the Yellow River a perched river channel. The riverbed is higher than the level of the dike by more than 10 m (Xie 1999; Yu 2002), which increases the frequency of dike breaching when flooding occurs. According to historical documents, flood events such as breaching and channel changes have occurred in the lower Yellow River more than one thousand times over a period of more than four thousand years (Shen et al. 1935). Breaching occurred up to three times in one year during the late Ming and early Qing dynasties in the middle of the seventeenth century (Chen et al. 2012). Once the dike of the Yellow River breaches, it can cause inundation of the part of or the whole Huang-Huai-Hai plain (Figure 1), resulting in large-scale crop failures that affect hundreds of thousands or even millions of people. The ancient capital city of Kaifeng in Henan Province is located at the apex of an alluvial fan and suffers from the most severe flood disasters. Seven floods have inundated Kaifeng city since the fourth century B.C. (Qiu 1999).

2.2 Data sources

The information regarding natural disaster impacts, relief measures and social responses in the capital of Zhili (currently Hebei, Beijing and Tianjin) and Kaifeng come primarily from *Relief chronicles authorized by emperor in XinYou* (Qing 1802) and *Flood description in Bian Liang* (Li et al. 2006). History records in Qing dynasty Rivers and canals annal (Zhao 1927), which recorded channel changes, dike breaching, government response and treatment related to the Yellow River during the Qing dynasty.

Relief chronicles authorized by emperor in Xin You is a historical text about official famine policy written by individuals appointed by the Jiaqing emperor (Shao 2006); it is included in *The Chinese famine policy book (series 1, volume 2)* (Li and Xia 2003). The book consists of 38 volumes containing more than 200,000 words. It is edited in chronological order according to the Jiaqing emperor's edicts and ministers' reports from volume 1 to volume 38. All of the Yongding River flood situations and relief processes from July 12, 1801 to September 9, 1802 are recorded. Therefore, these records provide a clear sequence and progression of events.

Flood description in Bian Liang includes a specific description of the flood in Kaifeng during the summer of 1841. A daily diary totaling eight months is provided from August 2, 1841 to March 27, 1842. At the end of the book, reports and imperial edicts about this flood in *Continuous Water Jin Jian* (China Institute of Water Resources and Hydropower Research 2004) and *Actual Annals of Qing Dynasty True Records of Emperor Xuanzongcheng in Qing Dynasty* (Zhao 1986) are also attached. In addition, a time series of governmental flood relief efforts during the Qing Dynasty are sourced by Huang (2013).

2.3 Methodology

To analyze mechanism of social response during the two floods, the capital region Zhili (present-day Hebei Province, Beijing City and Tianjin City) and Kaifeng city in Henan province were used as examples, and detailed information on social assistance, disaster response during the flood and protection after the disaster was examined and classified using historical document analysis methods (Figure 2). Then, a time series of recovery processes in response to these two floods was reconstructed using a the sequence of 10 days later, 1 month later, 3 month later and one year later (Figure 3). Response and recovery processes are classified into different areas including production, life, food and prices, water conservancy, tax, administration, and culture. Finally, government response measures and disaster impacts at different stages of the flood were compared and analyzed.

3. RESULTS

3.1 The two flooding events and the resulting hazards

Heavy summer rainfall occurring in two stages (July 10–16 and July 24–August 1) is the main cause of flooding in the Yongding River in 1801. The rainstorm in the first period led to the saturation of soil water, rising water level in the river, and river bank erosion and breach, which caused the heavy flood. After the rainstorm stopped, the flood abated but did not recede. On 25 July, heavy rain poured down again and torrents scoured the channel and road, seriously aggravating the effects of the flood disaster. After 2 August, the rain ended, and water level in the river gradually returned to normal. The flood of the Yongding River basin in 1801 was distributed predominantly along the Hai River, South Canal, Ziya River, Daqing River, Yongding River, North Canal, JiYun River and Luan River, affecting 160 counties in Shanxi, Inner Mongolia, Hebei, Beijing and Tianjin. Among them, Shijingshan, Wanping, Daxing, Changxindian, and Liangxiang, which were near the river's breaching point and along the place where the flood flowed, were the hardest-hit areas (Ye et al. 2014).

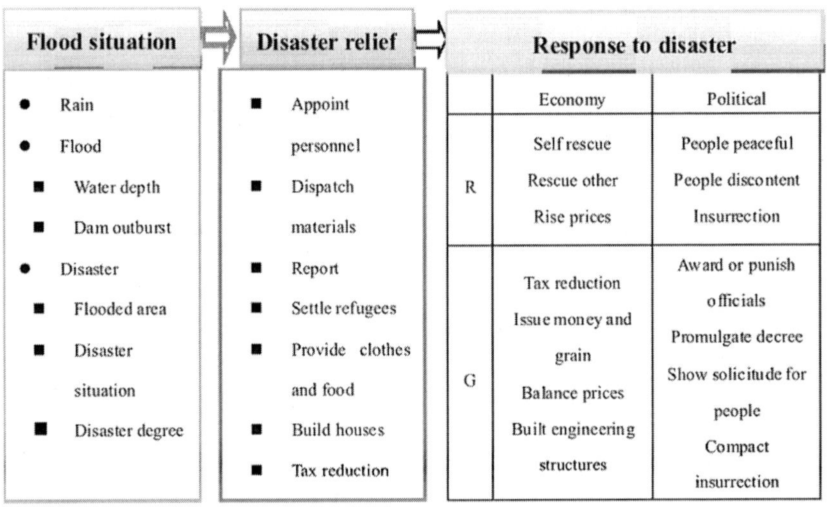

Figure 2. Classification of information on disaster relief and response to disaster (*R* Republic *G* Government)

The rising flow of the upper reaches of the Yellow River between July 23rd and 31st and the strong rainfall that followed from August 3rd to 22nd contributed to Kaifeng, Henan Province being surrounded by floodwaters of 1841 for 68 days in 1841. The most serious flood was in June of the lunar calendar, from 23 July to 1 August. Many states and counties along the Yellow River were seriously affected. In the records, the flood disaster in Kaifeng began with the water rushing into Kaifeng City on 2 August 1841, and ended with the closing of the breached dam on 19 March 1842. The flood of the Yellow River basin in 1841 affected almost 108 counties in

Shanxi, Shaanxi, Inner Mongolia, Ningxia, Shandong, Henan and Anhui. The most serious impacts occurred in lunar June, from July 23rd to August 1st. The River in Zhangjiawan breached, and Kaifeng city was surrounded by the flood. In late lunar June, the lower reaches of the Yellow River were cutoff, and the flood then flowed into the Huai River, causing breaching. This flood impacted 61 counties in Shanxi, Shaanxi, Henan, Anhui and other provinces.

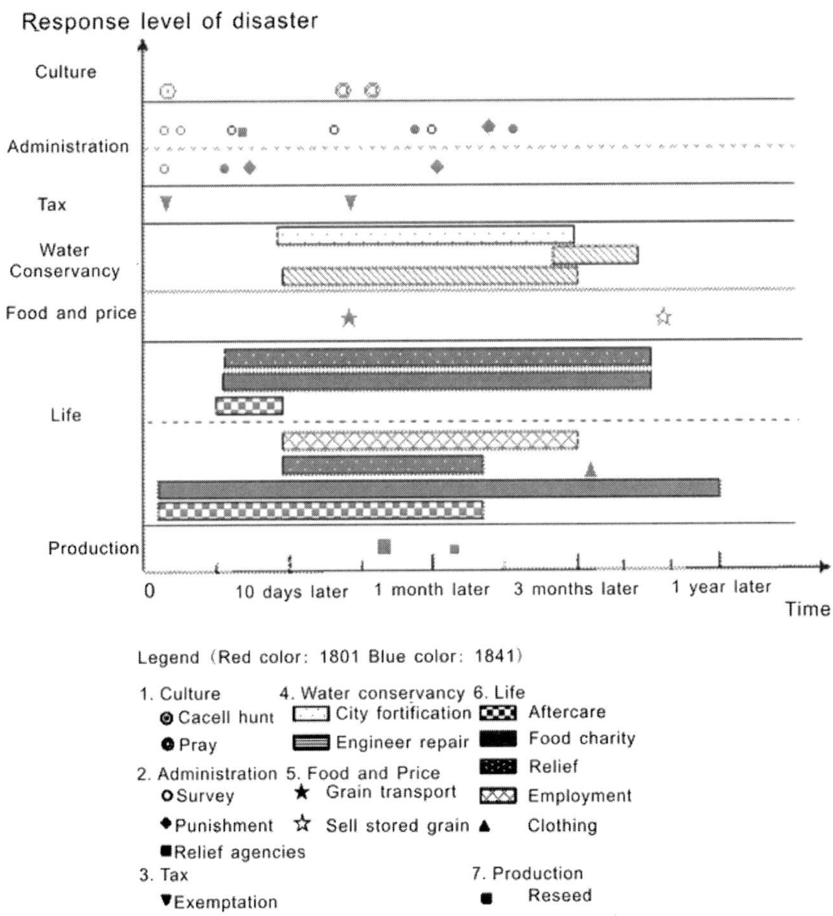

Figure 3. Comparison of the response and recovery measures and effects of the two floods

3.2 The relief and response speed to the two foods

Flood relief occurred differently for each event. In 1801, flood relief was led by the government, especially the central government. In 1841, local gentries played an important role in the early stages of food relief, from the first rescue to the management of the relief factory. The rescue work in the late period was taken over by engineer soldiers. At the same time, local officials served as relief directors throughout the flood event.

For example, in terms of surveying and reporting, the emperor of Jiaqing sent ministers to the disaster area from the west, east, north and south roads on the 7th day (July 17th) after the flood in 1801. The ministers had to guide local officials to carefully survey the flood and produce detailed and accurate reports. In addition to the breaching of the Yongding River, the extent of damage to cropland was assessed and reported with villages as the basic unit. In 1841, disaster surveying and reporting was complex. Central and local government officers and local gentries, especially local officers, surveyed the disaster situations. The Daoguang emperor instructed Wen Chong, the director of the east river channel, Niu Jian, the inspector of Henan province and the scholar Wang Ding in succession to survey the disaster situations.

The speed of the response varied. In 1801, the emperor ordered a thorough survey after 7 days; in 1841, the emperor did not send officials to supervise for 18 days, and officials arrived in Kaifeng after 40 days. In 1801, central officials were divided into four groups and ordered to assess and survey the disaster area 7 days after the flood. In 1841, however, complete surveying began very late, and the disaster situation was not reported until three months later.

3.3 The impact and mechanisms of the two floods

First, agricultural production was a key process in the flood response chain in both Zhili and Kaifeng, and the two floods both influenced society by impacting agriculture, food, materials and other aspects. Second, the level of the impact and subsequent related processes of the two floods were relatively consistent. A series of consequences generated by the floods ranged from reduced agricultural production and damage to water conservancy establishments to rising prices, vagrants and refugees, and robbery. Finally, the two floods both affected social stability to some extent, especially without timely national assistance, although there were also some differences in the social impact between the two floods.

3.4 Response and recovery measures

Response and recovery measures were different in the two floods due to differences in the objects affected (Figure 3). In 1801, the flood affected farmland and created many refugees. Governmental response actions involved in all aspects of society, including production, housing, food prices, tax, and water conservancy and administration, and even the emperor's personal behavior and culture. The government successfully organized a series of relief measures, including a set of rice porridge factories, tax exemptions, crop reseeding, engineer employment provisions, and relief substitution. National food relief during the flood season and extending into the following year effectively guaranteed the lives and social stability of the people. In 1841, the government made it a priority to protect Kaifeng city from the flood. The response measures were relatively simple, focusing mainly on providing shelter and food for victims. During the flood, food prices increased and robberies occurred even in the daytime; however, the government could do nothing but issue announcements to address it.

3.5 Flood mitigation infrastructure

The government organized the repair of water utilities after the floods. In 1841, the work mainly focused on flood emergency rescue, making urgent repair on the dam around Kaifeng city; however, in 1801, the government carried out more comprehensive disaster prevention measures such as channel dredging. In the flood of 1801, after completing repair work, the government employed the poor in digging the Tonghui River, repairing the Yongding River, South Canal, North Canal, and Hun River and dredging channels to prevent and mitigate impact of floods in the future.

3.6 Differences in administrative regimes between the two floods

Differences in the impact and response processes to the two floods reflected different administrations. In 1801, the central and local governments were unified, and there was strict and impartial award and punishment principle. Moreover, the government investigated and punished corrupt officials reported by victims on many occasions.

In 1841, problems delayed the construction of the dike outside the city. For example, officials blamed each other because of differing views on flood management, officers became corrupted and cut corners during the dike construction, engineer soldiers and gentries did not like working with each other and soldiers fought with citizens; all of these problems hampered the progress of the relief work.

All of above, although it is still unclear if the 1801 Yongding River Flood was of equivalent magnitude hydrologically and in some quantitative terms of societal impact to that of the Yellow River of 1841, we assume they were all heavy hydrological floods resulted by rainstorm and extensive region was seriously affected according to the abundant documental descriptions on hydrologic process and affected disaster areas. By the above comparison of these two floods, it confirmed the central conclusion of this study, ineffective government response to the flood disaster resulted in substantially more adverse impacts to the affected populations.

4. DISCUSSIONS

The climatic background of the two floods is essentially the same. The climate began to cool during the nineteenth century, and the frequency of flood and drought disasters increased. Climate instability increases the burden of social disaster response. There was a clear contrast, however, between the two floods in terms of magnitude of the flood, the disaster relief and the degree of attention drawn at a national level due to differences in regional and political status.

4.1 The role of climate variability

The two floods occurred in the first half of the nineteenth century, when the climate was cold. The average winter temperature was more than 1 °C lower than the present temperature in eastern China (Ge et al. 2002). During this period, there were also significant changes in precipitation. There were more flood and drought disasters in the first half of the nineteenth century than that in the eighteenth century.

According to the records of precipitation and flood disaster processes, the likely cause of both floods was the same—continuous heavy rainfall in the summer. The disaster processes, however, were different. In 1801, heavy rainfall was the determinant of the flood disaster. The Yongding River rose from the beginning of the heavy precipitation until the rain ended, and the flood gradually receded approximately one month later. In 1841, however, Kaifeng suffered for approximately 10 days after the flood in the lower reaches of the Yellow River. Heavy rainfall in Shanxi and Shaanxi led to rising water in the upper reaches of the Yellow River and its tributaries, which was the main cause of the flooding in Kaifeng, and consequently

Kaifeng was surrounded by the flood for more than one month after the heavy rain ended.

Overall, the climate began to cool down during the nineteenth century, and the frequency of flood and drought disasters rose. Continuous heavy rainfall in the summer frequently leads to river rose and flood waters surrounding the cities along the river. All climate instability increases the burden of social disaster response.

4.2 National strength reflected by the strength of governmental food relief

The flood in 1801 occurred during the transition from the "Kangxi-Qianlong Kingdom Flourishing Age" (1681–1796) to the "Jiaqing-Daoguang Kingdom Decline Age", when the territory and economic strength was at its peak in the Qing dynasty on the verge of decline. The flood in 1841 occurred on the eve of the Opium War. Financial shortages, official corruption, army laxity and domestic and foreign crises brought about the rapid decline of this prosperous feudal state.

The strength of governmental food relief for these disasters is also a reflection of national strength. The quantity of national food relief refers to total weight of governmental food transfer which unit is dan (1 dan equals 28 kg in the Qing Dynasty). National food relief strength is one of indexes of showing disaster relief effort, which is calculated by quantity of national food relief divided by the disaster index.

In the Qing dynasty, the quantity and strength of governmental food transfer in the north China plain was highest from 1740 to 1799; the annual average national food relief was 356,000 dans, and the average governmental food transfer strength index was 17.66. From 1800 to 1869, however, food relief was declining, and the annual average governmental food transfer was only 70,400 dans with an average national food relief strength index of 2.91.

The Qing government spent 1135,000 dans of grains on drought relief in 1785, but the amount of food relief decreased in the 1790s, with an annual average value of 287,400 dans. The amount dropped dramatically in the early 1800s, reaching minimum values of 18,800 and 10,000 dans in the 1830 and 1840s, respectively. During the 1830–1840s, the government did not organize powerful, effective food relief for the most severe floods and droughts, with the exception of the floods in 1822 and 1823, when the government offered 1161,000 dans of relief grains (Huang 2013). In conclusion, the national government in 1801 was stronger than in 1841; the Qing government had maintained a more comprehensive disaster response effort in 1801.

4.3 Different political statuses reflected by disaster response and recovery measures

The differing political statuses of disaster areas also led to differences in disaster response and recovery measures. As the political center of the Qing dynasty, Zhili (including present-day Hebei Province, Beijing City and Tianjin City) was a particular priority of the government and received the greatest and fastest food relief. The amount of food obtained by Zhili accounted for more than 50% of the total amount in the north China plain of the Qing dynasty. As the second-level administrative center, Henan province received less attention and food, although the northern zone served as a food distribution area in Henan. In 1801, tax exemption and food relief were still important disaster relief strategies. States and counties in the capital area enjoyed tax concessions according to the degree of damage. The total amount of flood relief food was 900,000 dans, mainly from Caoliang (rice transported to the capital by water). A small amount of food was bought from eastern Henan and northeastern China (Huang 2013). The government paid over 1,000,000 liang (1 liang equals 50 g) of silver for the Yongding River riverbank engineering in this year, which was the largest relief effort of the Qing Dynasty (Wang 2007). The scope and strength of the national relief in 1841 were much lower than those in 1801.

Generally speaking, the two floods had different national strength backgrounds. Different national disaster relief efforts were enacted to varying degrees, and the floods received different levels of national attention due to differences in regional and political status. These are all possible impact factors causing different responses during the two floods.

4.4 Significance of this research for contemporary flood hazard

Various flood relief measures and disaster management regimes can be used as a reference for responding to flood events. First, when flooding occurs, the government can organize a series of relief measures, including providing food and clothes, tax exemptions, crop reseeding, engineer employment provisions, and relief substitutions. Second, the impact conveying process and influence degree of the floods were relatively consistent. Agricultural production was a key process in the flood response chain, and the two floods both influenced society by affecting agriculture, food, materials and other aspects. Thus, contemporary flood hazard mitigation should follow the order of flood response chain, from measures on production, food, economy to society. All levels of relief measures should be implemented, including food relief, agriculture production, housing, food prices, tax, water conservancy and administration. Especially, governmental food relief and support of money, seed or technology for agriculture production in modern flood hazard mitigation are very necessary. Third, central and local governments should be unified and cooperate with each other, and officers should be rewarded or punished according to their performance in coping with flood hazard. The strict principle of reward and punishment should be implemented as a policy by legislation.

5. CONCLUSIONS

This study compared two floods that both occurred during the transition from warm to cold climate patterns during the Little Ice Age. The disaster responses processes were compared. It confirmed: the disaster relief systems had a significant effect on the consequences of the floods. The response and recovery processes of the Zhili flood in 1801 and the Kaifeng flood in 1841 were different.

1) The central government assumed a lead position, from flood surveying to relief processes, in 1801; local officers were coordinated and guided by the central government. In 1841, local government played an important role, and local gentries also contributed to surveying and relief under the supervision of the central government.

2) The response and recovery measures in 1801 were diverse and effective and operated on many levels; the strategies included forming rice porridge factories, creating tax exemptions, reseeding croplands and providing employment to the public. In 1841, the government prioritized preventing Kaifeng city from being surrounded by flooding. The government focused on providing shelter and food for victims.

3) National food assistance during the flood season in 1801 and even in the following year effectively guaranteed peoples' lives and social stability. Unfortunately, serious public security hazards such as daytime robberies still appeared in 1841.

4) From the aspect of water conservancy, the government carried out inspections and comprehensive disaster prevention measures such as dredging channels after the flood in 1801. In 1841, however, the work mainly focused on flood emergency rescue.

5) The flood in 1801 was controlled by a central authority. Materials transportation, disaster monitoring and punishment work were well executed during the flood. The authorities effectively controlled refugees and sustained a relatively stable regime. In 1841, the officials were corrupt and officers in different ranks blamed each other.

The climatic background of the two floods is essentially the same. However, there was a clear contrast between the two floods regarding the background of national strength, the stage of national disaster relief and degree of national attention due to differences in regional and political status. The flood in 1801 occurred when the prosperity of the Kangxi-Qianlong kingdom had just ended, and 20 years after the turning point of climate cooling. The flood in 1841 occurred when society had declined and was approximately 30 years from flourishing. These are all possible impact factors in the different responses during the two floods.

ACKNOWLEDGEMENTS

Authors would like to acknowledge the Springer Nature Author Services (https://secure.authorservices.springernature.com/) to edit this paper completely.

REFERENCES

1. Chen YZ, Syvitski JPM, Gao S et al (2012) Socio-economic impacts on flooding: a 4000-year history of the Yellow River, China. J Hum Environ 41:682–698
2. China Institute of Water Resources and Hydropower Research (2004) Continuous Water Jin Jian (1820–1911). Hubei Peoples Publishing House, Wuhan
3. David DZ, Harry F, Lee WC et al (2011) The causality analysis of climate change and large-scale human crisis. PNAS 10:17296–17301
4. Fang XQ, Xiao LB, Wei ZD (2013) Social impacts of the climatic shift around the turn of the 19th century on the North China Plain. Sci China Earth Sci 56:1044–1058
5. Fang XQ, Zheng JY, Ge QS (2014) Historical climate change impact-response processes under the framework of food security in China. Sci Geogr Sin 34:1291–1298
6. Ge QS (2011) Climate change in China. Science Press, Beijing
7. Ge QS, Zheng JY, Man ZM et al (2002) Reconstruction and analysis on the series of winter-half-year temperature changes over the past 2000 years in the eastern China. Earth Sci Front 9:169–181
8. Hao ZX, Zheng JY, Ge QS (2003) Climate change and harvest in Xi'an since 1736: the high-resolution data derived from the archives in the Qing Dynasty. Acta Geogr Sin 58:735–742
9. Huang H (2013) Temporal and spatial variation of relief grain transferring to flood and drought refugees in North China Plain in Qing dynasty. Beijing Normal University, Beijing
10. IPCC (2014) Working Group II report: impacts, adaptation and vulnerability. http://www.ipcc.ch
11. Li BZ (1999) Climate change and several times of drastic fluctuations of Chinese historical population. Popul Res 23:15–19
12. Li WH, Xia MF (2003) The Chinese famine policy book, series 1, vol 2. Beijing Ancient Books Publishing House, Beijing
13. Li JW, Wang SZ, Li TB (2006) Flood description in BianLiang. Henan University Press, Zhengzhou
14. Man ZM, Ge QS, Zhang PY (2000) Case studies on the impact of climatic changes on the farming-pastoral transitional zone in historical period. Geogr Res 19:141–147

15. Management Office of Yongding River in Beijing (2002) Floods and droughts in Yong-ding River. China Water & Power Press, Beijing
16. Medina-Elizalde M, Rohling EJ (2012) Collapse of Classic Maya civilization related to modest reduction in precipitation. Science 335:956
17. Qing G (1802) Relief chronicles authorized by emperor in Xin You (unpublished data)
18. Qiu G (1999) Archaeology on Kaifeng city in Song dynasty. J Hist Sci 6:105–109
19. Ren ME (2006) Sediment discharge of the Yellow River, China: past, present and future—a synthesis. Adv Earth Sci 21:551–563
20. Richard SJ, Wagner TS (2010) Climate change and violent conflict in Europe over the last millennium. Clim Change 99:65–79
21. Shao YZ (2006) Description of famine history books in China. J Huaibei Coal Ind Teach Coll 27:16–22
22. Shen Y, Zhao SX, Zheng DL (1935) Chronology of Yellow River. Military Commission and Records Commission, Nanjing
23. Wang XL (2007) About flood in Zhili area and government relief in the sixth and seventh year of the reign of emperor Jiaqing. Chin Soc Hist Rev 8(00):305–316
24. Weiss H, Courty MA, Wetterstrom W et al (1993) The genesis and collapse of third millennium north Mesopotamian civilization. Science 261:995–1004
25. Xiao LB, Fang XQ, Huang H, Wei ZD (2011) Changes in the ways of social response to floods and droughts in North China Plain during 1780~1819. J Catastrophol 26:83–102
26. Xiao LB, Huang H, Wei ZD (2012) Comparison of government relief food scheduling and social consequences during droughts of 1743–1744 and 1876-1878 AD in north China. J Catastrophol 27:101–106
27. Xie JH (1999) Present elevated situation of the lower Yellow River and preliminary discussion on its regulation. J Sedim Res 1:7–11
28. Ye Y, Fang XQ, Ge QS et al (2004) Response and adaptation to climate change indicated by the relationship between revolt and drought-flood in Shandong province during middle and late Qing Dynasty. Sci Geogr Sin 24:680–686
29. Ye Y, Fang XQ, Mohammad AK (2012) Migration and reclamation in Northeast China in response to climatic disasters in North China during the past 300 years. Reg Environ Change 12:193–206
30. Ye Y, Xu YF, Liang K et al (2014) Response to the Flood of Yong-ding River in 1801. J Chin Hist Geogr 29:13–19
31. Yu LS (2002) The Huanghe (Yellow) River: a review of its development, characteristics, and future management issues. Cont Shelf Res 22:389–403
32. Zhang D, Zhan ZY, Lin CS et al (2004) Climate change and war, social disturbance and dynasties change in China. Sci Bull 49:2468–2474
33. Zhang D, Jim C, Lin C et al (2005) Climate change, social unrest and dynastic transition in ancient China. Chin Sci Bull 50:137–144
34. Zhao EZ (1927) History records in Qing dynasty Rivers and canals annal (unpublished data)
35. Zhao EZ (1986) Actual annals of Qing Dynasty true records in Emperor Xuanzongcheng of Qing Dynasty. China Press, Beijing
36. Zou YL (1995) The movement of northern farming-grazing transitional zone and climate change during the Ming and Qing Dynasty. Fudan J 1:25–33

CHAPTER 16

Flood Hazard Mapping of Lower Indus Basin Using Multi-Criteria Analysis

Saba Zehra*, Sheeba Afsar

Department of Geography, University of Karachi, Karachi, Pakistan

ABSTRACT

Flooding has been one of the recurring occurred natural disasters that induce detrimental impacts on humans, property and environment. Frequent floods is a severe issue and a complex natural phenomenon in Pakistan with respect to population affected, environmental degradations, and socio-economic and property damages. The Super Flood, which hit Sindh in 2010, has turned out to be a wakeup call and has underlined the overwhelming challenge of natural calamities, as 2010 flood and the preceding flood in 2011 caused a huge loss to life, property and land use. These floods resulted in disruption of power, telecommunication, and water utilities in many districts of Pakistan, including 22 districts of Sindh. These floods call for risk assessment and hazard mapping of Lower Indus Basin flowing in the Sindh Province as such areas were also inundated in 2010 flood, which were not flooded in the past in this manner. This primary focus of this paper is the use of Multi-criteria Evaluation (MCE) methods in integration with the Geographical Information System (GIS) for the analysis of areas prone to flood. This research demonstrated how GIS tools can be used to produce map of flood vulnerable areas using MCE techniques. Slope, Aspect, Curvature, Soil, and Distance from Drainage, Land use, Precipitation, Flow Direction, and Flow Accumulation are taken as the causative factors for flooding in Lower Indus Basin. Analytical Hierarchy Process-AHP was used for the calculation of weights of all these factors. Finally, a flood hazard Map of Lower Indus Basin was generated which delineates the flood prone areas in the Sindh province along Indus River Basin that could be inundated by potential flooding in future. It is aimed that flood hazard mapping and risk assessment using open source geographic information system can serve as a handy tool for the development of land-use strategies so as to decrease the impact from flooding.

Keywords: Flood Risk Assessment, Flood Predictive Modeling, Flood Hazard Map, Geographic Information System

1. INTRODUCTION

Flooding has been one of the recurring occurred natural disasters that inflict irrecoverable and detrimental impacts on humans, property and environment [1] . According to Domeneghetti et al., (2013), natural disasters are the major rationales of irreparable losses all over the world [2] . Hussain et al. (2011) emphasize that natural disasters irrespective of any type induce considerable damages to both properties and people [3] . Almost every year, mainly during the monsoon seasons, Pakistan encounters flood events which cause substantial losses to livelihood, properties, settlement, roadways, livestock and agricultural lands. Particularly, recent flooding in the years 2010 & 2011 has caused considerable damages to Pakistan's economy. Heavy flooding in the Indus, Kabul, and Swat Rivers resulted in the life loss of a large number of people along with considerable damage to settlements, infrastructure and agriculture along with river embankments. If an early warning system and appropriate mitigation plan would have been in place then these consequences could have been minimized or decreased. There is an utmost need for accurate flood modeling and risk assessment for the reduction and mitigation of disastrous impacts of floods. Various hydrological approaches have been proposed by researchers. These methods include traditional approaches like rainfall-runoff modeling techniques and contemporary approaches like data-driven techniques for generating the accurate flood models. Tehrany et al. (2014) put forward that substantial contributions have been made by geographic information system (GIS) and Remote sensing (RS) techniques in modeling and prediction of flooding events [4] . RS and GIS techniques enable rapid hydrological data collection and analysis for the sake of flood susceptibility and hazard mapping. The major objective of this study is to generate flood risk and vulnerability map of lower Indus Basin on the basis of land use, hydrological and geomorphological features. GIS is extensively utilized in this paper for the integration of several layers which depict rainfall, land-use, elevation and slope, and soil.

2. STUDY AREA

Majority of the area of Sindh province falls in the Lower Indus basin that isgeographically located between latitudes from 66°E and longitudes 23°N and 23°31'N (see Figure 1). The lower Indus Basin is one of the tectono-stratigraphic provinces of Pakistan [5] . The soil of lower Indus plain is very fertile and crops can be easily grown where ever there is a proper water supply; however, Thar Desert situated in the eastern part of basin is quite infertile because of water scarcity. A large network of canals irrigates northern and central part of Sindh. In the southern part of Sindh, Indus River gets closer to Arabian Sea forming a deltaic region. During the monsoon, a considerable part of lower Indus plain is flooded with water causing devastation.

The reason for choosing the area is that in flooding events of the year 2010 & 2011, one of the hardest hit areas was Sindh province as about 22 out of 23 districts of Sindh were affected by floods [6] . After receiving heavy rainfall in the monsoons, the Sindh province faced destructive flooding that's why it was picked out as an appropriate application site for flood susceptibility mapping and risk.

3. METHODOLOGY

The first step for the flood risk assessment and modeling is creation of a spatial database [7] . The database consists of various factors relevant to floods. GIS tools and techniques were used for the transformation of the considered factors into a grid spatial database. Digital Elevation Model (DEM) of 3 arc resolution of the study area was acquired through http://srtm.csi.cgiar.org/ [8].

The DEM of the study area was used for calculating the flow direction, flow accumulation, slope, hill shade and aspect of the area. Slope curvature was also calculated by means of the ESRI routine in Arc View. Land cover map of the study area was classified using a Landsat 8 image through unsupervised classification technique and base maps in Erdas Imagine 2014. The land cover map has been classified into different classes.

The Meteorological data (precipitation) of stations in lower Indus Basin was collected from the Pakistan Meteorological Department. IDW technique was used for the interpolation of precipitation data for the whole study area over last twenty years. In order to have a general understanding and visual interpretation of the area, previous disaster history and satellite images (quick bird of resolution 0.6 m) of the area were also acquired. The demographic data of the area was determined through secondary sources like Census Reports of different districts.

In the present study a combination of different data sets such as morphologic, topographic, satellite imagery, and census data (2001) obtained from concerned government agencies (Table 1) have been used to compute a composite index of flood risk based on multi-parametric analysis.

The Multi Criteria Evaluation (MCE) approach through Rank Sum Method is used to identify the flood risk areas after assigned weights to each factor taken into consideration as per their sensitivity [9] . While conducted the present study following methodology schematically shown in Figure 2 is adopted. GIS application is fully used for reproducing, analyzing and integrated spatial data to prepare a flood risk map which not only defines the susceptibility of each settlement to inundation but also provides means for assessment of flood risk in terms of loss of life, crop land and property.

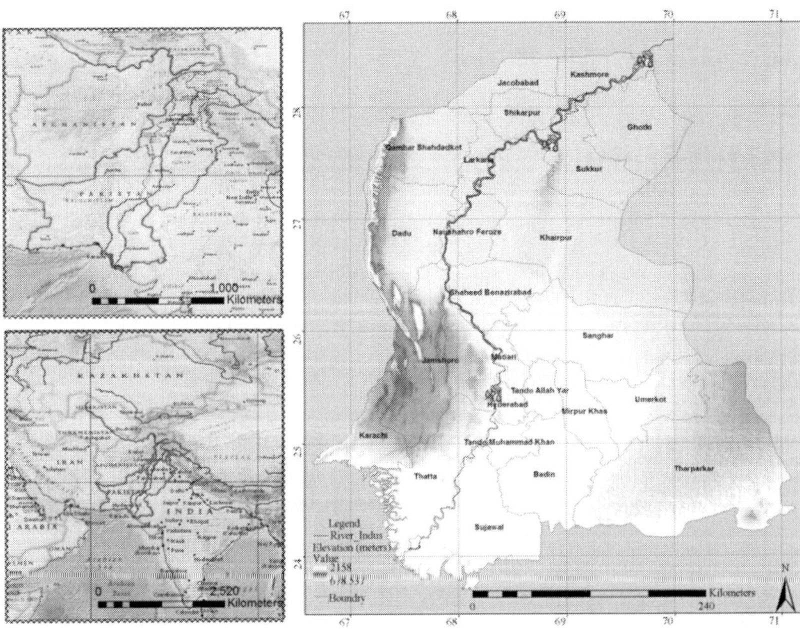

Figure 1. Map of study area.

Table 1. Spatial data used in the study.

Landsat 8 Image Acquired Date: November, 2014	http://glcfapp.glcf.umd.edu
Digital Elevation Model	http://srtm.csi.cgiar.org
Slope and Drainage Density	DEM
Population Data	Census Report
Rainfall (Precipitation)	Pakistan Meteorological Department

4. RESULTS & DISCUSSION

Multi-criteria evaluation (MCE) methods have been applied in several studies. Since 80 per cent of data used by decision makers is related geographically [10] , Geographical Information System (GIS) may provide more and better information about decision making situations. GIS allows the decision maker to identify a list meeting a predefined set of criteria with the overlay process [11] . In order to create flood risk maps ranking method is used and each factor is weighted according to the estimated significance for causing flooding.

1) Elevation: The 90 m SRTM DEM was converted into elevation raster layers using the ArcGIS conversion tool. On the basis of the susceptibility to flooding, elevation has been classified into five classes. Areas with the lowest elevation are very much susceptible to flooding, thus ranked 1, which is less than 77 m. Following the very high hazard class, there was a class high (77 - 232 m) ranked 2, class moderate (232 - 534 m) ranked 3, class low (534 - 1078 m) ranked 4 and class very low ranked 5 (>1078 m) (Figure 3).

2) Soil: The soil map of Sindh was converted into raster and then reclassified into five classes on the basis of their flood generating capacity. The soil type having very high flood generating capacity was ranked as 1, high ranked as class 2, moderate ranked as class 3, low ranked as class 4 and very low ranked as class 1. Below are the soil map and reclassified map of soil (Figure 4).

3) Slope of the Lower Indus Basin: The 90 m SRTM DEM was processed in ArcGIS to determine the slope of the AOI. Youssef et al. (2011) put forward that slope is regarded as the surface indicator to identify flood susceptible areas as it plays an important role in finding out vertical percolation and surface runoff velocity and, hence have an impact on flood susceptibility [12] . The lower slope value implies a flatter terrain, whereas the higher the slope value will be, the steeper will be the terrain, and higher will be the runoff. There will be a higher probability in areas with the lesser slope degree, much water will be received in

case of runoff. Slope has been classified into 5 classes on the basis of its susceptibility to flooding (Figure 5).

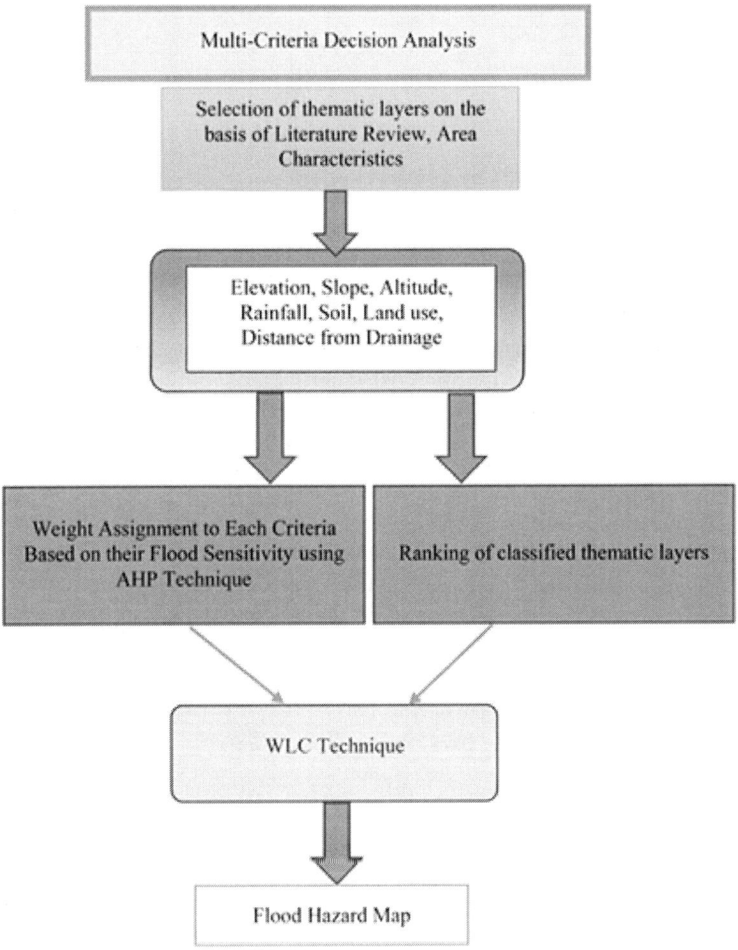

Figure 2. Methodological framework.

4) Rainfall (Precipitation): According to Musa et al. (2000), the availability of precipitation is regarded as a major source of recharge [13] . Adiat et al. (2012) [14] put forward that there is a significant impact of rainfall on the flood risk and the efficiency of MCDA. Monthly rainfall data of eight meteorological stations within the Sindh province for a period of 15 years (i.e., 1998-2012) were obtained from the Meteorological

Department. Monthly data were then converted to annual mean. The average annual rainfall of selected stations of Sindh of the years 1998-2012 was interpolated using the IDW (inverse distance weighted) technique. The raster layer was classified into five classes. It is indicated from the acquired annual mean rainfall pattern that there is low rainfall in the upper Sindh while there is high precipitation in the southeast peripheries, and moderate rainfall in the western lowlands of the river basin. In the classification process, area with higher rainfall (mm) is ranked 1 as it would be very highly affected by flood. Following the very highly susceptible to flooding class, there is a class high (191 - 223 mm/year), which is ranked as class 2, moderate (162 - 191 mm/year) ranked as class 3, low (136 - 162 mm/year) ranked as class 4 and very low ranked as class 5 (<136 mm/year) (Figure 6).

5) Land Cover: The Landsat 8 image was obtained from http://glcf.umd.edu/data/landsat/. A composite was made and then the area was mosaicked, clipped area and processed in Erdas Imagine 2014. The unsupervised land use classification was run and the image was classified into different classes of Water, Forest, Agriculture, Barren Land, Mangroves, Swamp and Settlement. Based on the flood generating characteristics of the land use type, water as very high (class 1) cultivated land was assigned as high flooding (class 2), crop/vegetation land as high (class 4), open to closed vegetation/grass land as moderate (class 3), shrub land as low (class 2) and forest land as very low (class 1) as shown in Figure 7.

After the reclassification of all parameters, the pairwise comparison matrix of the flood hazard inducing parameters was performed in the AHP in ArcGIS to compute the weights. Finally a composite risk map of flood was prepared using the Raster Calculator using the below equation:

$$\text{“Reclass\_Slope”} \times 0.3366 + \text{“Reclass\_Soil”} \times 0.3221 + \text{“Reclass\_Elevation”} \times 0.2154 + \text{“Reclass\_Rainfall”} \times 0.0743 + \text{“Reclass\_Landuse”} \times 0.0516$$

It can be seen from the above map that the riverine areas are the most vulnerable to flood as floodwaters will hit the riverine areas along both banks of the Indus River. River Indus after receiving heavy precipitation and snowmelt can possibly cause floods in the northern (like Kashmore, Shikarpur, Larkana, Naushero Feroze, Shaheed Benazirabad) and southern parts (Jamshoro, Thatta, Hyderabad, Matiari, Tando Muhammad Khan) of the Sindh province. The flood risks can be mitigated through development of land-use zoning guidelines and their effective implementation for reducing the magnitude of flooding to a certain extent. People should not be restricted to build settlements in the flood plains through legislation (Figure 8).

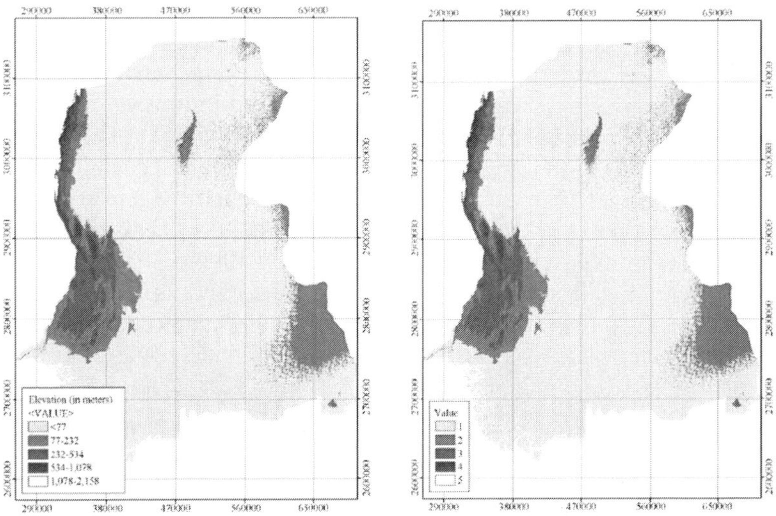

(a) (b)

Figure 3. (a) & (b): Elevation map & reclassified map.

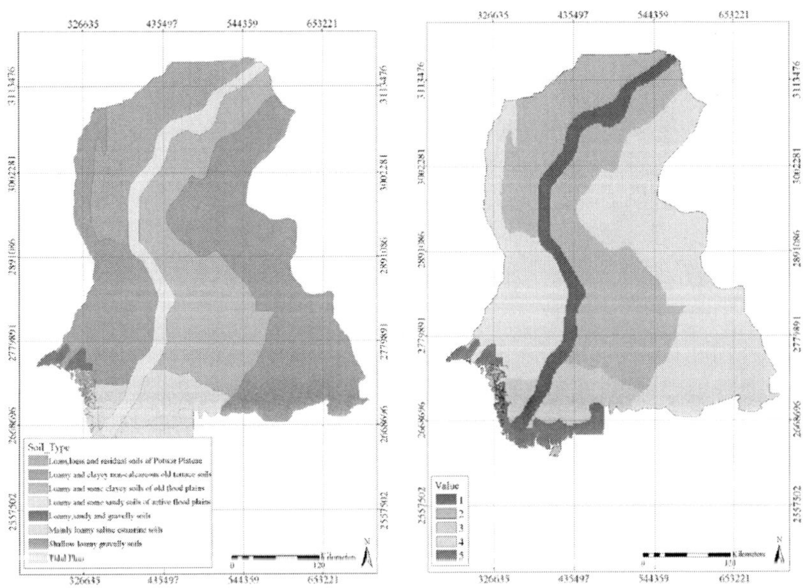

(a) (b)

Figure 4. (a) & (b): Soil map & reclassified map.

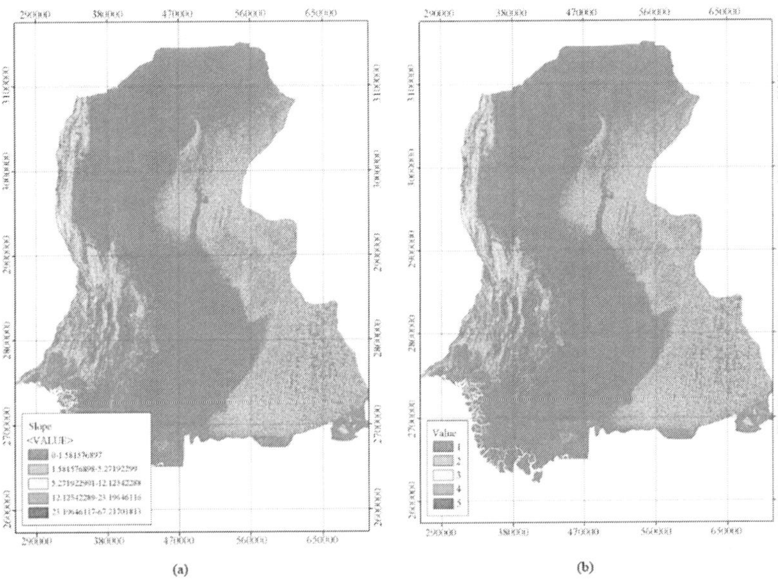

Figure 5. (a) & (b): Slope map & reclassified map.

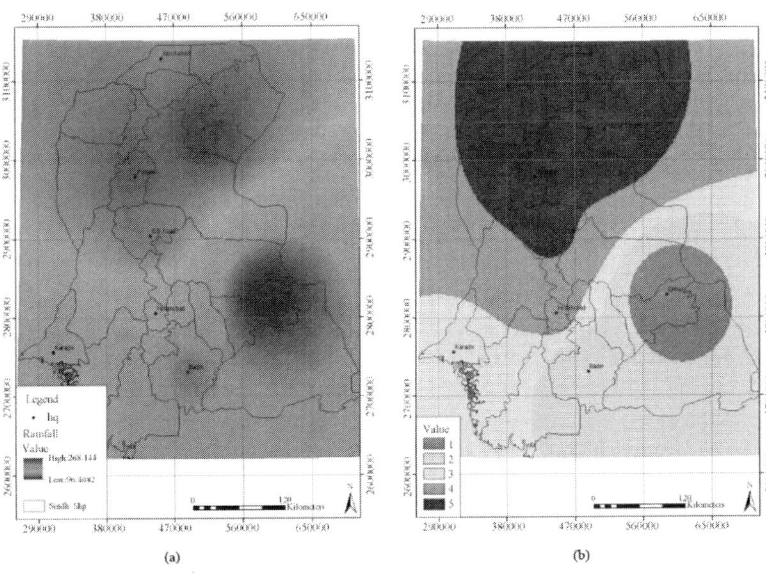

Figure 6. (a) & (b): rainfall map & reclassified map.

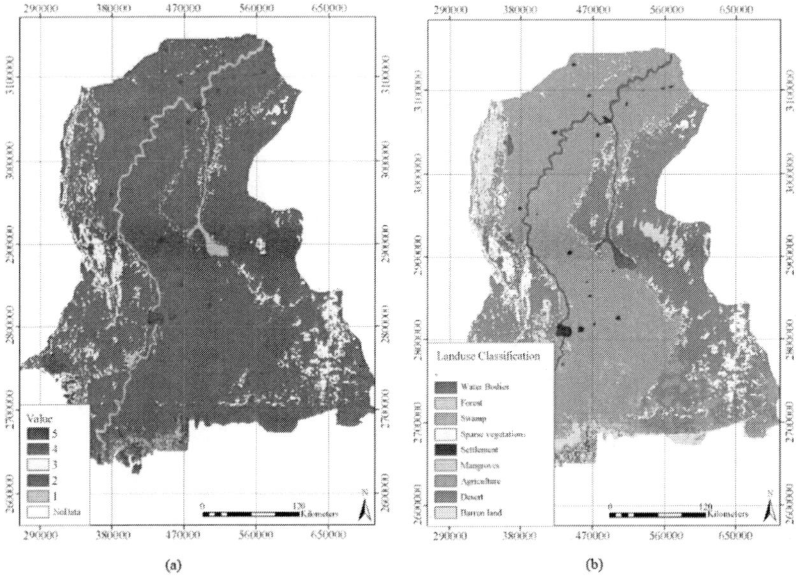

Figure 7. (a) & (b): land use map & reclassified map.

5. CONCLUSION

This paper presented a predictive model of flood of Sindh using AHP along with geographical information system. This paper initially constructed a spatial database of digital elevation model (DEM), geological and soil map, land cover map, topographical map and historical data of flood events for the evaluation of factors associated with flood vulnerability analysis. Then, this paper applied AHP technique to prepare flood vulnerability map. The flood hazard map delineated the flood prone areas that could be inundated by substantial flooding. To conclude, detailed information is needed for analysis of comprehensive flood risk analysis; such information could include hydrologic statistics, field conditions, & characteristics of flood-defense structures in order to determine the severity & extent of the flood impact of on the specific areas on the basis of probability-based result. In this study, with the methodology proposed, the integration of AHP and GIS can provide an initial assessment of the flood risk in the certain area. Moreover, GIS and AHP could provide the decision makers a more meaningful, interactive & detailed flood risk assessment. It is anticipated that this study could be helpful for the local community & flood managers to comprehend the factors causing flood plains inundation.

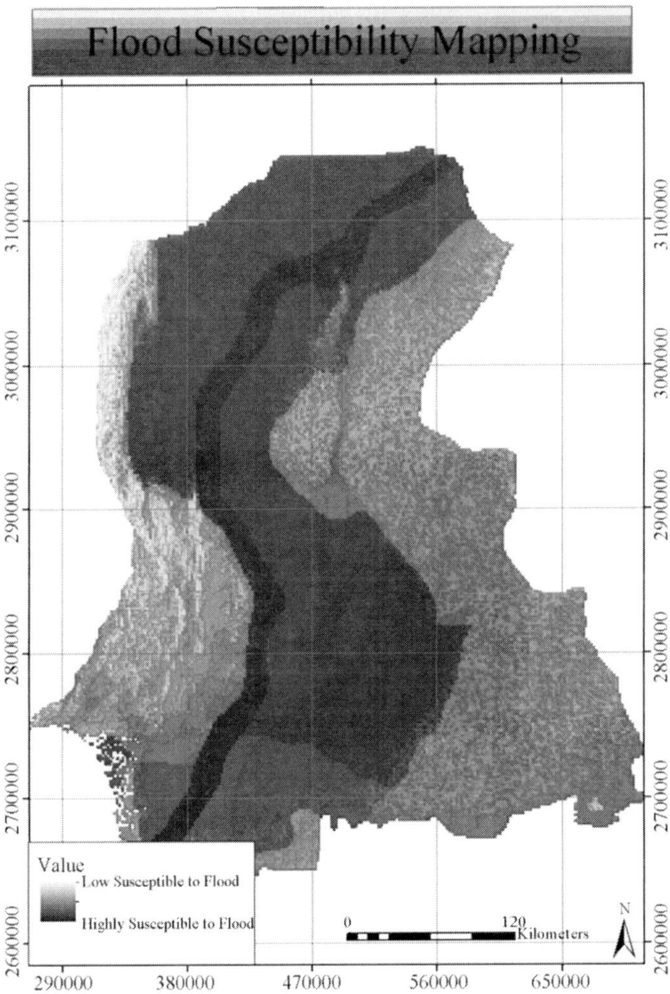

Figure 8. Final flood susceptibility map.

REFERENCES

1. Khan, B., Iqbal, M.J. and Yosufzai, M.A.K. (2011) Flood Risk Assessment of River Indus of Pakistan. Arabian Journal of Geosciences, 4, 115-122.
2. Domeneghetti, A., Vorogushyn, S., Castellarin, A., Merz, B. and Brath, A. (2013) Probabilistic Flood Hazard Mapping: Effects of Uncertain Boundary Conditions. Hydrology and Earth System Sciences, 17, 3127-3140.
3. Hussain, A., Weisaeth, L. and Heir, T. (2011) Psychiatric Disorders and Functional Impairment among Disaster Victims after Exposure to a Natural Disaster: A Population Based Study. Journal of Affective Disorders, 128, 135-141.

4. Tehrany, M.S., Pradhan, B. and Jebur, M.N. (2014) Flood Susceptibility Mapping Using a Novel Ensemble Weights-of-Evidence and Support Vector Machine Models in GIS. Journal of Hydrology, 512, 332-343.
5. Khan, N., Konaté, A. and Zhu, P. (2013) Integrated Geophysical Study of the Lower Indus Platform Basin Area of Pakistan, International Journal of Geosciences, 4, 1242-1247.
6. Memon, N. (2012) Malevolent Floods of Pakistan, Islamabad. Strengthening Participatory Organization, p. 8.
7. Zerger, A. and Wealands, S. (2004) Beyond Modelling: Linking Models with GIS for Flood Risk Management. Natural Hazards, 33, 191-208.
8. SRTM, C.C., 90m DEM Digital Elevation Database. http://srtm.csi.cgiar.org/
9. Papaioannou, G., Vasiliades, L. and Loukas, A. (2015) Multi-Criteria Analysis Framework for Potential Flood Prone Areas Mapping. Water Resources Management, 29, 399-418.
10. Malczewski, J. (1999) GIS and Multicriteria Decision Analysis. John Wiley & Sons, Hoboken.
11. Heywood, I., Oliver, J. and Tomlinson, S. (1995) Building an Exploratory Multi-Criteria Modelling Environment for Spatial Decision Support. Innovations in GIS, 2, 127-136. [Citation Time(s):1]
12. Youssef, A.M., Pradhan, B. and Hassan, A.M. (2011) Flash Flood Risk Estimation along the St. Katherine Road, Southern Sinai, Egypt Using GIS Based Morphometry and Satellite Imagery. Environmental Earth Sciences, 62, 611-623.
13. Musa, K.A., Akhir, J.M. and Abdullah, I. (2000) Groundwater Prediction Potential Zones in Langat Basin Using the Integration of Remote Sensing & GIS. The 21st Asian Conference on Remote Sensing.
14. Adiat, K.A.N., Nawawi, M.N.M. and Abdullah, K. (2012) Assessing the Accuracy of GIS-Based Elementary Multi Criteria Decision Analysis as a Spatial Prediction Tool—A Case of Predicting Potential Zones of Sustainable Groundwater Resources. Journal of Hydrology, 440-441, 75-89.

CHAPTER 17

Flood Management in California

Jay R. Lund

Department of Civil and Environmental Engineering, University of California, Davis, CA 95616, USA

ABSTRACT

California's development and success have been shaped by its ability to manage floods. This management has varied over the history of California's economic and political development and continues in various forms today. California will always have flood problems. A range of options are available to aid in flood management problems and have been used over time. These options can be contrasted with flood management elsewhere and the types of options used to manage other types of hazards in California, such as earthquakes, wildfires, and droughts. In the future, flood management in California will require greater reliance on local funding and leadership, reflecting diminished federal and state funding, with more effective state and federal guidance. Effective flood management will also tend to integrate flood management with actions to achieve environmental and other water supply objectives, both to gain revenues from a broader range of beneficiaries as well as to make more efficient use of land and water in a state where both are often scarce.

Keywords: California; flood; floodplain; management

1. INTRODUCTION

California is a land of many extremes. One extreme is floods. From its earliest days of statehood, California has had to contend with and eventually manage frequent and extreme floods. Flood management did not come easily to California, and remains a major issue for some areas of the state. This paper reviews the array of options available for flood management, the nature of floods in California, the history and current condition of flood management in California, as well as long term problems and solutions. Flood management is contrasted with flood management elsewhere and with the management of other hazards in California. Finally, some ideas for improving management and conclusions are presented.

2. FLOOD MANAGEMENT OPTIONS

A wide range of options can be discussed for flood management, and classified in various ways. Here, these options are classified by when they can be applied, in preparation for a flood, during a flood, and in flood recovery (Table 1).

Options also are classified as actions which protect land uses from flooding ("Protection" actions) or reduce the vulnerability of land to flooding ("Vulnerability reduction," reducing damage and loss of life consequences when a flood occurs). An effective and cost-effective flood management program usually will combine these options suitably for local conditions. The history and future of flood management in California is the story of how these options have been employed and combined over time.

3. FLOODS IN CALIFORNIA

The frequency, magnitude, and duration of floods are driven by local climate and basin characteristics. California is a large and diverse state with somewhat different flood regimes in different regions. In each area, floods have been managed somewhat differently. Also, the hydrologic variability of floods in each area served somewhat different ecological functions before the economic development of California's lands and rivers [1]. Most flood problems in California tend to have different characters, depending on local regions, as summarized in Table 2.

Table 1. Major flood management portfolio options.

Preparatory actions	
Protection	**Vulnerability reduction (reduced damage and casualty potential)**
Levees	Relocation of vulnerable human activities
Flood walls and doors	Floodplain zoning and building codes
Closed conduits	Floodproofing–raising structures, sacrificial first floor, watertight doors, and flood vents
Channel improvements and flood corridors	Flood warning and evacuation systems
Reservoirs	Flood insurance and reinsurance
Bypasses	Flood risk disclosure
Sacrificial flooding	Public and policymaker education
Flood easements (bypasses and designated\ flood areas)	Flood preparation and training exercises
Local detention basins, drainage, and pumps	Floodplain mapping, gaging, data collection and availability
Regular inspections, assessments, and maintenance	Community engagement and multi-hazard planning
Response actions	
Protection	**Vulnerability reduction**
Levee and flood wall monitoring (structures and seepage)	Warnings, evacuation calls, and emergency mobilization
Flood fighting–sandbagging, sheet pile installation, wave wash protection, splash cap installation, ring levee construction, relief cut, pumping, and breach closure and capping	High water staking
Flood door closure and gate operation	
Reservoir operation–including coordinated operations, rule curve operations and encroachment, flash board installation, and spillway surcharging	
Reconstruction and repair of flood infrastructure	Flood damage assessment–flood infrastructure surveys, system performance, damage, response costs Flood insurance and reinsurance Reconstruction and repair Relocation or reconstruction to reduce future flood vulnerability

For all of California, most flood-generating storms arrive in the winter with predominantly orographic precipitation. Particularly in northern and central California, snowfall and snowmelt are important. Colder storms tend to have more precipitation fall as snow, and cause less flooding than comparable precipitation from rainier warmer storms. However, the accumulation of snowpack from earlier cold storms can contribute greater runoff from following warmer storms [2].

At the end of the winter and into the spring (particularly March and April, but sometimes into July), snowmelt floods can arise in northern California. Flow peaks for snowmelt floods tend to be less than the larger rain-generated floods, but have much larger volumes and durations of high flows.

Table 2. Major flood types in California.

Lands below sea level
Sacramento-San Joaquin Delta: Large areas of deeply subsided diked lands, mostly agricultural, but with important statewide water supply and ecosystem implications; immediate flood causes can range from high river flows, earthquakes, burrowing animals, and ship grounding, exacerbated by long term land subsidence and sea level rise. Managed mostly by local levee maintenance districts with significant dependence on state funding.
Major River Flooding
Sacramento Valley, San Joaquin Valley, major north coast rivers: Flooding from major rainstorms (particularly "atmospheric rivers") or snowmelt. Floods can cover large areas to substantial depths and cause large amounts of property damage and life loss. Flood protection mostly by levees, bypasses, and upstream reservoirs. Flood warning and evacuation systems are usually effective in reducing loss of life, except where levee breaches cause rapid deep flooding. Largest potential damage for Sacramento metropolitan area, Marysville/Yuba City area, Stockton, and about 27 smaller river towns. Managed by regional water and flood control projects involving local, state, and federal governments.
Rapid River flooding
Most common in Southern California and alluvial fans: Local floods caused by intense rainstorms on steep watersheds, which can also include substantial debris flows. Property damage can be large over a small area, and loss-of-life risks are larger due to difficulties with flood warning. Managed by local governments (particularly using land use authority) with involvement from federal and state governments.
Local rainfall flooding
Everywhere: Flooding of small areas to shallow depths caused by local rainfall or rapid snowmelt. Inconvenience can be frequent, but property damage is usually small and loss of life is relatively rare. Managed mostly by landowners and local governments.
Dam failures
Dam failures are infrequent in California, but there is always some residual risk, particularly given the large earthquakes which can occur in California. Roughly 1,250 sizable non-federal dams are regulated by the California Department of Water Resources Division of Dam Safety.
Coastal flooding
Coastal storms and tsunamis are a risk in low-lying coastal areas. These risks will increase with sea level rise and continued coastal development. Managed mostly by coastal hardening and warning and evacuation systems.

Most significant floods in California are caused by "atmospheric rivers" from the Pacific Ocean, which include warmer and more intense "pineapple express" storms, in California parlance, where a stream of warm atmospheric moisture precipitates for several concentrated days over the mountains of California. [3,4,5] (Figure 1) Historically, large examples of these storms flooded much of the Sacramento Valley, with floods as large as 17,000 cms (600,000 cfs) near Sacramento. The size and flatness of the Sacramento and San Joaquin Valleys, with high tributary mountain watersheds, can bring large amounts of water quickly to slow-draining populated and agricultural areas, making them especially flood-prone.

North Coast rivers, having less snowpack and more direct access to Pacific Ocean moisture, are similarly exposed to major floods. However, their smaller drainage areas and low population densities make damages there, while locally significant, less important from a statewide perspective.

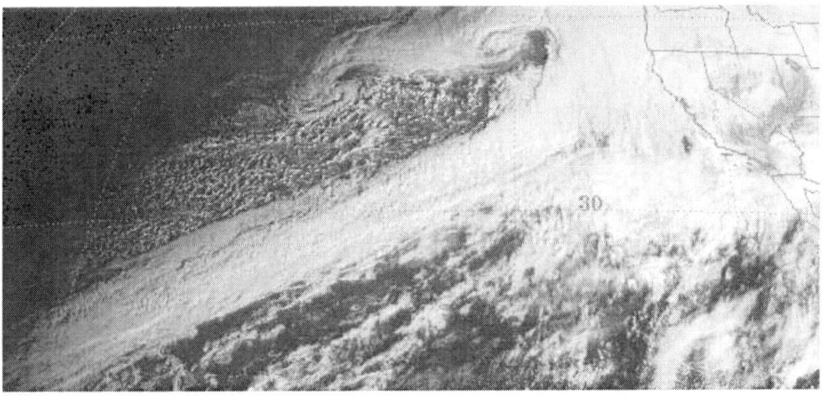

Figure 1. "Atmospheric Rivers" bring major floods to California, 31 December 1996, 1800 UTC, GOES-9 image [6].

In southern California, steeper and smaller watersheds close to urban settlements cause floods to develop rapidly for shorter periods of time. Debris flows and mud slides accompany floods in some steeper areas. Warnings and evacuations are more difficult. Damage reduction is more reliant on land use planning and large channel capacities, along with some detention storage, particularly for debris flows [7].

A unique form of flooding occurs in the subsided islands of the Sacramento-San Joaquin Delta, which often lie several meters below sea level and are protected by relatively weak levees. Although many floods are caused by high water from storm-generated inflows, island failures also can arise from higher than usual tides, animal burrowing, earthquakes, ship grounding, and other causes. Continued land subsidence and sea level rise will worsen flooding problems in this area [8,9].

Dam failures are rare in California, but still pose potentially significant residual risks, especially given the state's susceptibility to major earthquakes. (Residual risk is the risk remaining after flood management activities have been undertaken.) The largest dam failure in California was the St. Francis dam collapse in 1928, killing hundreds of people. Smaller failures or partial failures have occurred at Lower San Fernando Dam (1971, earthquake), Calaveras Dam (1918, in construction), San Luis

Dam (1981, drawdown), and elsewhere, requiring repairs or reductions in capacity. Potential inundation areas for major dam failures in the densely populated San Francisco Bay Area are illustrated in Figure 2.

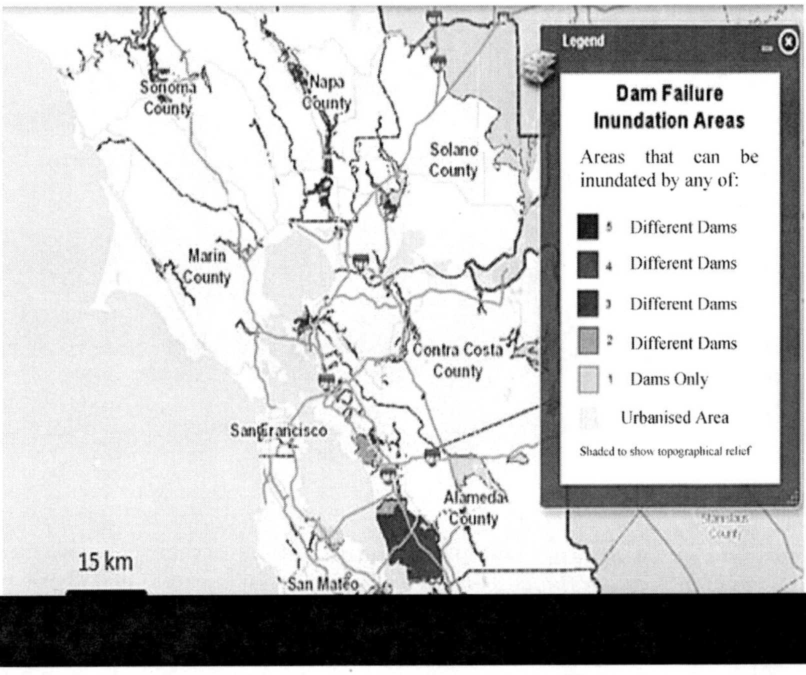

[redacted]

Figure 2. Dam failure inundation areas for the San Francisco Bay Area [10].

Coastal flooding occasionally occurs from storms and tsunamis. Tsunami and coastal flood warning systems are in place. However, coastal flooding is expected to increase with sea level rise and continued development in coastal areas, posing both economic and environmental difficulties.

4. MANAGING FLOODS IN CALIFORNIA

Early European settlers to California faced abrupt climate change, relative to the more humid non-Mediterranean climate they came from on the eastern seaboard of the United States. Instead of a climate with predominantly summer precipitation and major hurricane-driven floods, they came to an overwhelmingly winter precipitation climate with major winter floods from winter "atmospheric river" events. As the settlers came to understand the flood-prone nature of California's semi-arid climate, they were driven to create local institutions for flood management [11]. California's ever-changing economy and society have also changed the nature and importance of California's flood problems with time.

Floods are always most acutely felt locally. Flood management in California began with unaided flood protection efforts by local landowners. Local flood efforts were then aided by legislation enabling formation and governance of local levee or reclamation districts in 1868, allowing groups of landowners to more efficiently build more effective levees for a district, rather than a single property. These efforts reduced seasonal flooding in most years. However, as the number of districts grew, often on opposite sides of the river, violent interactions sometimes arose among neighboring levee districts during floods. Larger floods would often overwhelm many levee districts.

Following major floods in 1907, it was widely recognized that the system of local levees had created a regional fragility to large storms, with potential to stymie urban and agricultural development in the Sacramento Valley. State and federal authorities, after years of study, developed a set of flood bypasses, channel expansions, and stronger project levees to provide greater relief from large floods and more systematic coordination of levees and channels. The bypass system was built largely during the 1920s under federal and state authority, with local levee districts remaining in charge of levee maintenance. The bypass system restored some of the flood channel capacity, eliminated by earlier levee construction, by building large seasonal channels with flood easements purchased from farmers (who conduct seasonal farming in the bypasses). These bypass channels are separated from the main stem of the Sacramento River by low weirs which overtop with high water (Figure 3).

A major problem for flood management in the Sacramento Valley during the early 1900s was the immense quantity of debris and sediment from earlier gold mining in and near the rivers. To mobilize these sediments, which had greatly raised local river beds, levees were placed close to the river to induce scour. While this placement of levees was effective in mobilizing sediment, now that most accessible mining sediment has flushed from the system, the close placement of river levees has caused problems of levee erosion and increased the cost of making improvements to levees as higher levels of protection are desired for more valuable and populated land uses in protected floodplains.

With the development of major water supply storage reservoirs, starting in the 1940s through the early 1980s, reservoir flood storage was added to the flood management system. Almost all major reservoirs have a reserved seasonal flood management pool to reduce peak flows from storms. This flood capacity was paid for and regulated by the federal government.

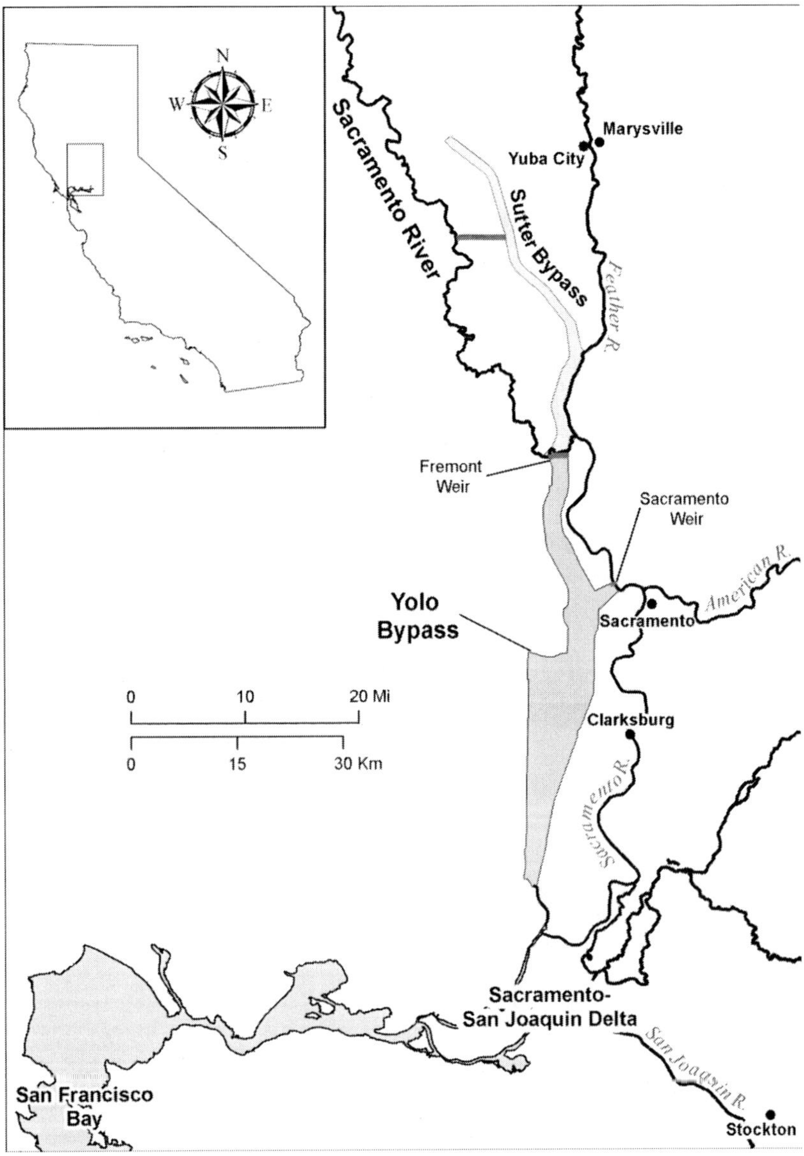

Figure 3. Sacramento River Bypass system [14].

Because California's flood season falls outside of the major water use and irrigation season, reservoir operations are better able to both moderate floods and supply water, compared to reservoirs in climates with wet summers, such as China or the Eastern U.S. With climate warming, the loss of snowpack will force reservoirs to operate more for floods, which will be more rain-fed and less snow-melt. Although a warmer climate will be inconvenient and cause modest costs, if well managed, climate change alone seems unlikely to cause catastrophe [2,12]. Shifting drought water storage from surface reservoirs to aquifers can also make substantially more storage capacity available for winter flood management [13]. This shift will require integrated reoperation of many flood and water supply system elements.

5. CURRENT FLOOD PROBLEMS AND RISKS

As summarized in Table 2, small frequent floods are fairly well managed in California, in terms of economic risk. Residual risks are large in some places, but California's flood problems are more of a collection of local flood problems, sometimes with regional management implications, than a core threat to California's society and economy, as they arguably were a century ago [14,15,16,17,18,19].

Despite the tremendous successes of flood management in California, California will always remain prone to major floods and flooding. The extremes of California's climate are unlikely to moderate. Economically, it is infeasible and undesirable to prevent all floods, as the cost of preventing floods from all storms and extreme conditions comes to exceed the damages incurred, particularly in an expected value and risk-based sense. Also, as a levee-dominated flood management system, much of the Central Valley will always retain a substantial residual risk of levee breaches and the major damages and potential loss of life that such sudden and deep floods entail.

Flood potential remains significant and even limiting for several major urban areas of California, as well as more numerous small towns. The Sacramento metropolitan area, including the cities of Sacramento and West Sacramento, remains one of the most flood-prone urban areas of the country. The Marysville-Yuba City and lower reaches of the Feather River are also inherently flood-prone. Parts of Stockton, Napa, Victorville, and areas of Santa Clara and Ventura counties also are flood-prone, often with unique circumstances. Some urban areas of southern California also have potential for major urban flooding [18,19]. Smaller former river towns, such as Clarksburg, as well as some small Delta towns on subsided or low-lying lands (such as Bethel Island) are more difficult flood-prone areas, with little economic activity to support flood protection actions. Recent legal rulings (the "Paterno decision") which found the state government of California liable for damages from flooding in areas protected by state project levees have shifted flood responsibilities toward state government, and have led to increased state activity in flood management.

Greater potential and opportunities remain for the management of more frequent floods for environmental purposes [20,21]. Frequent seasonal floodplain inundation was historically important for many of California's native fish species as well as migratory waterfowl. Today, roughly 95% of native wetlands in the state have been leveed and drained, mostly for agriculture as well as for urban uses and salt production in some areas. Re-connecting floodplains to rivers to allow for seasonal flooding is a major environmental objective which will take decades to realize. Over such a long period of time, there will be opportunities to re-configure and expand parts of the levee system by setting them back from the river or improving the system of bypasses [14,22]. Such changes in the levee and bypass system provide opportunities to improve environmental performance, as well as to increase flood channel capacities and reduce scour of levees, reducing economic risk from floods. Alas, most environmental benefits will come from small frequent floods, whereas most flood protection benefits will come from avoiding very large infrequent floods. Environmental and flood benefits will also often require more complex planning due to likely greater losses of water stored for agricultural and urban water supplies during the summer.

The management of floods in California is going through major transitions. With the decline in federal budgets and capabilities, leadership has shifted back to local and state governments. Local leadership has been particularly effective. Following the 1986 flood, which nearly overwhelmed levees for Sacramento, the flood protection level for the city was downgraded from one in 500 years to about once in 60 years (annual exceedence probability from 0.002 to 0.017, or 0.2% to 1.7%) [22]. In 1989, several local governments in the Sacramento area, east of the river, formed SAFCA (Sacramento Area Flood Control Agency) to better coordinate local, state, and federal actions for flood protection in the member area. SAFCA has been very effective in marshalling federal, state, and local resources to improve levee reliability, expand flood channel capacity, improve the outlet capacity for Folsom Dam, and reduce development potential in rural areas on the western side of the Sacramento River.

Following decades of local, state, and federal neglect, attention brought by the 2005 Katrina flood of New Orleans, California voters approved two general obligation bonds in 2006 with $5 billion to support flood management. California also passed legislation requiring urban areas of the Central Valley to eventually adopt 200-year levels of flood protection for major river floods (0.5% annual exceedence probability). This has rejuvenated state flood management efforts and investments, particularly levee improvements, although roughly $20 billion more is estimated to be needed for flood system improvements. After the flood bonds are all expended, flood problems and risks will remain, and it will be important for the state to establish a sustainable finance model for local and state support of flood operation, maintenance, and assessment. State government has largely supplanted federal leadership in flood policy in California [22].

6. CONTRASTS

Flood management in California can be compared with other parts of the world. Flood management in the Sacramento Valley has significant parallels with management of the Mississippi River, which was another major national flood hazard area in the first half of the 1900s. On both systems, the U.S. Army Corps of Engineers was active in designing major basinwide flood management systems, particularly the combined use of flood bypasses and levees [11]. Bypasses have been very successful for both systems, however the shorter duration of floods in California give reservoirs a greater role in reducing peak flows. In the Netherlands, where major river floods also tend to have long durations, upstream reservoirs, largely outside of Dutch territory, are not of major importance, and dikes, levees, and bypasses are important [23]. In China, historical floods arguably shaped early Chinese civilization, but are now largely managed, in most years, with a system of levees, reservoirs, and retention areas. In all these regions, floods have historically shaped the region's land use, economy, and society. While floods have been largely "tamed" for these regions, substantial residual risk remains, along with challenges for managing floods for ever-changing economic and societal objectives. In these cases, major expenses for construction of flood management infrastructure has been supported by national governments, reflecting the national importance of flood control at the times when these projects were constructed. Operations and maintenance is largely left to local and state agencies. Flood damages are almost non-existent in the Netherlands, due to their much higher economically-designed risk safety standards (up to one in 10,000 years for urban areas) [23]. Flood damages on the Mississippi River system are largely paid by land users through flood insurance and uncompensated damages. The U.S. East Coast is subject to major hurricane flooding from rains and storm surges over very large areas. Dunes and occasional urban storm barriers provide some protection, but most flood risk reduction is from reducing the vulnerability of coastal communities to flooding by land use actions, building codes (particularly elevated structures), and strong flood warning and evacuation systems.

California's management of floods also can be compared with the management of other hazards in California, such as earthquakes, fires, and droughts. Major earthquake hazards, which vary throughout the state, are managed primarily by using building and construction codes and earthquake design to make structures resistant to all but the largest earthquakes. Some structures carry earthquake insurance. So most earthquake-related costs are therefore borne by building owners, both in construction costs and insurance, as well as damage (although considerable federal and state relief and recovery support also becomes available). Wildfires are another major hazard in California. Wildfires occur in most years and can damage large exurban and even suburban areas, such as the Oakland Hill fire of 1991 ($3.5 billion in damage). Most damage is covered by homeowners' insurance, and rural homeowners often pay fees to support fire protection services. Effective local evacuation programs have eliminated most wildfire-related deaths (fire damages commonly exceed $1 billion/year) [24]. Droughts in California are mostly managed by water agencies and utilities. Seasonal droughts have been largely eliminated through the use of surface reservoir and aquifer water storage, mostly by local agencies and water users as well as through major reservoir projects. Today, major drought losses occur from large over-year droughts, and many are concerned with additional perhaps permanent losses with climate change [13]. Disaster relief aid is minimal for droughts in California, with most expenses for accommodating and managing for droughts being borne by water users themselves. Given the low frequency of major flooding in California, it is important that floodplain emergency management be brought into other local and state emergency management activities, such as for fires, drought, and earthquakes (which rarely occur at the same time or season). Federal disaster response efforts are largely centralized with the U.S. Army Corps of Engineers and Federal Emergency Management Agency (FEMA).

7. SOME IDEAS FOR THE FUTURE

For California, the future of flood management seems likely to be as tied to restoring ecosystem functions as to traditional economic risk and public safety. Managing economic risks and public safety will take the form mostly of major investments in stronger levees to protect urban areas. However, there are also large opportunities to simultaneously reduce economic risks (especially for large floods) and improve environmental performance by re-connecting some floodplain lands and bypass systems with rivers to allow more seasonal flooding. This reconnection and recreation of seasonal floodplain wetlands has substantial environmental potential for native fish and waterfowl, and if done carefully, can also improve bypass and river channel capacities while reducing scour on mainstem river levees. Similarly, there will be interactions between flood and water supply functions of California's water management system [22].

Climate change, particularly climate warming, will be an issue for flood operation of reservoirs. Greater winter peaks and volumes are likely, with reductions in later snowmelt. Many reservoirs seem capable of regulating such floods, sometimes with modifications to operating rules, although longer periods of high flows may increase levee scour [2,12]. The combination of continued urban growth and climate change are likely to make larger and perhaps set-back levees more desirable [25,26]. Discussion of climate change raises another point for flood management more generally, that we are likely to never have a great deal of certainty regarding the frequency distribution of flood peaks, durations, and volumes [27,28]. Nevertheless, probabilistic risk analysis remains the most rational framework for understanding and evaluating flood problems and solutions [22].

Flood management in California, and elsewhere, is always hampered by a lack of stable financial resources. Federal, state, and local resources are often committed in the immediate aftermath of floods, especially for recovery, but operation, maintenance, and planning are often neglected for decades without major floods. Federal and state budget problems are likely to worsen funding problems for flood management. Without sustained funding for flood management, land use management is often driven by private and local revenue potential from new floodplain development and water management is driven by demands for water supply, both of which can work to increase flood risks.

Many ideas can be explored for better managing floods in California [16,29]. Some of these include:

- More portfolio-based approaches to flood management, with greater coordination in the planning and operation of preparatory, response, and recovery actions to increase protection and reduce vulnerability (Table 1) [22].
- Establish floodplain land use and insurance policies which come into effect only after some years, or a decade, to make the transition to new policies easier in political and economic terms.

- Impose a state-collected fee on land uses in the flood plain to cover flood damages, essentially insuring the state for its liability for flood damages in areas covered by state flood control projects. Such a fee might also take the form of a flood damage bond on new development (insuring the additional risk from new urban development in floodplains).
- Better coordinate local, state, and federal actions to take advantage of the comparative advantages of each level of action and institution [22].

Flood management in California's Central Valley is currently being examined under the ongoing State of California FloodSAFE effort [14]. In the coming months and years, there will be improvements in our abilities to characterize and model floods, flood damages, and flood management and policies.

If accompanied by political leadership, such technological advances should greatly improve flood safety and reduce the ecological impacts of flood management.

8. Conclusions

California has had considerable success in managing floods. Large, mostly agricultural, parts of California are now roughly suitably protected from floods, in a risk analysis sense. However, several large urban areas remain at large risk of flooding, including the state capital, and there is considerable urban development pressure in many deep floodplains behind fairly weak levees.

Much of the state's water supply would also be at risk from flooding of subsided lowlands in the Sacramento-San Joaquin Delta. So, despite significant successes, a large and growing residual flood risk remains. Fortunately, the state and many local governments now understand these risks better and have embarked on a major effort to improve flood protection and management. Time will tell if these efforts are fruitful.

ACKNOWLEDGMENTS

Nathan Burley and Christy Jones, two anonymous reviewers, and two editors are thanked for suggestions on earlier drafts. Anna Fryjoff-Hung is thanked for her cartography.

REFERENCES

1. Moyle, P.B.; Bennett, W.A.; Fleenor, W.E.; Lund, J.R. Habitat variability and complexity in the upper San Francisco estuary. San Francisco Estuary Watershed Sci. 2010, 8, 1–24.
2. Willis, A.D.; Lund, J.R.; Townsley, E.S.; Faber, B. Climate change and flood operations in the Sacramento Basin, California. San Francisco Estuary Watershed Sci. 2011, 9, 1–18.
3. Dettinger, M. Fifty-Two Years of "Pineapple-Express" Storms across the West Coast of North America; PIER Project Report; Prepared for California Energy Commission, Public Interest Energy Research Programme; 2004 December; CEC-500-2005-004.
4. Dettinger, M.; Ralph, F.; Das, T.; Neiman, P.; Cayan, D. Atmospheric rivers, floods and the water resources of California. Water 2011, 3, 445–478.
5. Ralph, F.M.; Neiman, P.J.; Wick, G.A.; Gutman, S.I.; Dettinger, M.D.; Cayan, D.R.; White, A.B. Flooding on California's Russian River: Role of atmospheric rivers. Geophys. Res. Lett. 2006, 33.
6. GOES Data; National Climate Data Center: Asheville NC, USA, 2012. Available online: http://www.ncdc.noaa.gov(accessed on 11 February 2012).
7. Rantz, S.E. Urban Sprawl and Flooding in Southern California; USGS Circular 601-B, U.S. Geological Survey: Menlo Park, CA, USA, 1970.
8. Lund, J.; Hanak, E.; Fleenor, W.; Bennett, W.; Howitt, R.; Mount, J.; Moyle, P. Comparing Futures for the Sacramento-San Joaquin Delta; University of California Press: Berkeley, CA, USA, 2010.
9. Suddeth, R.; Mount, J.F.; Lund, J.R. Levee decisions and sustainability for the Sacramento San Joaquin Delta. San Francisco Estuary Watershed Sci. 2010, 8, 1–23.
10. Earthquake and Hazards Program. Dam Failure Inundation; Association of Bay Area Governments: St. Oakland CA, USA, 2011. Available online: http://quake.abag.ca.gov/dam-failure/ (accessed on 11 February 2012).
11. Kelley, R. Battling the Inland Sea; University of California Press: Berkeley, CA, USA, 1989.
12. Connel-Buck, C.R.; Medellín-Azuara, J.; Lund, J.R.; Madani, K. Adapting California's water system to warm vs.warm-dry climates. Clim. Change 2011, 109, 133–149.
13. Tanaka, S.K.; Zhu, T.; Lund, J.R.; Howitt, R.E.; Jenkins, M.W.; Pulido, M.A.; Tauber, M.; Ritzema, R.S.; Ferreira, I.C. Climate warming and water management adaptation for California. Clim. Change 2006, 76, 361–387.
14. California Department of Water Resources (DWR). 2012 Central Valley Flood Protection Plan; DWR: Sacramento, CA, USA, 2011. Available online: http://www.water.ca. gov/floodsafe/ (accessed on 11 February 2012).
15. California Department of Water Resources (DWR). Responding to California's Flood Crisis; DWR: Sacramento, CA, USA, 2005. Available online: http://www.water. ca.gov/pubs/flood/flood_warnings__responding_to_california%27s_flood_crisis/011005floodwarnings.pdf (accessed on 11 February 2012).
16. U.S. Army Corps of Engineers (USACE), Post-Flood Assessment for 1983, 1986, 1995, and 1997 Central Valley, California; USACE, Sacramento District: Sacramento, CA, USA, 1999.
17. California Floodplain Management Task Force, California Floodplain Management Report: Recommendations of the California Floodplain Management Task Force (Galloway Report); California Department of Water Resources: Sacramento, CA, USA, 2002.
18. Alluvial Fan Task Force, Alluvial Fan Task Force: Findings and Recommendations Report; California Department of Water Resources: Sacramento, CA, USA, 2010.
19. Alluvial Fan Task Force, The Integrated Approach for Sustainable Development on Alluvial Fans; California Department of Water Resources: Sacramento, CA, USA, 2010.
20. Jeffres, C.A.; Opperman, J.J.; Moyle, P.B. Ephemeral floodplain habitats provide best growth conditions for juvenile Chinook salmon in a California river. Environ. Biol. Fishes 2008, 83, 449–458.
21. Opperman, J.J.; Galloway, G.E.; Fargione, J.; Mount, J.F.; Richter, B.D.; Secchi, S. Sustainable floodplains through large-scale reconnection to rivers. Science 2009, 326, 1487–1488.

22. Hanak, E.; Lund, J.; Dinar, A.; Gray, B.; Howitt, R.; Mount, J.; Moyle, P.; Thompson, B. Managing California's Water: From Conflict to Reconciliation; Public Policy Institute of California: San Francisco, CA, USA, 2011; pp. 1–500.
23. Woodall, D.L.; Lund, J.R. Dutch flood policy innovations for California. J. Contemp. Water Res. Educ. 2009, 141, 45–59.
24. Ross, T.; Lott, N. A Climatology of 1980–2003 Extreme Weather and Climate Events; National Climatic Data Center Technical Report No. 2003-01; NOAA: Asheville, NC, USA, 2003.
25. Zhu, T.; Lund, J.R. Up or out?-Economic-engineering theory of flood levee height and setback. J. Water Resour. Plan. Manag. ASCE 2009, 135.
26. Zhu, T.; Lund, J.R.; Jenkins, M.W.; Marques, G.F.; Ritzema, R.S. Climate change, urbanization, and optimal long-term floodplain protection. Water Resour. Res. 2007, 43.
27. Klemes, V. Sensitivity of water resource systems to climatic variability. In Common Sense and Other Heresies: Selected Papers on Hydrology and Water Resources Engineering; Klemes, V., Ed.; Canadian Water Resources Association: Cambridge, ON, Canada, 2000.
28. Klemes, V. Design implications of climate change. In Common Sense and Other Heresies: Selected Papers on Hydrology and Water Resources Engineering; Klemes, V., Ed.; Canadian Water Resources Association: Cambridge, ON, Canada, 2000.
29. Burby, R.J. Flood insurance and floodplain management: The U.S. experience. Environ. Hazards 2001, 3, 111–122.

CHAPTER 18

Framework of National Non-Structural Measures for Flash Flood Disaster Prevention in China

Dongya Sun [1], Dawei Zhang [1,2,*] and Xiaotao Cheng [2]

[1]*Research Center on Flood and Drought Disaster Reduction of the MWR, Beijing 100038, China*

[2]*State Key Laboratory of Simulation and Regulation of Water Cycle in River Basin, China Institute of Water Resources and Hydropower Research, Beijing 100038, China*

ABSTRACT

: In recent years, disasters caused by flash floods with many casualties have occurred frequently in China. In order to effectively prevent flash flood disasters, the State Council approved the *National Flash Flood Control Planning*(NFFCP) in 2006. In this planning, non-structural measures are recommended as the first step to be adopted in the prevention of flash floods, debris flow and landslide disasters caused by heavy rainfall. In order to effectively build up a comprehensive non-structural measure system for flash flood disaster prevention, the Ministry of Water Resources has asked the local authorities to conduct studies in 103 pilot counties aiming to explore practical schemes. Based on the experience from these pilot counties, the construction of non-structural measures on flash flood disaster prevention for 1836 counties was officially initiated in 2010. The contents of non-structural measures for flash flood disaster prevention at county level include eight aspects.

This paper will systematically illustrate the main contents of this national project for the construction of non-structural measures for flash flood disaster prevention in China. This work could provide a reference for other countries and regions in terms of flash flood prevention.

Keywords: flash flood disaster prevention; non-structural measures; monitoring; warning; China

1. INTRODUCTION

The flash floods referred to occur after heavy local rainfalls, often in mountainous areas. They are characterized by a quick rise of water levels causing a threat to the lives of those exposed. The time available to predict flash floods is limited. Flash floods often induce landslides and debris flows leading to severe damage and casualties in local areas. Due to the complex process of flash flood formation, the complicated characteristics of river basins in mountainous and hilly areas and the uncertainty of short-duration heavy rainfalls, flash flood prevention is a worldwide problem posing a great challenge. Many countries and regions attach great importance to the work of flash flood control. In the United States, flash flood warnings and watches are issued by local NWS Weather Forecast Offices (WFOs), which base the information on the comparison of rainfall amounts with flash flood guidance (FFG) values. FFG procedures were first formed in the 1990s. FFG was established by regional River Forecast Centers (RFC) and provided to local WFOs for flash flood forecasting [1]. In the European Union, a FP6 project titled FLASH ran from 2006–2009 in an attempt to improve the knowledge of historic flash floods in the Mediterranean region, while trying to improve forecasts of heavy precipitation events [2].

Compared to advanced countries, the research on flash flood disasters started relatively late in China. In recent decades, along with the rapid social and economic development as well as increasing losses caused by flash flood disasters, the issue of flash flood disasters has attracted wide attention in the whole society. Some scholars carried out thematic studies on this topic and the prevention work of flash flood disaster has progressed step by step. Particularly since 2000—with the development of the national economy and infrastructure construction and due to increasingly serious disasters caused by flash floods —China has increased inputs into the research on flash flood disasters and their prevention. After some endeavors in recent years, a set of methodologies on flash flood disaster prevention featuring Chinese characteristics has been established. Construction of non-structural measures on flash flood disaster prevention was carried out in 1836 counties nationwide.

This paper first introduces the basic situation of flash flood disasters in China, and then focuses on the Chinese government's efforts to establish non-structural measures regarding flash flood disaster prevention in recent years.

2. BASIC SITUATION OF FLASH FLOOD DISASTERS IN CHINA

Mountainous and hilly areas account for two thirds of the land area in China, among which the prevention area of flash flood disasters reaches 4.63 million km^2 (see Figure 1), involving a population of 560 million. The priority prevention area equals 970,000 km^2, affecting a population of 130 million. The priority prevention area can be subdivided into a first-grade prevention area and a second-grade prevention area according to the occurrence frequency of flash floods, geology and topographical conditions and the socio-economic situation.

In recent years, the characteristics of flood disaster losses have changed in China. The most outstanding change is that the percentage of casualties caused by flash floods remains high and holds an increasing trend, which has become a significant issue of the flood control and disaster mitigation task. In 2010 China witnessed particularly serious flash flood disasters, with 3887 people dead or missing. Representative events include: Massive debris flows that occurred on 7 August in Zhouqu, Gansu Province, leading to 1,500 people dead and 265 people missing; massive debris flows that occurred on August 18 in Puladi Town, Gongshan County, Yunnan Province, leading to 92 people dead or missing.

According to the statistical data, the proportion of deaths caused by flash flood disasters to those caused by flood disasters from 2004 to 2010 were 78%, 84%, 71%, 75%, 80%, 80% and 90%, respectively (shown in Figure 2). Flash flood disasters have become the major reason for casualties and have severely restricted the social and economic development in mountainous and hilly areas. It is imperative to strengthen the preventive work of flash flood disasters [3].

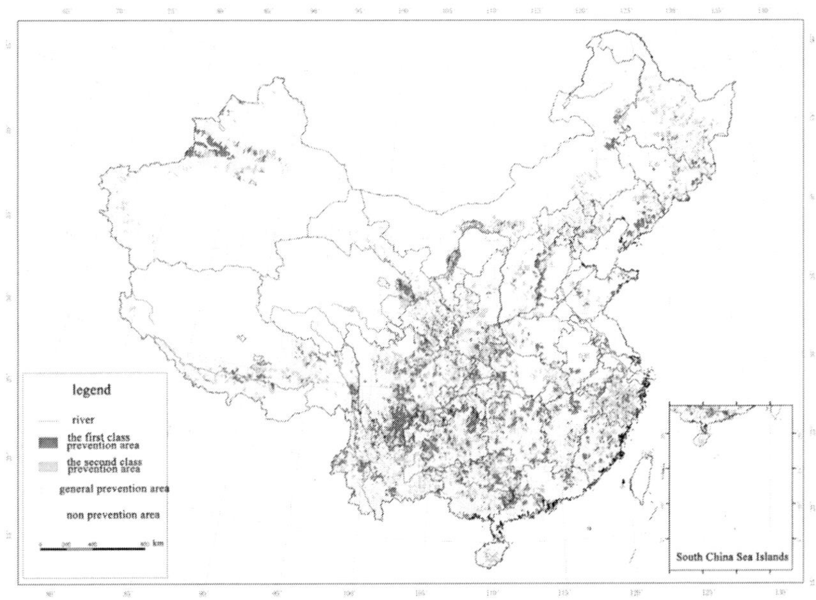

Figure 1. Zoning map of prevention area of flash flood disasters in China.

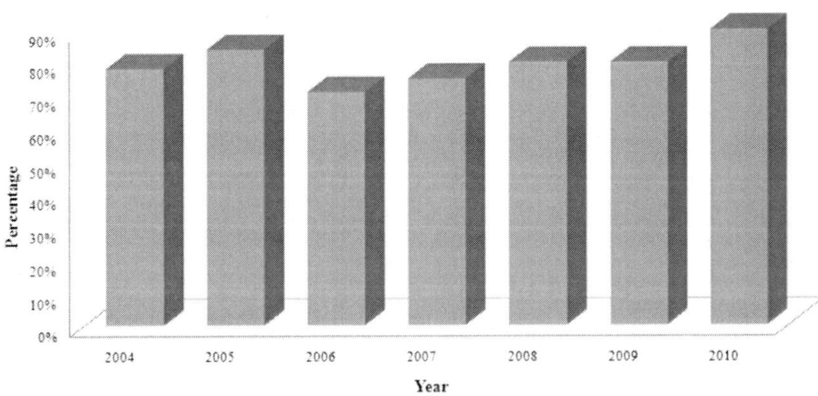

Figure 2. Proportion of deaths caused by flash flood disasters compared to those caused by flood disasters.

3. NATIONAL FLASH FLOOD CONTROL PLANNING (NFFCP) AND PILOT PROJECT

The Chinese government has attached great importance to the issue of flash flood disasters. In 2005, the Ministry of Water Resources, together with the Ministry of Land and Resources, China Meteorological Administration, the former Ministry of Construction and the former State Environmental Administration, compiled the *National Flash Flood Control Planning*(NFFCP)[4], which is the first thematic plan aiming at flash flood disaster prevention. Covering 1836 counties of 29 provinces, autonomous regions and municipalities, the planning adheres to the principle that non-structural measures are given priority while planning structural measures. A structural measure is seen as any physical construction to reduce or avoid possible impacts of hazards, or any application of engineering techniques to achieve hazard-resistance and resilience in structures or systems (for example building levee, embankment, *etc.*). Non-structural measures are defined as any measures not involving physical construction that use knowledge, practice or agreement to reduce risks and impacts, in particular through policies and laws, public awareness raising, training, education *etc.* With the guidance of this principle, a disaster prevention system is established including monitoring, communication, forecasting and warning. In October of 2006, the State Council officially approved the NFFCP and stipulated that—with the short-term planning projects as the priority and timely personnel evacuation and casualty reduction as the objectives—the construction of pilot projects should be sped up in priority prevention areas of flash flood disasters by adopting effective non-structural measures such as rainfall monitoring and construction of warning systems. On top of that, knowledge dissemination and management should be strengthened, the general public's awareness on disaster prevention and mitigation should be improved, and the masses' capacity of self-prevention, self-rescue and mutual assistance should be promoted to set up a monitoring and prevention system enforced by the masses themselves.

In order to actively explore effective ways and means for flash flood disaster prevention, and to accumulate experience for the implementation of the NFFCP, the Chinese government set aside special funds in the budget for pilot projects focusing on non-structural measures for flash flood disaster prevention in 103 counties of 30 provinces (autonomous regions and municipalities) in 2009.

The pilot projects of flash flood disaster prevention mainly focused on non-structural measures, such as monitoring and warning facilities, warning platforms, evacuation plans, organizational systems, knowledge dissemination, training and exercises. These above-mentioned non-structural measures could be further elaborated as follows:

- define the safety areas and hazardous areas;
- determine the critical rainfall(water level) that will trigger the occurrence of flash floods;
- adapt the monitoring and warning equipments;
- comprehensively equip the counties with the necessary warning facilities;
- develop the warning platform of flash flood disasters at the county level by using GIS and data bases as well as large scale electronic maps;
- establish the hierarchical responsibility system consisting of five levels: county, town, village, production group and household. Clarify the duty of each level and form a defense system in which the monitoring and prevention activities are enforced by the masses;
- formulate preparedness plan to guide the disaster prevention and avoidance activities;
- carry out knowledge dissemination, training and exercise activities to promote the awareness and capacity of the grass-roots cadres and masses on disaster prevention and emergency response.

Based on the actual situation, all the pilot counties of the 30 provinces have successfully accomplished the task with explorative efforts and positive innovation. A number of valuable experiences and practices are summarized below:

- Intensify the construction of monitoring sites and strengthen information sharing. Establish a rainfall and water level monitoring network that basically covers all villages and construction groups within the pilot regions by combining automatic telemetry and simple artificial observation. Provide timely and accurate rainfall and stream water level information and realize the information auto-writing into database.
- Adopt various measures to distribute warning messages timely and instantly. Through adopting traditional means and developing new ones, each pilot county worked out many ways and means to distribute and send warning messages in order to ensure that they were received by each town, village, construction group and each person in the flash flood disaster prone areas.

- Conduct dynamic monitoring and analysis of warning messages timely in order to effectively improve the commanding and decision-making capacities of defending flash flood disasters in the grass-roots areas. Each pilot county established a platform for warning of flash flood disasters. The telemetry rainfall and water level information can be automatically written into the database, and the warning message can be released automatically or manually through the platform.
- Improve organizational structure and assign responsibility to make sure that the prevention work of flash flood disasters was effectively and orderly carried out. Each pilot county formed a relatively perfect responsibility network of flash flood disaster prevention in the grass-roots areas by establishing the responsibility system of flash flood disaster prevention, setting up the five hierarchy responsibility system that covers the level of county, town, village, production group and household, clarifying the organizational structure and responsibilities of each level, and forming a wonderful defense system in which the monitoring and prevention activities were enforced .
- Optimize the plan system of flash flood disaster prevention in the grass-roots areas. Through overall investigation and research, each pilot county formulated or optimized the preparedness for flash flood disasters at the county, town and village levels in accordance with guidelines issued by the Office of State Flood Control and Drought Relief Headquarters (SFDH), which provided powerful guidance for conducting the work of flash flood disaster prevention.
- Strengthen knowledge dissemination, training and exercise activities to promote the masses' awareness of disaster prevention and avoidance, and their capacity on self-rescue and mutual assistance. By carrying out various forms of publicity, training and exercise activities at different levels, the knowledge of flash flood disaster prevention was rooted in the people's heart.
- Lay a sound technical foundation for the extension and popularization of the pilot projects to a large scale. The Ministry of Water Resources developed series of technical documents about project construction and management. Moreover, the Water Conservancy and Flood Control Department also formulated a series of systems and methods regarding project construction and management while carrying out pilot projects on different levels. On that basis, a group of intellectuals were involved that hold a good understanding of both the monitoring and warning system of flash flood disasters and project construction management. On top of that,

many monitoring and warning facility manufactures and software development companies have accumulated experience, which will provide sound technical support on extending the pilot projects to a large scale.

- Use the pilot projects to vigorously drive the project construction and implementation on a large scale. The pilot projects created a good demonstrative effect and laid a sound fundament for constructing non-structural measures for flash flood disaster prevention in the whole country.

4. DEVELOPMENT OF NON-STRUCTURAL MEASURES FOR FLASH FLOOD DISASTER PREVENTION AT COUNTY LEVEL IN CHINA

On the basis of the pilot project construction, the Ministry of Water Resources, together with the Ministry of Finance, the Ministry of Land and Resources and China Meteorological Administrative, organized video conference in November 2010 to officially initiate the construction of non-structural measures for flash flood disaster prevention at county level nationwide.

With the purpose of improving the capacity of flash flood disaster prevention, minimizing personnel casualties and property loss as well as avoiding the occurrence of massive injury and death particularly, it was planned to initially establish the non-structural measures system of disaster prevention that mainly focuses on the monitoring and warning aspects in 1,836 counties that had suffered from flash flood disasters within the past three years.

Regarding a county as the basic unit for the construction of non-structural measures on flash flood disaster prevention, the flood control department also gradually established corresponding systems on central, provincial and municipal levels in order to receive the monitoring and warning information of flash flood disasters timely and to provide support for decision making.

The overall structure is shown in Figure 3.

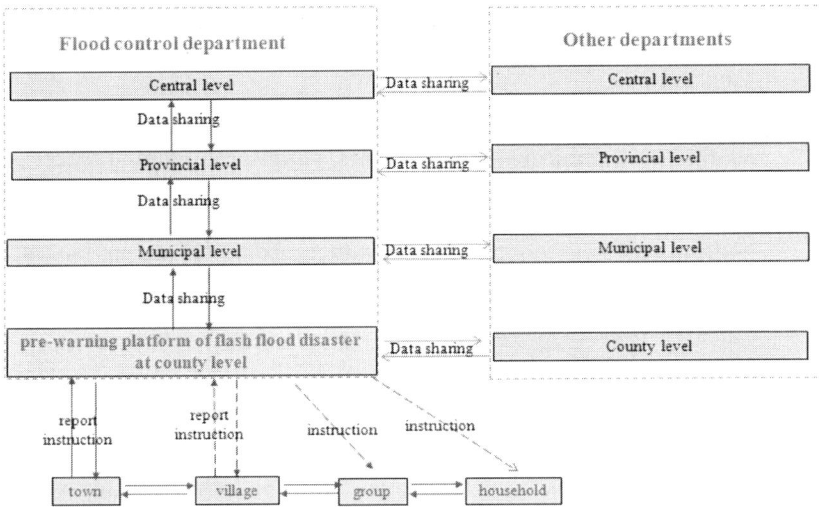

Figure 3. Overall structure of non-structural measures construction for flash flood disaster prevention nationwide.

The major content of the non-structural measures construction for flash flood disaster prevention is as follows:

- Comprehensive investigation. Investigate the natural, social and economic situation, the population distribution situation, the types of flash flood disasters, historical flash flood disasters, the populations threatened by flash flood disasters and the distribution of major economic facilities in all small river basins.
- Risk assessment and risk area delineation. Based on the investigation results, define hazardous areas and safety areas within the flash flood disaster prevention area. The delineated area was then plotted on the maps used to prepare by considering the villages and small river basins as the basic unit.
- Determination of preparedness parameters and their magnitudes, such as critical rainfall and water levels. Based on historical records of rainfall and flash flood disasters, determine preparedness parameters and their magnitudes (e.g. critical rainfall and water levels) for each small river basin and village with consideration of topography, vegetation and soil type. This kind of threshold setup is very difficult. At a primary stage, the critical values are determined by experience based on historical disaster events or numerical analyses using distributed hydrological models. This could be reviewed and improved during practical implementation along with

the accumulation of monitoring data in small watersheds in long time series.

- Construction of monitoring systems including automatic rainfall and water level monitoring instruments, simple rainfall and water level measurement devices used by local residents. At the same time, improve the monitoring properties by making full use of the existing monitoring instruments built by meteorological departments and hydrological departments.
- Construction of warning system platform at county level. The platform has the function of basic information inquiry, real-time inquiry of hydrological data from the monitoring system, hydrological forecasting, warning message dissemination and system management. The platform software will assess the monitored rainfall or water level data and compare this with pre-determined critical values corresponding to different warning grades. If the critical values are exceeded, text messages will be automatically sent to orderly individuals or/and the broadcast equipments in villages by using GPRS/GSM communication networks. The broadcast equipment can transform the text messages into sound signals and broadcast them to the public. It is also operational that the orderly individuals call the broadcast equipment and their phone call can be broadcasted to the public.
- Equip with necessary warning facilities. In addition to traditional ways such as hand cranking alarm, gong and whistle, warning facilities also include telephone, fax, text message, radio broadcast, television, *etc*.
- Establish a community-based defense system for disaster prevention in which the man-observed rainfall or water level monitoring and prevention activities are enforced by the masses themselves. Set up the five hierarchy responsibility system that covers the level of county, town, village, production group and household. Improve the commanding organization at town village level and clarify the responsible personnel and corresponding duties at different levels.
- Formulate the preparedness of flash flood disaster in accordance with the *Guideline for the Compilation of Flash Flood Disaster Prevention Plan* issued by the SFDH [5]. Carry out knowledge dissemination, training and exercise activities.

The investment for each county was six million Renminbi (RMB) shared by the central government and local government. In 2010 and 2011, the central government issued 3.8 billion subsidies to support the development of non-structural measures on the flash flood disaster prevention in 1,100 counties. In 2012, the remaining 736 counties will complete the construction of non-structural measures.

The SFDH as the project management agency actively took all efforts to promote the implementation of the project. For example, it formulated several technical guidances, such as the *Technical Requirements on the Development of Warning System for the Flash Flood Disaster Prevention at County Level* [6], and the *Guideline for Compiling Implementation scheme of Non-structural Measures Construction for the Flash Flood Disaster Prevention at County Level* [7]. The SFDH also organized the inspection meeting and developed the *National Management Method on Non-structural Measures Construction for the Flash Flood Disaster Prevention at County Level* [8]. Currently, the SFDH is setting up a website of national flash flood disaster prevention (www.qgshzh.com) to introduce the most current project situation [9].

5. PRELIMINARY ACHIEVEMENTS—A CASE STUDY

Currently, the projects developed in 2010 have been progressing smoothly and are in full implementation throughout the country. The projects developed in 2011 will be finished before the flood season in 2012. In this year's flood season, Hunan Province, Yunnan Province, Jiangxi Province and Hubei Province distributed warning information timely by using non-structural measures systems recently completed or under development when fighting against flash flood disasters. The masses in the threatened areas were evacuated in advance and significant benefits were achieved. Following, a case study of Shennongjia forest region was given.

The Shennongjia forest region with an area of 3,253 km^2 is located in the northwestern corner of Sichuan and Hubei Province. There are a total of 317 streams in the region and the confluence area of 15 streams exceeds 50 km^2. All the streams originate in the territory of the district and they are all mountainous rivers. There are eight towns and 66 villages in this region.

Flash flood disasters have always been severe in this area. Therefore, the non-structural measures construction for flash flood defense had been initiated. The non-structural measures mainly include the construction of monitoring and warning systems, formulating preparedness plans, establishing a community-based defense system, flash flood control knowledge dissemination, training, *etc.* Specifically, besides the warning system platform, 77 simple rainfall measurement devices, 21 automatic rainfall monitoring instruments and three automatic water level monitoring instruments were installed. Detailed preparedness plans for flash flood disasters were established for eight towns and 66 villages. Various methods were used for knowledge dissemination and training such as meetings, broadcast, TV, website, paper, billboards, brochures, wall charts, CDs, cards, *etc.* Using these means, the awareness and capacity of withstanding risk of the local residents were improved. Everyone is now aware of the serious consequences potentially caused by flash floods and knows when and how to evacuate to a safe place.

The completed monitoring and warning system played an important role in the flood season of 2011. From 3:00 am on August 22 to 8:00 am on August 23, the Shennongjia forest region was hit by a heavy rainstorm. With the three-hour rainfall exceeding 100mm, the total rainfall was 270 mm, being the maximum of the past 58 years since the construction of an existing hydrological station. The storm centre was located in the southern area of the forest region that covers four towns, *i.e.* Muyu Town, Hongping Town, Jiuhu Town and Songluo Town. The real-time rainfall data of automatic rainfall monitoring instruments were transmitted to the warning platform through a GSM/GPRS network. Three hours after the beginning of rainfall, the cumulative rainfall exceeded the presupposed critical rainfall. Then the warning beep and red flashing signals were automatically launched by the platform. The local flood control department instantly sent the warning messages using the platform and at the same time notified the responsible personnel of towns in the hazardous areas by telephone. Because of the timely warning, the potentially affected 79,427 people of 28,631 families in 327 villages were evacuated in advance following the preparedness plan. Due to these efforts, no person was injured or died. For the residents suffering from flash flood disasters, the system of non-structural measures for flash flood disaster prevention has become a real 'umbrella for life protection'.

6. CONCLUSIONS

Flash flood disasters occur frequently over a broad area in China. Due to the features of sudden occurrence and severe damage as well as difficulties in warning and prevention, flash floods lead to significant damages to people's life safety, infrastructure and eco-environment. The Chinese government attaches great importance to flash flood disaster prevention. In 2006, the first thematic plan aiming at the flash flood prevention was issued. Later, the pilot projects regarding flash flood disaster prevention were carried out. Based on the exploration, a set of effective technical systems and management systems for the flash flood disaster prevention were established. With casualty minimization as the primary target, the work of flash flood disaster prevention mainly focused on non-structural measures with the structural measures as supplementary. Based on the experience distilled from the pilot projects, the construction of non-structural measures on the flash flood disaster prevention was carried out comprehensively in 1836 counties. This work is organized and financed by the central government, esp. the Ministry of Water Resources and the Ministry of Finance. It is implemented by local governments. It is one of the activities of overall flood risk management of China. Currently, the completed projects have played a tremendous role in disaster prevention.

In parallel with the development of non-structural measures against flash flood disasters, and in accordance with the decision of the Chinese government, the Ministry of Water Resources and the Ministry of Land and Resources re-formulated the *Thematic Plan on the Geological Hazard Control Caused by Flash Flood* based on the former NFFCP [10].Structural measures, non-structural measures and personnel resettlement will be implemented comprehensively and the capacity of flash flood disaster prevention will be improved comprehensively.

ACKNOWLEDGMENTS

The authors are grateful to Qiu Ruitian, Shang Quanmin and Huang Xianlong, who worked in SFDH, for their guidance to write this paper. Great thanks is also given to GuanYuhui and Wang Yanwei for their help in translating this article. Additional thanks is given to three anonymous reviewers for their comments and suggestions.

REFERENCES

1. Carpenter, T.M.; Sperfslage, J.A.; Georgakakos, K.P.; et al. National threshold runoff estimation utilizing GIS in support of operational flash flood warning systems. J. Hydrol. 1999, 224, 21–44.
2. Price, C.; Yair, Y.; Mugnai, A.; Lagouvardos, K.; Llasat, M.C.; Michaelides, S. FLASH: A new EU project related to Mediterranean flash floods. In Geophysical Research Abstracts, Vienna, Austria, 2007.
3. Huang, X.L.; Chu, M.H.; Zuo, J.C.; Yang, Y.X. Strengthen the development of non-structural measures on the flash flood disaster prevention in China. China Flood Drought Manage. 2010, 20, 4–6.
4. Editorial Group of NFFCP. National Flash Flood Control Planning; NFFCP: Beijing, China, 2006.

5. Office of State Flood Control and Drought Relief Headquarters. Guideline for the Compilation of Flash Flood Disaster Prevention Plan; SFDH: Beijing, China, 2009.
6. Office of State Flood Control and Drought Relief Headquarters. Technical Requirements on the Development of Warning System for the Flash Flood Disaster Prevention at County Level; SFDH: Beijing, China, 2010.
7. Office of State Flood Control and Drought Relief Headquarters. Guideline for Compiling Implementation scheme of Non-structural Measures Construction for the Flash Flood Disaster Prevention at County Level; SFDH: Beijing, China, 2010.
8. Office of State Flood Control and Drought Relief Headquarters. National Management Method on Non-structural Measures Construction for the Flash Flood Disaster Prevention at County Level; SFDH: Beijing, China, 2011.
9. Office of State Flood Control and Drought Relief Headquarters. Beijing, China, 2010. Available online: http://www.qgshzh.com (accessed on 19 March 2012).
10. United Editorial Group. Thematic Plan on the Geological Hazard Control Caused by Flash Flood. Beijing, China, 2010.

CHAPTER 19

Managing the Limbe Floods: Considerations for Disaster Risk Reduction in Cameroon

Roland Ndille1, Johannes A. Belle2

1. Faculty of ArtsUniversity of BueaBueaCameroon

2. Disaster Management Training and Education Centre for AfricaUniversity of the Free StateBloemfonteinSouth Africa

ABSTRACT

Cameroon has recorded disasters, both natural and man-made, in the past. The coastal city of Limbe and many other geographical locations are potential sites for future disasters. On the basis of interviews and related primary and secondary literature, this article looks at past occurrences of floods in Limbe, particularly those of 2001, and how the state and local authorities managed the challenge. Through this, it attempts an evaluation of state policies and the institutional structure for disaster risk reduction (DRR) and management in the country. The study reveals that Cameroon does not have a national disaster management structure and a national platform for DRR. Disaster related activities are lumped together with other civil protection responsibilities of the Ministry of Territorial Administration and Decentralization (MINATD) despite the country's commitment to the Hyogo Framework for Action 2005–2015. Government interventions are more reactive than proactive although its strategy to disaster management indicates both disaster preparedness and emergency intervention and rehabilitation. The study concludes that the present strategy is bureaucratic, financially burdensome, and does not really achieve the essential goals of DRR in saving lives and reducing vulnerability. For DRR to be effective in Cameroon, a permanent disaster management unit with regional and local structures should be put in place independent of other branches of the Department of Disaster and Emergency Services of MINATD and with each level exercising a significant degree of operative autonomy. A national policy and framework with emphasis on DRR should be adopted in the country.

Keywords: Cameroon, Disaster risk management, Disaster risk reduction, Flood Landslide

1. INTRODUCTION

A disaster is a serious disruption of the functioning of a community or a society involving widespread human, material, economic, or environmental losses and impacts that exceed the ability of the affected community to cope using its own resources (Wisner et al. 2004; Birkman 2006; UNISDR 2009; Coppola 2011). The number and seriousness of disasters in terms of people affected and property lost have been on the increase around the world since the 1960s when reliable data started to be recorded (DFID 2006). Between 2000 and 2012, disasters killed 1.2 million people and affected an annual average of 268 million people. Disasters also caused damages worth USD 1.7 trillion between 2002 and 2012 (Guha-Sapir et al. 2013; UNISDR 2013).

Preparedness and relief activities have often shown the extent of the effects of disasters. That is why local, regional, national, and international organizations have often been on the alert, ensuring that effective and well prepared disaster management plans are made available in vulnerable areas for prevention, preparedness, rapid and effective relief, and recovery. Cameroon has experienced a number of natural and man-made disasters over the years and the approach to solving the crises by bringing relief to the affected population has left many people uncertain of the availability and approach to disaster management in the country. A paucity of research literature on disaster management in Cameroon adds to the problem. An in-depth study of the history of disasters and disaster management in Cameroon has become imperative. The aim of this study is to contribute to the historiography of disaster management in Cameroon and to create awareness of the necessity to regularly question and assess policies and actions in managing disasters.

2. STUDY AREA

Cameroon is a democratic republic with a surface area of 475,000 km^2. Recent estimates (2012) put the population at about 22 million people (African Development Bank 2013). Although the estimates indicate a moderate population density of 46 people per km^2, the rural–urban population divide shows that 59.9 % of the population lives in urban areas while 40.1 % lives in rural areas (UNISDR-Africa 2010). The population density is thus uneven and particularly intense in the urban areas. This exacerbates the quest for basic services such water, electricity, and other social amenities like appropriate housing, invariably leading to the development of slums and ghettos thereby increasing vulnerability to disasters. It is estimated that between 70 % and 80 % of the urban population lives in slums with access to portable water limited to only 37.5 % (Nussbaum 2010). This situation is made even worse by the fact that almost a half of the country's population lives below the poverty line and unemployment currently stands at 30 % (CIA World Factbook 2013). The lack of money and the need to sustain livelihood by any means ultimately lead to the lack of a culture of safety and disaster prevention among this portion of the population.

In 2012, the country's annual growth rate was 4.1 % and the GDP per capita stood at USD 2,400. The national budget balance grew in deficit from 2.7 % in 2011 to 3.5 % in 2012 of the GDP and inflation stood above 3 % (African Development Bank 2013). All these in addition to the fact that the country over the past 10 years has featured prominently amongst the most corrupt nations in the world (Transparency International 2013) emphasize the difficulty on the part of the government and other local stakeholders of finding resources for disaster management and putting them into effective use.

Cameroon is divided into 10 administrative regions—two (the North-West and South-West) are part of the former British administered territory of Southern Cameroons and the eight others are part of the former French administered territory of Cameroun. The country can also be divided into a Christian south and Muslim north, and forested south and savanna north. There are more than 260 ethnic groups, each speaking a different dialect (Neba 1990). The country has abundant natural resources ranging from fertile soils for agriculture, oil and gas, minerals such as gold, bauxite, and tin, to timber and other non-timber forest products. Since the 1990s, due to the extraction of oil off the coast of Limbe in the Gulf of Guinea, the country's economic emphasis to some extent has shifted from agriculture to the extraction of oil and gas but revenues from the exploitation of these resources, and from oil in particular, have not been sufficiently channelled into structural investments in infrastructure and the productive sectors (African Development Bank 2013).

The city of Limbe is located along the coastal area of Fako Division, South-West Region of Cameroon (Figure1). The indigenous people include the Bakweri, Isubu, and Creoles (Matute 1988). It has beautiful coastal beaches, historic monuments, a botanic garden, and a wildlife center. It is not only an international tourist destination but the major petroleum and agricultural hub of Cameroon. The lone oil refinery is located in Limbe and the Cameroon Development Corporation's (CDC) banana, palm, and rubber plantations are spread over a majority of the most favorable lands for settlement around the city. The Corporation also has its headquarters in Limbe. The city's economic importance therefore cannot be overemphasized and this continues to account for the high influx of economic migrants who then exert pressure on the city's already scarce natural and man-made resources. With a surface area of 545 km^2, and a population currently estimated at 120,000 inhabitants (Mosoke 2012), Limbe has one of the highest population densities in Cameroon with 220 people per km^2.

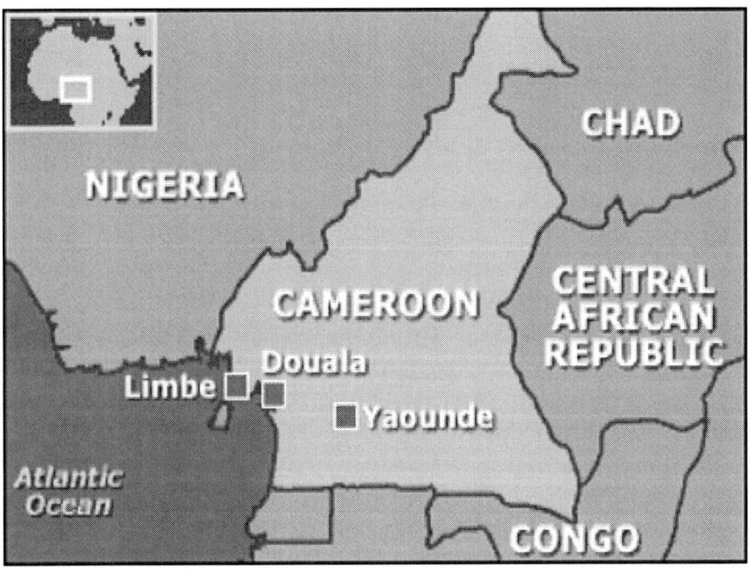

Figure1. Geographic location of Limbe in Cameroon. Adapted from Kometa (2012)

The city is characterized by a low-lying coastal plain, rising to a chain of horseshoe shaped hills with slopes of varying intensities towards the northeast and east, with the highest points reaching 362 m above sea level (Njabe and Fobang 2006). The area exhibits the characteristics of the Gulf of Guinea tropical equatorial climate of hot, moist, and dry conditions (Folack and Gana 2003). Only about 10 miles from Dibuncha, the second wettest place in the world after Cherrapunji in India (UNU-EHS and UFS-DiMTEC 2010), Limbe experiences very heavy torrential rains in the long rainy season (March–October) with the highest average monthly precipitation of about 700 mm recorded in June, July, and August (Neba 1990).

A drive through the city indicates an acute spatial disorganization with a mixture of planned and unplanned settlements, which are a reflection of a relative scarcity of suitable land for settlement as a result of rapid population growth and a lack of any effective land use development plan. This is because, possibilities of expansion are limited by the hilly nature of the eastern side, the Atlantic ocean to the south, the Mt. Cameroon lava flow route to the west, and the plantations of the CDC to the north. The imminent construction of a deep seaport facility and a naval shipyard has not only exacerbated population influx but has further intensified land speculation. Within the city small streams flow into larger drainage beds that converge into two main rivers (Limbe and Jenguele) that empty into the Atlantic Ocean (Figure2). These rivers frequently overflow their banks in the rainy season causing floods in the low-lying areas that are just 1–2 m above sea level. These are ideally the best parts for settlement but unfortunately there are no properly conceived drainage channels throughout the city. The situation is aggravated by settlement very close to the coastline at sea level with very poorly constructed, mostly wooden buildings. The hills that surround the city are made up of loose ferralitic and volcanic soils that easily disintegrate when they absorb a lot of water (Nwankiti 1983).

Poverty affects over 85 % of the Limbe population (Awum et al. 2001). That is why over 70 % of the population of Limbe lives in make-shift temporal structures constructed on floodplains, too close to the coastline, or on the hillslopes (areas declared as not suitable for habitation) and under slum-like conditions—a situation of high vulnerability to disaster (Mosoke 2012). The intense seismic activity from Mt. Cameroon located in the nearby Buea Subdivision and the threat of lava flows into the city have made matters worse. The last eruption of Mt. Cameroon in 1999 emitted hot magma that burnt down a large section of a palm plantation and the closest magma solidified just a few kilometers from the city. All these, make Limbe a city highly vulnerable to disasters.

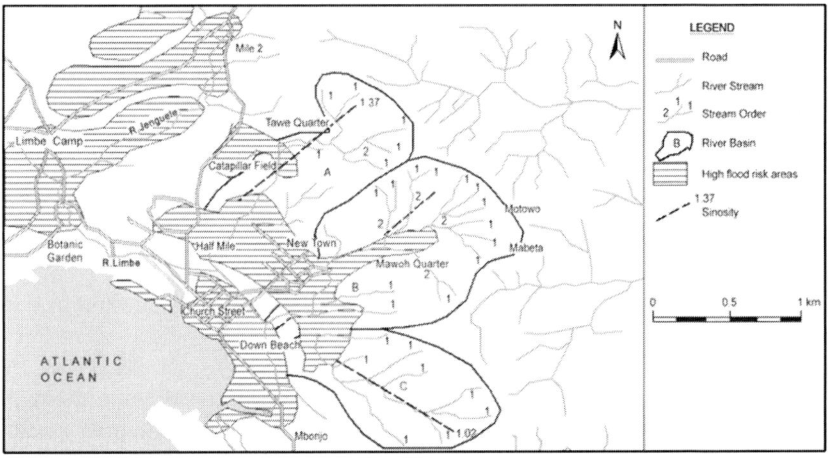

Figure 2. High flood risk areas in Limbe, Cameroon. Adapted from Kometa (2012)

3. METHODOLOGY

This study is qualitative and makes use of a blend of relevant historical inquiry and social science data gathering tools and presentation approaches. The researchers have relied heavily on oral accounts of hazard related experiences and data from primary and secondary written sources. Interviews were conducted in the neighborhoods affected by floods and landslides between February and June 2012, with affected persons and other eyewitnesses as well as some staff of the Limbe City Council, social workers, and some members of the 2001 Local Crisis Commission for the 2001 flood and landslide disasters in Limbe. A total of 50 participants were interviewed by convenience sampling on a one-to-one, non-directive basis (Maree 2009). The data were analyzed by grouping the responses into thematic areas and presented in a narrative format.

Archival data from committee reports and legislations on disasters were obtained from the National Archives in Buea (NAB), National Archives in Yaoundé (NAY), the Ministry of Territorial Administration and Decentralization (MINATD), and the Limbe City (former Urban) Council (LCC). These materials gave the researchers a view of the past and present policies on disasters in the country. Published books and internet sources provided us with general, theoretical, and conceptual knowledge of the field of study. For a first-hand impression of the floods, landslides, and drainage situation in the city during the rainy season and to enable us corroborate the oral and written data, we also carried out field observations in the city in June, July, and August 2013. We recognize, however, that major sources of error for research reporting could arise from the authentication of sources and from the subjective bias of the authors. We therefore ensured that all respondents confirmed their statements by sending them feedback from the interviews before using them in the study. To enhance objective research reporting a wide variety of informants and documents were consulted.

The hypothesis is that the Cameroon government has not set up an efficient disaster risk reduction (DRR) framework and that current strategies are very inefficient. The objectives are to analyze the government's handling of contemporary flood and landslide situations in Limbe; to show the inefficiency of the current strategies; and to highlight the need for a reconsideration of government strategies with the ultimate goal of establishing an efficient proactive national DRR framework that is completely autonomous from the current reactive civil protection structures of the MINATD.

4. CONTEMPORARY FLOOD DISASTER EVENTS IN LIMBE AND THE GOVERNMENT'S RESPONSE

Despite the awareness that some of the low-lying areas in Limbe are not habitable, many people in the last two decades have adopted a resigned attitude toward this and also have encroached on the steep hillslopes known for their catastrophic landslides (Mosoke 2012). The consequences have often been grievous. Each year, between the months of June and August, often characterized by week-long torrential rains, the city is submerged in water, above a meter deep in some quarters, causing serious distress to the local population and great embarrassment to the municipal authorities and local government officials. Residents on the slopes find the hills collapsing on their houses. According to statistics, each year about 100 homes are flooded and an average of five landslides occurs in the city (Lyonga 2012). In some years the situation became so critical that it required national emergency alerts—such as in 1990, 2000, 2001, and 2007. Emphasis in this analysis is on the 2001 disaster because of its devastating nature and the fact that it received significant government attention from which state disaster management strategies could best be appreciated.

On 26 and 27 June 2001 the city was engulfed in heavy rains that led to a combination of floods and landslides and resulted in more than 30 landslide scars, several tensional cracks (Lambi et al. 2002), 154 completely destroyed houses, and rendered over 233 people homeless. About 93 people died and over 197 cases were reported for post-disaster related illnesses and wounds of which 82 were reported at the Limbe District Hospital alone (Andin 2012). A significant number of livestock was lost and many people were rendered permanently or partially unemployed. The floods and landslides took a great toll on community infrastructure such as roads, water and electricity supply lines, communication systems, schools, hospitals, and churches, which were either completely or partially shattered. All losses were estimated at about three million US dollars (Limbe Urban Council 2001). A comparative summary of the damages caused by floods and related landslides in Limbe since 1990 is presented in Table 1.

Table 1. Impact of floods and landslides in Limbe, Cameroon

Item	1990	2000	2001	2007	Total
Displaced households	32	79	233	63	407
Deaths (human)	10	14	93	12	129
Reported health cases	19	20	197	62	298
Buildings destroyed	16	33	154	83	286
Animals/fowls lost	206	147	835	244	1,432
People rendered unemployed	51	107	438	79	675
Roads and bridges damaged	7	5	19	4	35
Electricity/water installations damaged	15	8	54	6	83

Limbe Urban Council (2000, 2001) and Andin (2012)

When disasters occur as they have in Limbe, it is expected that relief and rehabilitation would be swift. In 2001, the Senior Divisional Officer for Fako Division through the Governor of the South-West Region informed the various administrative levels of the situation as policy demanded. Through Prefectoral Order No. 94/2001 of 27 June 2001, the Senior Divisional Officer set up a crisis commission that comprised of several subcommittees. The Crisis Commission included the Delegate of Town Planning and Housing, Fako; the Divisional Delegate for Mines, Water and Energy; a representative of the Divisional Officer for Limbe; the Divisional

Delegate of Public Works; the Mount Cameroon Project[1]; the Limbe Botanic Garden; the Delegate of Urban Affairs, Limbe; the Delegate of Environment and Forestry; the representative of the CDC; the local Red Cross officials; the representatives of the Bimbia and Bonadikombo Development Councils; Forces of Law and Order; and the Department of Geology at the University of Buea (Limbe Urban Council 2001).

The Crisis Commission rallied a team of 725 persons, of which 230 were volunteers from humanitarian and charitable organizations. Financial resources of USD 65,000 equivalent and material resources of 15 trucks, two caterpillars, 14 assorted vehicles, and many others began to arrive Saker Baptist College, which was transformed into an emergency center as the students had gone on holidays 2 days before. Relief workers began to move destitute survivors from the affected areas to the college. Truckloads of bags of rice, beans, potatoes, onions, vegetables, oil, milk, meat, fish, bread, basic drugs, mattresses, clothes, bed sheets, pillowcases, tents, coffins, and so on continued to flow in from all directions as the affected persons were being moved and the dead buried (Lyonga 2012). The Minister of Territorial Administration and Decentralisation, who is responsible for civil protection, arrived Limbe a few days after.

Apart from immediately assisting victims of the floods and landslides, to prevent post-disaster crises such as the spread of epidemics like cholera, the Crisis Commission was expected to determine ways of rehabilitating and reconstructing the Limbe Municipality; identify the risk zones; define unhabitable areas; propose strategies for drainage in the city; and make general recommendations to minimize potential risks. After the floodwaters had receded, the Commission went around the city to identify the affected areas and assess the level of damage in a disaster assessment exercise. After the assessment tour they began paying compensation to the victims and launched rehabilitation and resettlement works.

Damaged roads, water channels, and bridges that had collapsed and completely cut off some neighborhoods from the rest of the city were gradually reconstructed. Rehabilitation also involved the dredging and widening of the stream channels and the construction of reinforced concrete embankments on the streams. Tarred roads heavily destroyed by floodwaters were also rehabilitated while gravel was thrown on the earth roads (Lyonga 2012).These huge financial engagements notwithstanding, the government disbursed another USD 1.5 million to identify risk zones and evacuate the occupants. About 150 ha of land were provided by the Limbe Urban Council for resettlement to this effect in the Bonadikombo area on the northern outskirts of the city. By the beginning of 2002 the government had begun to construct low-cost houses in this area. The last activities from the recommendations of the 2001 Crisis Commission were carried out by the Limbe Urban Council in 2005. In collaboration with the

Divisional Officer of Limbe the Council began demolishing houses and structures within the municipality built on waterways. These structures were said to have been obstructing the flow of rainwater causing floods during the rainy seasons. Structures constructed on hillslopes were also targeted (Mosoke 2012). These efforts notwithstanding, the floods of 2007 with relatively serious impacts as well came to prove to the government that there is much more to be done in disaster management than existing relief and rehabilitation strategies can accomplish.

5. ASSESSING THE LEGAL AND INSTITUTIONAL FRAMEWORKS FOR DISASTER MANAGEMENT IN CAMEROON

Until September 1986, when the Lake Nyos gas emission disaster struck Cameroon, the country had not thought of a disaster management policy or strategy. The magnitude of this disaster in which 1,785 people were killed (Ayanji 2004) demonstrated to the government that it was time to act. The government passed Law No. 86-16 of 6 December 1986 to reorganize civil protection in Cameroon (Republic of Cameroon 1986). Despite this law, it was only after 10 years following the 1996 measles outbreak with over 53 lives lost in Wum in the northwest region not far from the scene of the 1986 disaster and the cholera epidemic in northern Cameroon, which began in 1992 and by 1996 had killed over 440 people (Ayanji 2004) that the state began to act concretely. The President of the Republic signed Decree No. 96/054 of 12 March 1996 to determine the composition and duties of the National Council for Civil Protection (Republic of Cameroon 1996) and followed this up 2 years after by Decree No. 98-31 of 9 March 1998 to determine the organization of emergency and relief plans (ex post disaster measures) (Republic of Cameroon 1998a); and Law No. 98-15 of 14 July 1998 relating to establishments classified as dangerous, unhealthy, or obnoxious (Republic of Cameroon 1998b).

These, alongside Decree No. 2004/99 of 26 April 2004 to reorganize the MINATD (Republic of Cameroon 2004), Decree No. 2004/320 of 8 December 2004 and Decree No. 2005/104 of 13 April 2005 (Bang 2013), have since been the operating legal frameworks for civil protection under which any disaster management effort is carried out. While Decree No. 2004/320 made civil protection the second most important function of MINATD, Decree No. 2005/104 transformed the Department of Civil Protection in MINATD to the Department of Disaster and Emergency Services (DDES) and gave it the responsibility to organize and coordinate civil protection activities throughout the country for natural and human induced disasters. Table 2summarizes the structure of disaster response in Cameroon.

Table 2. Structure of disaster response in Cameroon

Government organ	Duties
PRESIDENCY OF THE REPUBLIC (The National Commission on Civil Protection, NCCP)	Defines policies, sets up the NCCP; harnesses national and international relief efforts
MINATD (DDES)	National supervisory authority of relief efforts; custodian of national disaster and emergency fund
REGIONAL GOVERNORS (Economic and Social Service; Regional Delegation of Social Affairs)	Coordinates disaster relief at the regional level
SENIOR DIVISIONAL OFFICERS (Divisional Delegation of Social Affairs)	Supervisory authority at the divisional level, sets up local crisis commission with the approval of the Regional Governor
Subdivisional Officers/Municipal Councils/Local NGOs, and so on	Implement relief efforts

By virtue of this legal and institutional framework, the President of the Republic, assisted by the National Council for Civil Protection, defines policies of disaster risk management, and such policies are implemented by MINATD, through the DDES (formerly the Department of Civil Protection). This national structure is further supported by the regional governors as well as the regional and divisional services of the specialized ministries such as the Ministry of Environment and Nature Protection and the Ministry of Social Affairs. This structure shows a "top down" approach to disaster management with the effect that implementation is bound to be delayed by bureaucratic and administrative bottlenecks. From the

Presidency to the ministerial departments, down to the regional governors and delegations, and further down to the divisional officers and divisional delegations, and ending with the local civil and municipal administrators, subdivisional officers and mayors, it takes a significant amount of time to have disaster management related decisions to be made or for relief efforts to reach the concerned.

In the event of a disaster the competent divisional/subdivisional civil administrator with reliable and irrefutable information on the nature and scope of the disaster informs the Regional Governor, who also informs the Minister of Territorial Administration and Decentralization. S(he) also relays the Prime Minister, who then informs the President. Authorization then leaves the Presidency following the same channel before the local administrative officer concerned can send out an alert; initiate emergency relief activities; mobilize the necessary human, material, and financial resources; convene a crisis commission; and inform the general public.

The convened crisis commission then assumes the responsibility of coordinating public relief action; directing relief operations; preparing rescue facilities; forwarding relief to the disaster area; determining the assistance and relief needs of victims; assessing the immediate and after effects of the disaster; and managing all the human, material, and financial resources provided for the circumstance by the government authorities, public, or private international organizations and, in general, all gifts and legacies.

From the above description and with the experience of the 2001 Limbe floods and landslides, it can be observed that practically, the disaster management strategy of the Cameroon government is centered on providing post-disaster relief and rehabilitation rather than on being proactive and preventive. Where the government strategy has shown a leaning towards some disaster preparedness efforts, they have remained theoretical than practical. For instance although Law No. 86-16 of 1986 to reorganize civil protection provides for the setup of a national observatory for disasters with local branches to permanently monitor disaster-prone areas, this organ has not yet been set up 28 years after this law was enacted. Also, in 2002, the government set aside March 1 to be observed yearly as the National Day for Civil Protection and for more than 10 years has observed the International Day for Disaster Reduction on 9 October. The events on these days, contrary to expectation, make no significant impact in terms of disaster awareness and preparedness as they are often characterized by speeches by government officials instead of grassroots sensitization campaigns.

In Limbe in 2012, of the 50 people sampled for this study, only 21 indicated that they were aware that the area where they lived was a disaster-prone area despite their regular experiences with floods and landslides. Forty-eight indicated that they have never heard of a talk on Limbe as a disaster-prone city while only seven indicated that they have been met by a local NGO worker, municipal or government official on a door to door sensitization campaign related to disaster prevention. Five of these seven however revealed that such an authority only came to serve them a demolition notice and not for any disaster risk education. There is also an annual report on the state of civil protection in Cameroon but the extent of its dissemination to zones and people concerned remains questionable. None of the 50 participants has ever heard, seen, or read a portion of this document. These revelations clearly demonstrate the lack of focus of the government strategy towards DRR in the country despite the fact that partners such as the United Nations Development Programme (UNDP), the United Nations Office for the Coordination of Humanitarian Affairs (UNOCHA), and ProVention Consortium have expressed their willingness to cooperate in this domain. In terms of national risk identification, assessment, monitoring, and early warning, a general study was conducted in 1999 in hazard-prone areas (UNISDR-Africa 2010) but this study, from the level of awareness in Limbe, either passed the people by or begs for a second phase.

Judging from the level of casualties, material, psycho-social, and infrastructural damages (most often irreparable) that disasters inflict on individuals, communities, and governments, reactive post-disaster strategies such as those currently in place in Cameroon should be discouraged as they do not meet the objectives of DRR of saving lives and reducing vulnerability. The administrative bureaucracy involved in sounding a disaster alert makes intervention slow, stressful, and increases the level of casualties and damages. Although the current civil protection structure seems decentralized by involving municipalities and divisional as well as subdivisional delegations of the concerned ministries, the devolution of authority to lower administrative levels is limited in practice (Bang 2013). For example, Law No. 2004/018 of 22 July 2004 stipulates that municipal councils are responsible for such tasks as organizing and coordinating relief operations (Bang 2013). But they must await authorization and approval from higher administrative levels, the immediate one being the Senior Divisional Officer. As the local authorities await authorization to come down the chain of command, the disaster continues to claim victims.

Experience from the Limbe disasters has demonstrated that post-disaster activities such as compensation and rehabilitation are hardly successfully carried out. The 2001 Local Crisis Commission was dissolved sometime in 2002 (Andin 2012) when it was not certain that it had fulfilled its terms of reference and without a proper auditing of its accounts. Besides, it has become very difficult to hold commission members accountable after the commissions have been dissolved. The Commission was made up of civil servants, some of them novices in disaster management and a majority must have been transferred to other localities considering the fact that such officials hardly stay in one station for up to 5 years. The fact that disaster field operations are managed by the Local Crisis Commission with lack of accountability discourages other agencies such as municipal authorities and NGOs from investing in DRR activities.

The finances for the implementation of the post-disaster management system come from the calamity budgetary head of the DDES in MINATD. This stands at USD 1 billion annually (UNISDR-Africa 2010). Often due to bureaucracy the finances are not disbursed on time although there is common knowledge that the first 24 h following a disaster are critical for search and rescue operations, which require a responsive disbursement of funds. Apart from that, there are no standards and norms of accountability for the use of relief resources. Where donations are made, as they were made in Limbe, the amounts collected are often poorly accounted for by local commissions. In most cases they are mismanaged. The 2001 Crisis Commission and the DDES have to date not informed the Cameroon public on the total amount of public and private financial and material donations received during the 2001 Limbe disaster and how they used the funds. Many of the victims still express feelings of disappointment with the 2001 Crisis Commission for mismanaging disaster relief resources. The situation in Limbe is not different from that of the 1986 Lake Nyos gas disaster as many of the victims are still waiting compensation from the government (Bang 2013). Cameroon was ranked the sixth most corrupt country in the world by Transparency International in 2004 and 144 out of 177 in 2013 (Transparency International 2013). The insurance system in Cameroon is very underdeveloped as well, implying that the government bears the brunt of rehabilitation and reconstructions, because private sector involvement in risk transfer is limited. Moreover, the roles of NGOs and the private sector are not indicated in the existing legislation. This limits their involvement to relief work when they are made part of crisis commissions.

6. THE FUTURE: FROM DISASTER RESPONSE TO DISASTER RISK REDUCTION

At present, the overall coordination of the disaster management system in Cameroon lies with the DDES at the MINATD. This is the only nodal agency for disaster management in Cameroon. The intervention strategy of the DDES is expected to focus on three main complementary approaches. Before disasters, it is expected to promote disaster preparedness, prevention, and mitigation mainly through information, awareness-raising, sensitization, and education and other activities that would promote a culture of safety and readiness in the community—hence the establishment of contingency plans during the predisaster phase (Bang 2013). It is expected to initiate an emergency intervention plan during the crisis phase and coordinate the implementation of emergency response programs (sectoral contingency plans) for efficient and effective disaster management. Such activities include the on-the-spot coordination of emergency responses, search and rescue operations, and the supervision of the activities of local crisis commission. After disasters they are charged with performing activities relating to compensation and rehabilitation of affected populations, disaster and hazard risk assessment, reconstruction through the crisis commission, and the communication of risk zones. The approach so far indicates a marked disregard for the first phase, which unfortunately should be the one emphasized as it involves DRR.

It is important to emphasize that not all disasters, particularly natural disasters, can be prevented, but the risk of loss of life, injury, and damage to property can be mitigated with good evacuation plans, land-use planning, environmental management, and building codes or standards (Wisner et al. 2004; DFID 2006; UNISDR 2013). Cameroon should therefore adopt policies designed to provide permanent protection from disasters; any disaster-prone area should be equipped with disaster warning mechanisms; and the people need to be educated constantly on how best to carry out livelihood activities in such places. In January 2005, 168 governments including Cameroon adopted the Hyogo Framework for Action (HFA) 2005–2015 (UNISDR 2005), a 10-year global plan for natural DRR. It offers vulnerable communities such as Limbe guiding principles, priorities for action, and practical means for achieving disaster resilience.

Almost 9 years after the Hyogo Framework was adopted, very few of its recommendations are being applied in disaster-prone areas like Limbe in Cameroon. An examination of the extent of disaster preparedness in Limbe confirms this conclusion. After the June 2001 floods and landslides, the 2001 Local Crisis Commission identified the alignment of 30 landslide scars around the Mabeta neighborhood (Limbe Urban Council 2001). The people were informed of this after the disaster and some left the area but a majority has continued to live there. Since then, nothing has been done to inform the people of the dangers inherent in living around this area. Tremors from nearby and active Mt. Cameroon volcano can seriously affect this area and even reach the sea, where a tsunami could occur and hit the city.

In terms of disaster awareness and preparedness, it is observed that the level of disaster education in Limbe is very low. A survey carried out in the multi-hazards, disaster-prone areas of Limbe in 2010 (UNU-EHS and UFS-DiMTEC 2010) revealed that very little had been done to prepare the people for any impending disasters and to warn them to evacuate the areas before imminent disasters strike. Two years after, our sampling of the situation as indicated above reveal an even more alarming level of ignorance and unpreparedness. Limbe has no meteorological station to monitor and communicate rainfall, wind speed, and sea level to the people. The city remains highly vulnerable to disasters, as very few people have moved out of the disaster-prone areas of the city especially Mabeta, Motowo, Limbe Camp, Down Beach, and Church Street in the southern part of the city. The evacuation of these persons can only be achieved by a sincere government effort beginning with the local inhabitants being sensitized on dangers of living in such places. This can be achieved through local and international cooperation and partnerships.

The government of Cameroon must empower local communities and authorities to manage and reduce disaster risk through appropriate legislation; by providing access to the necessary information and resources; and through the employment of professionally trained man-power to effectively implement DRR policies. These basic tenets of DRR have yet to be realized in Cameroon as a whole and Limbe in particular. In a city with over 85 % of the population affected by poverty, it is almost an impossibility to talk of the average inhabitant of the disaster-prone quarters buying a piece of land in a designated disaster-free area like Bonadikombo and constructing an average house of two bedrooms. Land speculation raised the cost of land to USD 7,000 for a 240 m^2 plot and USD 10,000 for a 400 m^2 plot (Awum et al. 2001). The cost of over USD 20 for 1 m^2 of land is very high for the urban poor. The local municipality therefore must make available low cost housing facilities in disaster-free areas as a measure of risk reduction.

The poor road network and the absence of proper drainage facilities account for the poor harmony in rain water collection and the high level of flooding that the city registers in the rainy season. The municipality must make efforts towards infrastructural provision especially with regards to roads, gutters, and drains in Mabeta, New Town, Motowo, and Church Street and implement clean up campaigns to ensure that these structures are cleaned up and maintained on a regular basis. The indiscriminate building of houses in violation of town planning rules and regulations must be checked. Those built too close to the Jenguele River must be demolished and no more houses should be built along floodplains or hillslopes with landslide histories, especially on the Mabeta hills. Neither should any other type of landed property be located in disaster-prone areas. The invasion and occupation of vacant risky sites should be stopped; streams such as the Jenguele should be maintained and their channels kept wide enough to facilitate the quick evacuation of run-off waters through monthly clean-up campaigns. Projects for the upgrading of occupied floodplains and corresponding resettlement schemes should be prepared. Those that were started in 2001 in Bonadikombo should be completed and resettlement carried out there.

Measures should be taken to increase the vegetation cover in the city, especially around steep hillslopes to help reduce the speed of run-off water after rainfalls. Deep-rooted trees should be planted on steep slopes and hillsides. The deforestation of slopes and steep hills around Limbe should be checked. There should be awareness and sensitization campaigns in Limbe, advising people on the negative consequences of dumping waste in drains as this eventually blocks their flow paths. The seashore must be reinforced with solid stones or block walls to avoid tidal overflow, and the removal of coastal mangroves should be regulated and checked. If the tides had been high on 26 and 27 June 2001, Limbe could have suffered more deaths. Government should set up permanent local emergency relief teams that can be activated in the event of an impending disaster. The municipality should also set up its own disaster management unit or service and equip it with the necessary human and material resources aimed at DRR.

At the national level, a strong political will is needed in Cameroon to gradually move from disaster response to disaster prevention, mitigation, and preparedness. The DDES, as the nodal disaster management agency, must have full-time paid staff who are experts in disaster management at national, provincial, as well as local levels. Their department must be linked to the research institutions of the Ministry of Scientific and Technical Research and Innovation, which would enable the disaster management structures to benefit from advice in land use, planning, and mapping of high-risk areas. A permanent disaster management structure should be put in place independent of other branches of the DDES, with regional as well as local organs each with a certain degree of operational autonomy. The structure must incorporate specific roles for each level of the disaster management structure, local administrative authorities, economic operators, and the public so that all can be held accountable for action or inaction.

When such structures are established, their sources of financing must be clearly defined and budgeting must lay emphases on DRR activities such as research, education, training, and public sensitization/awareness creation. The role of the Ministry of Finance and the Budget (though it is a member of the National Council for Civil Protection) must be evident in the legislations in ensuring that crucial financial resources are budgeted by all the ministries involved in disaster management. The keeping of historical records with the DDES relating to disaster risks is poor, and the information on disasters is often incomplete, especially in terms of statistics. This is a major problem that hampers research in this domain. The DDES and all agencies involved in disaster management in the country must also provide for a documentation and archives unit to meet the demand for the collection of systematic and accurate records.

7. CONCLUSION

The study aimed at assessing the Cameroon government's policies and practices on disaster management, especially in the context of the 2001 Limbe floods and landslides. The study revealed that although these disasters are a yearly phenomenon in the city of Limbe, regularly taking a heavy toll on human life, infrastructure, and material and financial resources, the government continues to rely on post-disaster relief and response policies and practices that were put in place after the 1986 Lake Nyos gas disaster. Other subsequent disasters have occurred in the country, notably the explosion of gas and petrol tanks in Nsam (1996), the Lagdo dam failure (1998), the eruption of Mt. Cameroon in Buea (1999), the landslides at Bafaka Balue (1997) and Lebialem (2009), the Ndop and Logone floods (2012), and a number of recorded epidemics, and bush and market fires (Ayanji 2004). Despite these numerous disasters and the heavy financial burdens they have been for the government, the state has not so far recognized the need to reform the national disaster management system.

The study concludes that the reform of the disaster management system in Cameroon is a matter of urgency and the application of the Hyogo Framework for Action five priority areas should now be seen as an urgent priority before the next calamity takes its toll on the people and their assets. Current climate change and the increase in extreme weather events should act as an impetus for the national government to put the disaster management function in Cameroon in line with international best practices. For example, South Africa adopted a national policy and framework for disaster management that emphasizes DRR without neglecting efficient and effective response (Government of South Africa 2002, 2005). Cameroon could follow the same example with basic adjustments to suit local needs. The state must ensure that DRR is prioritized and that strong institutional bases are created for its implementation at the national, provincial, and local levels. Structures must be put in place all over the country to identify, assess, monitor, and respond to disaster risks. This would enhance early warning and preventive actions. The use of knowledge, innovation, and education to build a culture of safety and resilience at all levels (UNISDR 2005) must be guaranteed through seminars, workshops, door-to-door as well as radio and television sensitization. Most importantly, where possible, the underlying risk factors must be reduced through resettlement, sustainable livelihoods, and effective environmental management. Disaster preparedness must be strengthened for effective response at all levels. The government's role as a policy developer, implementer, and evaluator is urgently needed.

FOOTNOTES

1 An international nongovernment organization involved in research and the preservation of forest, wildlife, and environment around the Mt. Cameroon area that includes Limbe.

ACKNOWLEDGMENTS

We have benefited from the field assistance (especially interviews) of Solange Enoh Andin, former student and supervisee of Roland Ndille in the Department of History at the University of Buea between 2009 and 2012. Immense thanks.

REFERENCES

1. African Development Bank. 2013. African Development Bank African economic outlook: Cameroon. http://www.afdb.org/en/countries/central-africa/cameroon/cameroon-economic-outlook/. Accessed 12 Dec 2013.
2. Andin, S.E. 2012. Natural disasters and disaster management in the Limbe Municipality 1990–2007. Unpublished undergraduate long essay at the Department of History, University of Buea.
3. Awum, D., K. Bayie, and N.L. Fonda. 2001. *Mainstreaming potentials for sustainable development in the Limbe Urban Municipality*. Yaounde: PAFP.
4. Ayanji, E.N. 2004. A critical assessment of natural disaster management framework in Cameroon. Unpublished end of course research report at the Department of City Management and Urban Development, World Bank Institute, Rotterdam.
5. Bang, H.N. 2013. Governance of disaster risk reduction in Cameroon: The need to empower local government. *Journal of Disaster Risk Reduction Studies* 5(2): 1–10.
6. Birkman, J. (ed.). 2006. *Measuring vulnerability to natural hazards: Towards disaster resilient societies*. New York: United Nations University Press.
7. CIA World Factbook. 2013. Cameroon economy profile 2013. http://www.heritage. org/ index/country/cameroon. Accessed 12 Dec 2013.

8. Coppola, D.P. 2011. *Introduction to international disaster management*, 2nd ed. New York: Elsevier.
9. DFID (Department for International Development). 2006. *Reducing risk of disaster—Helping to achieve sustainable poverty reduction in a vulnerable world. A policy paper.* Geneva: DFID.
10. Folack, J., and W.F. Gana. 2003. Cameroon: Project brief on the integrated management of the Limbe-Kribi coastal area. 4th draft.
11. Government of South Africa. 2002. Disaster management act. Act No. 57 of 2002. Cape Town: Government Gazette.
12. Government of South Africa. 2005. *National disaster management framework.* Durban: Lexis Nexis.
13. Guha-Sapir, D., P. Hoyois, and R. Below. 2013. *Annual disaster statistical review 2012: The numbers and trends.* Brussels: Centre for Research on the Epidemiology of Disasters.
14. Kometa, S.S. 2012. Ensuring human safety in the disaster prone town of Limbe Cameroon. *Journal of Geography and Geology* 4(2): 156–165.
15. Lambi, C.M., S.S. Kometa, and L.F. Fombe. 2002. Environmental hazards and land use planning for sustainable development: The Limbe unstable coastal region, Cameroon. In *Instability: Planning and management; seeking sustainable solutions to ground movement problems*, ed. R.G. Macinnes, and J. Jakeways, 151–160. London: Thomas Telford.
16. Limbe Urban Council. 2000. Report of the technical sub-committee on the 2000 floods. Limbe: Limbe Urban Council.
17. Limbe Urban Council. 2001. Report of the Crisis Commission of the 2001 flood and landslide disasters in Limbe. Limbe: Limbe Urban Council.
18. Lyonga, Casimir Njime. 2012. Personal interview. Civil engineer at the Limbe City Council. Limbe, 17 May 2012.
19. Maree, K. 2009. *First steps in research*. Pretoria: Van Schaik.
20. Matute, D.L. 1988. *The socio-cultural legacies of the Bakweri of Cameroon*. Yaoundé: Ceper.
21. Mosoke, Sebastine. 2012. Personal interview. Town planner at the Limbe City Council. Limbe, 15 April 2012.
22. Neba, A. 1990. *Modern geography of the Republic of Cameroon*, 3rd ed. Bamenda: Neba.
23. Njabe, R.K., and R. Fobang. 2006. *Illustrated physical geography and map reading for Cameroon*, 3rd ed. Limbe: Sunway.
24. Nussbaum, L. 2010. More than just a roof over my head. Cameroon: Housing for people living with HIV AIDS. http://www.thebody.com/content/art58858.html. Accessed 29 Apr 2014.
25. Nwankiti, O.C. 1983. *Man and his environment.* London: Longman Group.
26. Republic of Cameroon. 1986. Law No. 86-16 of 6 December 1986. On the reorganization of civil protection in Cameroon. National Archives in Buea.
27. Republic of Cameroon. 1996. Decree No. 96/054 of 12 March 1996. To determine the composition and duties of the National Council for Civil Protection. National Archives in Buea.
28. Republic of Cameroon. 1998a. Decree No. 98-31 of 9 March 1998. To determine the organization of emergency and relief plans (ex post disaster measures). National Archives in Buea.
29. Republic of Cameroon. 1998b. Law No. 98-15 of 14 July 1998. Relating to establishments classified as dangerous, unhealthy or obnoxious. National Archives in Buea.
30. Republic of Cameroon. 2004. Decree No. 2004/99 of 26 April 2004. To reorganize the Ministry of Territorial Administration and Decentralization. MINATD.
31. Transparency International. 2013. *Corruption perception Index 2013.* Berlin: Transparency International.
32. UNISDR (United Nations International Strategy for Disaster Reduction). 2005. *Hyogo framework for action 2005–2015: Building the resilience of nations and communities to disasters.* Geneva: UNISDR.
33. UNISDR (United Nations International Strategy for Disaster Reduction). 2009. *Terminology.* Geneva: UNISDR.

34. UNISDR (United Nations International Strategy for Disaster Reduction). 2013. *Disaster impacts 2000–2012*. Geneva: UNISDR.
35. UNISDR-Africa (United Nations International Strategy for Disaster Reduction Regional Office for Africa). 2010. *Inventory of national coordination mechanisms, legal frameworks and national plans for disaster risk reduction in Africa*. Nairobi: ISDR.
36. UNU-EHS and UFS-DiMTEC (United Nations University Institute for Environment and Human Security and University of the Free State Disaster Management Training and Education Centre for Africa). 2010. Vulnerability assessment of floods in the Limbe District of Cameroon. Presented at the Ph.D. block course held from 26 May to 2 June 2010 at DiMTEC, UFS.
37. Wisner, B., P. Blaikie, T. Cannon, and I. Davis. 2004. *At risk: Natural hazards, people's vulnerability and disasters*, 2nd ed. London: Routledge.

INDEX

A

assessment tools, 210

B

bank protection, 193, 194

Bengal basin, 106

Block method, 1

Brisbane River, 139, 140, 141, 142, 143, 144, 148, 150, 155, 160, 161, 164, 165, 166

C

Cameroon, 31, 379, 380, 381, 382, 383, 384, 385, 386, 392, 393, 394, 395, 396, 397, 398, 399

China flood risk, 170

Climate change, 24, 48, 50, 90, 334, 335, 361, 363, 364

coastal flood risk, 77, 78, 80, 82, 83, 89, 284, 287, 313, 314

coastal flooding, 25, 52, 78, 80, 81, 91, 170, 283, 284, 285, 287, 288, 298, 302, 311, 312, 313, 354

Coping strategies, 320

Crowdsourcing, 94, 103

D

defence failure, 283, 285, 293, 295, 304, 308

dilemmas, 169, 170, 172, 173, 176, 186, 187

dredging, 194, 206, 220, 226, 320, 329, 333, 387

E

Environmental risk management, 1

F

Flood, 1, 3, 4, 5, 18, 23, 24, 28, 34, 40, 44, 45, 47, 48, 49, 50, 369, 371, 372, 374, 375, 377, 378, 380, 385

Flood Hazard Map, 338

Flood modeling, 24

flood plain, 44, 143, 193, 194, 195, 207, 247, 362

Flood Potential Map, 242

Flood risk, 1, 49, 95

Flood vulnerability assessment, 24

floodplain, 49, 106, 155, 162, 165, 169, 171, 172, 174, 310, 311, 315, 349, 358, 360, 361, 363, 364

G

Gumbel distributnion, 1

L

land use planning, 140, 160, 165, 212, 213, 216, 227, 260, 261, 270, 272, 274, 353, 398

Landslide risk, 210, 238, 239

Littoral tract, 106

Low-salinity flooding, 53, 54, 74

M

main channel, 194, 196, 198, 199, 200, 202, 203, 206, 207

Model Historic Floods, 121

Multi-Criteria Decision Analysis, 242

N

Non-Linear Flow, 121

O

Odra basin, 1

Oil recovery, 54, 63, 64, 65, 66, 76

P

pathways flood, 170

R

regional inundation modelling, 283

resilience, 27, 38, 40, 44, 50, 94, 99, 102, 122, 169, 170, 171, 172, 174, 176, 180, 183, 186, 187, 240, 279, 369, 393, 397, 398

risk reduction, 23, 26, 27, 28, 29, 30, 32, 36, 37, 40, 43, 44, 379, 380, 385, 394, 397,399

S

Social media, 94

T

The Yongding River, 320, 323, 330

V

Volunteered geographic information, 94, 103

W

Water logging, 106, 111

water release strategy, 140, 151, 154, 165

Wettability, 53, 54, 57

Y

Yellow River, 181, 189, 319, 320, 322, 324, 326, 330, 334, 335